T0309749

LASER SCANNING FOR THE ENVIRONMENTAL SCIENCES

Laser Scanning for the Environmental Sciences

EDITED BY

George L. Heritage and Andrew R.G. Large

A John Wiley & Sons, Ltd., Publication

This edition first published 2009, © 2009 by Blackwell Publishing Ltd

Blackwell Publishing was acquired by John Wiley & Sons in February 2007. Blackwell's publishing program has been merged with Wiley's global Scientific, Technical and Medical business to form Wiley-Blackwell.

Registered office
John Wiley & Sons Ltd, The Atrium, Southern Gate, Chichester,
West Sussex, PO19 8SQ, UK

Editorial offices
9600 Garsington Road, Oxford, OX4 2DQ, UK
The Atrium, Southern Gate, Chichester, West Sussex, PO19 8SQ, UK
111 River Street, Hoboken, NJ 07030-5774, USA

For details of our global editorial offices, for customer services and for information about how to apply for permission to reuse the copyright material in this book please see our website at www.wiley.com/wiley-blackwell

Library of Congress Cataloguing-in-Publication Data
Laser scanning for the environmental sciences / edited by George L. Heritage and Andrew R.G. Large
p. cm.
Includes bibliographical references and index.
ISBN 978-1-4051-5717-9 (hardcover : alk. paper)
1. Environmental monitoring—Remote sensing. I. Heritage, G.L. (George Leonard) II. Large, Andrew R.G.

GE45.R44L37 2009
526.9—dc22

2008046482

A catalogue record for this book is available from the British Library.

Set in 9/11.5 pt Trump Mediaeval by Newgen Imaging Systems (P) Ltd, Chennai, India
Printed and bound in Malaysia by Vivar Printing Sdn Bhd

1 2009

Contents

List of Contributors

GABRIEL AMABLE *Department of Geography, University of Cambridge, Downing Place, Cambridge, CB23EN, UK*

CHRIS BRUNSDON *Department of Geography, University of Leicester, Leicester, LE17RH, UK*

GUY BYRNE *CSIRO Division of Land and Water, Urrbrae, South Australia 5064, Australia*

WILLIAM E. CARTER *Department of Civil and Coastal Engineering, University of Florida, P.O. Box 116200, 216 Larsen Hall, Gainesville, FL 32611-6200 USA*

MARTIN E. CHARLTON *National Centre for Geocomputation, National University of Ireland, Maynooth, County Kildare, Ireland*

SEAMUS J. COVENEY *National Centre for Geocomputation, National University of Ireland, Maynooth, County Kildare, Ireland*

SIMON CRUTCHLEY *Senior Investigator, English Heritage, Kemble Drive, Swindon Wilts SN22GZ, UK*

F.M. DANSON *Centre for Environmental Systems Research, School of Environment and Life Sciences, University of Salford, Manchester M54WT, UK*

BERNARD DEVEREUX *Department of Geography, University of Cambridge, Downing Place, Cambridge, CB23EN, UK*

NEIL S. ENTWISTLE *School of Environment and Life Sciences, University of Salford, Manchester, M54WT, UK*

IAN C. FULLER *Geography Programme, School of People, Environment and Planning, Massey University, Private Bag 11–222, Palmerston North, New Zealand*

JOHN C. GALLANT *CSIRO Division of Land and Water, Urrbrae, South Australia 5064, Australia*

GEORGE L. HERITAGE *JBA Consulting, The Brew House, Wilderspool Park, Greenall's Avenue, Warrington, WA4 6HL, UK*

DAVID HETHERINGTON *Ove Arup and Partners, Central Square, Forth Street, Newcastle Upon Tyne, NE13PL, UK*

DAVID HODGETTS *School of Earth, Atmospheric and Environmental Sciences, The University of Manchester, Manchester, M139PL,UK*

OCTAVIAN IERCAN *Department of Remote Sensing and Landscape Information Systems, University of Freiburg, 79106 Freiburg, Germany*

B. KOETZ *ESA – ESRIN, EO Science, Application and Future Technologies Department, Via Galileo Galilei – Casella Postale 64, 00044 Frascati, Italy*

ANDREW R.G. LARGE *School of Geography, Politics and Sociology, Newcastle University, Newcastle upon Tyne, NE17RU, UK*

MICHAEL LIM *Department of Geography, Institute of Hazard and Risk Research, University of Durham, Science Laboratories, South Road, Durham, DH13LE, UK*

TIMOTHY MCCARTHY *National Centre for Geocomputation, National University of Ireland, Maynooth, County Kildare, Ireland*

DAVID J. MILAN *Department of Natural and Social Sciences, University of Gloucestershire, Cheltenham, GL504AZ, UK*

JON MILLS *School of Civil Engineering and Geosciences, Cassie Building, Newcastle University, Newcastle Upon Tyne, NE17RU, UK*

F. MORSDORF *Department of Geography, University of Zurich, Winterthurerstrasse, Zurich, Switzerland*

IAN C. OVERTON *CSIRO Division of Land and Water, Urrbrae, South Australia 5064, Australia*

DAVID PENTON *CSIRO Division of Land and Water, Urrbrae, South Australia 5064, Australia*

NICHOLAS ROSSER *Department of Geography, Institute of Hazard and Risk Research, University of Durham, Science Laboratories, South Road, Durham, DH13LE, UK*

RAMESH L. SHRESTHA *Department of Civil and Coastal Engineering, University of Florida, P.O. Box 116200, 216 Larsen Hall, Gainesville, FL 32611-6200, USA*

ANDERS SIGGINS *CSIRO Division of Sustainable Ecosystems, Clayton South, Victoria, Australia*

K. CLINT SLATTON *Department of Civil and Coastal Engineering, University of Florida, P.O. Box 116200, 216 Larsen Hall, Gainesville, FL 32611-6200, USA*
and
Department of Electrical and Computer Engineering, University of Florida, P.O. Box 116200, 216 Larsen Hall, Gainesville, FL 32611-6200, USA

MICHAEL J. STAREK *Department of Civil and Coastal Engineering, University of Florida, P.O. Box 116200, 216 Larsen Hall, Gainesville, FL 32611-6200, USA*

CHRISTOPH STRAUB *Department of Remote Sensing and Landscape Information Systems, University of Freiburg, 79106 Freiburg, Germany*

YUNGSHENG WANG *Department of Remote Sensing and Landscape Information Systems, University of Freiburg, 79106 Freiburg, Germany*

Preface

Since the use of compass and chain field survey, the acquisition of precise terrain information has been instrumental in advancing our knowledge and understanding of the natural and built environment. As time has passed, technological developments have facilitated faster and more efficient acquisition of ever more accurate field data. Laser ranging has been used extensively since the development of the ruby laser in the 1960s – the same time that the term LiDAR (Light Detection And Ranging) appeared in the geoscience literature. Over the last decade or so LiDAR mapping has gained acceptance as a rapid, accurate and adaptable method for three-dimensional surveying of the Earth. The principle advantages of the technology over other surveying methods are the provision of precise *x,y,z* measurement from the instrument, its high level of automation, rapidity of coverage and fast delivery time for often extremely large datasets. The inspiration for this book stemmed from one of the most recent and certainly most significant advances, the advent of portable, robust LiDAR instruments. The book has two purposes, the first is to introduce LiDAR to the reader keen to learn more about the new technology and the second is to stimulate research using LiDAR through a series of applied chapters across a range of fields in the Environmental Sciences. It brings together some of the leading proponents of airborne and terrestrial laser scanning from across the globe, showcasing some of the innovative and groundbreaking research being conducted using the recently developed instrumentation.

LiDAR is of growing importance to the world's environmental scientists as a cost effective way to acquire massive amounts of high resolution 3D digital data. Many new users are keen to participate in this surveying revolution and the widespread utility of LiDAR is reflected in the Applications section of the book. It should be emphasised at the outset that indiscriminate collection and use of LiDAR data is a danger that we face at the moment; data acquisition is remarkably easy and DEM construction quickly follows with data quantity overriding data quality as a driver. Many of the chapters in this volume warn against such practices and protocols are slowly emerging in an effort to standardize disparate practice and define benchmark data quality standards. Perhaps the most succinct way of expressing this is *via* a poem by Dr Keith Kirby which appeared in the *Bulletin* of the British Ecological Society (with due acknowledgement by Keith to Lewis Carroll) in August 2008; we reproduce it here by virtue of his kind permission with a couple of tiny tweaks:

Surveyors ancient and modern

'You are old, Father William', the team leader said
'And your PDA's never been used.
Yet your reports come in, in a couple of weeks,
So you see I'm a little confused.'

'At your age', said the sage, 'I was given some maps,
A Beagle, and sent to the woods.
The skills that I learnt in those midge-haunted days
Still help me deliver the goods.'

'Yes', came the e-mail, 'We've noted the fact
That you still go to visit the site.
When satellites pass over at least twice a day,
Are you sure that your methods are right?'

'LiDAR sensors see the woods and the trees,
Now even the lie of the ground.
My visit though explores the cause and effect,
A view of the site in the round.'

'But virtual visits emit less CO_2
That's more carbon mileage to spare,
To come to a meeting in London next week
On the corporate plan; so, be there!'

'It's a bugger I know, what you say is part true.
We could argue the toss half the night.
Let's resolve the dispute the traditional way.
A clipboard round the ear says I'm right.'

We began this book with two purposes in mind: first to introduce LiDAR to an interested readership and second to stimulate research into the use of LiDAR in the environmental sciences. LiDAR, both terrestrial and airborne, has the potential to help us understand the natural systems which have shaped, and continue to shape, our world. The experiences of the contributors to this volume reinforce our confidence in that potential, although all of us recognise that LiDAR is not a universal panacea in environmental data gathering. To ignore that potential would be to our ultimate disadvantage – never has our need to understand the physical world been greater, and we are beholden to use the best technologies to help us in our research. LiDAR, as we have collectively demonstrated, is one of those technologies. However, its usage within the Environmental Sciences is still in its infancy and there are many research areas which would greatly benefit from the acquisition of greater amounts of spatial data. As editors, we have made every effort to encapsulate the variety of fields in which the technology is being applied around the world and intend this text be innovative in examining in detail the issues and areas of application under the banner of laser scanning. However, we acknowledge that, due to the rapidly evolving nature of the subject, there are inevitably omissions. Urban, atmospheric and volcanic applications are gaps we, as editors, originally sought to fill. Despite this, we are confident that this book captures the state-of-the-art in airborne and ground-based laser scanning, and trust that it will provide a catalyst for future interdisciplinary research and collaboration, and a starting point for scientists, practitioners and end users of the science.

The aim, therefore, is that this book will encourage the use of LiDAR, stimulate collaboration and inspire high quality scientific research. We are looking at a potential revolution in our understanding of Earth systems provided that the technology is used correctly, data limitations are acknowledged and handled competently and data analyses are conducted in a rigorous and scientific manner. If so, it is hoped that publication of potentially paradigm-shifting research can follow.

George L. Heritage
Andrew R.G. Large

1 Laser Scanning – Evolution of the Discipline

ANDREW R.G. LARGE[1] AND GEORGE L. HERITAGE[2]

[1]School of Geography, Politics and Sociology, Newcastle University, Newcastle Upon Tyne, UK
[2]JBA Consulting, Greenall's Avenue, Warrington, UK

A BRIEF HISTORY OF LASER DEVELOPMENT FOR TOPOGRAPHIC SURVEYING

Since its introduction in the 1960s, the laser has assumed a central role in the accurate measuring of natural environments. The historical background to laser scanning began in 1958 when two scientists, Charles Townes and Arthur Schawlow, suggested the potential for a narrow beam of very intense monochromatic radiation travelling over large distances that could be precisely directed (Price & Uren 1989). The first solid-state ruby laser was developed in 1960 and emitted powerful pulses of collimated red light. The period 1962–68 saw basic development of laser technology ('laser' is an acronym for *Light Amplification by the Stimulated Emission of Radiation*), and was followed in the 1970s by a period of improvement in the reliability of the technique. It was not long before the potential for a narrow, straight, reflectable beam as a reference direction in alignment was recognised. Early surveying instruments were developed specifically for laboratory use, the first laser distance-measuring instrument appeared in 1966 and the first alignment laser was marketed from 1971 onwards (Price & Uren, 1989). Despite reliability issues (the first instruments only had an operating life of 1000 hours), commercial success followed and the 1970s saw a rapid uptake in the use of lasers in engineering surveying and the construction industry. Once the early systems were adapted into weather-proofed machines that were specifically designed for more rugged situations, environmental scientists rapidly took up the new technology, and the 1980s and 1990s saw a wide range of applications in a broad range of environmental systems. Initially there was commercial inertia: Price and Uren (1989) quote a survey of commercial operators in the UK and USA which showed that, less than two decades ago, only 5% of commercial contractors in the UK used lasers at some time in their work (the comparative figure for the US was 95%).

Today, laser-based instrumentation is standard in a wide range of applications, with laser surveying instruments falling into three categories: fixed beam, rotation beam and distance measurers. Since the end of the 20th century the pace of technological progress has been breathtaking and field scientists now have the ability to rapidly measure environmental systems virtually in their entirety.

THE THEODOLITE AND THE EDM

Prior to development of laser scanning instrumentation, the theodolite was the most versatile and extensively used of all surveying instruments. This versatility was due to the manner in which

Laser Scanning for the Environmental Sciences, 1st edition. Edited by G.L. Heritage and A.R.G. Large.
 ISBN 978-1-4051-5717-9

all theodolites performed two simple operations – measuring angles in horizontal and vertical planes (Ritchie *et al.*, 1988). To date, theodolites remain standard surveying tools due to their versatility, accuracy and ease of operation.

The development of Electromagnetic Distance Measuring (EDM) devices designed for accurate distance measurement had a great influence on the discipline and soon became one of the most widely used pieces of technology for surveying exercises based on either triangulation, traversing and radiation, all of which have accurate distance measurement at their heart. In essence, an EDM can rapidly record spot heights and distance measurements in the field in the laying out of baselines over much longer distances than were obtainable with conventional methods. Most EDMs use a near-visible light source, or electromagnetic beam in the form of a modulated sine wave. In earlier instruments, the signal generated by the EDM was reflected back by a prism, with distance calculated by measuring the phase shift between the outgoing and returned signals. Distances measured can only be less than the wavelength of the carrier wave. Given that the wavelengths of the usual light sources are very typically short (e.g. infra-red at about 0.0009 mm), EDMs use a carrier wave (near-visible radio, microwaves) which can be modulated to allow it to carry a more useful wavelength. For example, if the carrier wave is modulated to 10 m, the instrument is capable of measuring the phase shift of the modulated wave to an accuracy of 1–2 mm. The shorter the wavelength of the carrier wave, the higher the accuracy of distance measurement (but with a trade-off – the wavelength of the carrier controls the distance over which accurate measurements can be taken). Most modern EDM instruments use a near visible light source as this has the least cost and power requirement. More recent machines have dispensed with the need for prisms – a development which has again speeded up the rate of data acquisition.

Until recently, EDM proved an efficient and reliable way of collecting the data necessary to produce a DEM. Questions have been posed, however, as to the best way to represent features in the field (Chappel *et al.*, 2003; Brunsden: Chapter 5, this volume). Most workers interpolate elevations using an arbitrary mathematical surface fitted through some of the data to form a regular grid. This assumes, however, that data are spatially dependant, yet this is rarely determined or quantified (it may also introduce additional sources of error). While different sampling strategies can emphasise the effect of data redundancy on DEM generation it is undeniable that, the more data points used, the more representative the DEM will be of the 'real-world' situation.

Intensive mapping in large (>1 km^2) natural systems using traditional theodolite EDM may exert a high demand on operator time and cost, especially when including the survey of all detail at a scale larger than the geomorphological unit (gravel bar, riffle, dune slack, moraine etc.). Sub-unit morphology (e.g. chute channels, sediment lobes) may be ignored to rationalise time in the field or at the photogrammetric plotter; however these features may represent significant changes in sediment storage within the channel system, and comprise important habitat for biota within these dynamic systems as they are subject to intermediate levels of disturbance (Cornell, 1978). Airborne LiDAR is a thus a useful tool for the acquisition of such terrain data as, while methods like EDM and DGPS also have potential for rapid topographic data acquisition (e.g. Brasington *et al.*, 2000), areal extent remains a limiting factor for these other techniques. Laser scanning techniques have the ability to vastly increase the data collection ability and, theoretically, to improve the accuracy of DEM representations of a range of natural environments.

REVIEW OF PREVIOUS DEVELOPMENTS

From point sampling to data clouds – a sea change in laser surveying

LiDAR, variously termed in the literature as *Laser Induced Direction and Ranging* (Marks & Bates, 2000) or, more commonly, *Light Detection and Ranging* (e.g. Wehr & Lohr, 1999; Smith-Voysey, 2006), provides laser-based measurements of the distance between an aircraft carrying the sensor and the ground. The resulting measurements can be post-processed to provide a digital elevation model with a precision within 15 cm (Charlton *et al.*, 2003). LiDAR consequently has significant potential for generating high-resolution digital terrain surfaces accurately, representing complex natural and semi-natural environments incorporating morphological features at a range of scales (e.g. McHenry *et al.*, 1982; Jackson *et al.* 1988; Ritchie, 1995; Ritchie & Jackson 1989; Ritchie *et al.*, 1992a, 1992b, 1994, 1995; Krabill *et al.*, 1995; Wadhams, 1995; Gauldie *et al.*, 1996; Bissonnette *et al.*, 1997; Parson *et al.*, 1997; Innes & Koch, 1998; Irish & White, 1998; Geist *et al.*, 2003; Godin-Beckman *et al.*, 2003; Ancellet & Ravetta, 2003; Staley *et al.*, 2006). As such, LiDAR technology has huge attractions for the environmental scientist. Hodgetts (Chapter 11, this volume) summarises these as follows:

- The technology is characterised by very high speed data collection – for example the Riegl LMS420i terrestrial laser scanner can scan at up to 12,000 data points per second (later generation machines being faster), with each data point having *x,y,z* positional information, reflection intensity and colour provided by a calibrated high resolution digital camera. Speed of collection is, however, machine-dependant.
- Datasets, once collected, may be interrogated at a later date for information which was not the focus of the original project.
- There is a high degree of coverage – therefore the LiDAR data can be returned to at a later stage in order to look for other features which may have initially been missed in the field.
- Accurate spatial data can be easily collected.

The downside, it has to be said, is cost: typical entry costs for terrestrial laser scanning equipment is currently in the order of £100,000 (*circa* £70,000 for hardware and £30,000 for processing software). LiDAR costs are prohibitive and ownership of this equipment is restricted to consortia and governmental research councils (in the UK the main provider is the Airborne Remote Sensing Facility (ARSF) section of the government-funded Natural Environment Research Council). Despite these restrictions, the techniques have considerable advantages over conventional surveying techniques (see other chapters in this volume for numerous examples of how they have recently been widely applied in the environmental sciences). Certain fields have been slower to adopt the technology; use of LiDAR for hydrologic and hydraulic applications has, for example, been relatively limited (Hollaus *et al.*, 2005). Wealands *et al.* (2004) have discussed the usefulness of remotely sensed data for distributed hydrological models, while Brügelmann and Bollweg (2005) describe how roughness coefficients can be used to derive hydraulically relevant cover classes (see case study below for an expansion on this topic). Pereira and Wicherson (1999) describe the potential for airborne laser scanning for collecting relief information for use in river management, and in the UK the Environment Agency regularly uses LiDAR-derived flood surface predictions as a public education tool.

How does LiDAR work?

Position for any *x,y,z* point on the Earth's surface is generated from three sources: (i) the LiDAR sensor, (ii) the Inertial Navigation Unit (INU) of the aircraft and (iii) GPS. The LiDAR measurements must be corrected for the pitch, roll and yaw of the aircraft, and the GPS information allows the slant distances to be corrected and converted into a measurement of ground elevation relative to the WGS84 datum. The measurements are taken from side-to-side in a swath as the aircraft flies along its path (Baltsavias, 1999); those measurements at the centreline of the swath are more precise than those near the edge (Figure 1.1). Brinkman and O'Neill (2000) also observe that both horizontal and vertical precision depend on the flying height (where horizontal precision is 1/2000th of the flying height); horizontal precision will thus be 'accurate to 15 centimetres or better' when the flying height is at or below 1200 m. The standard altitude for airborne LiDAR acquisition is *circa* 3,500 m.

Airborne LiDAR sensors generally emit anything between 5000 and 50,000 laser pulses per second, although Smith-Voysey (2006) states a higher figure of 100,000 pulses per second, a figure claimed for machines such as the Optech ALTM3100EA

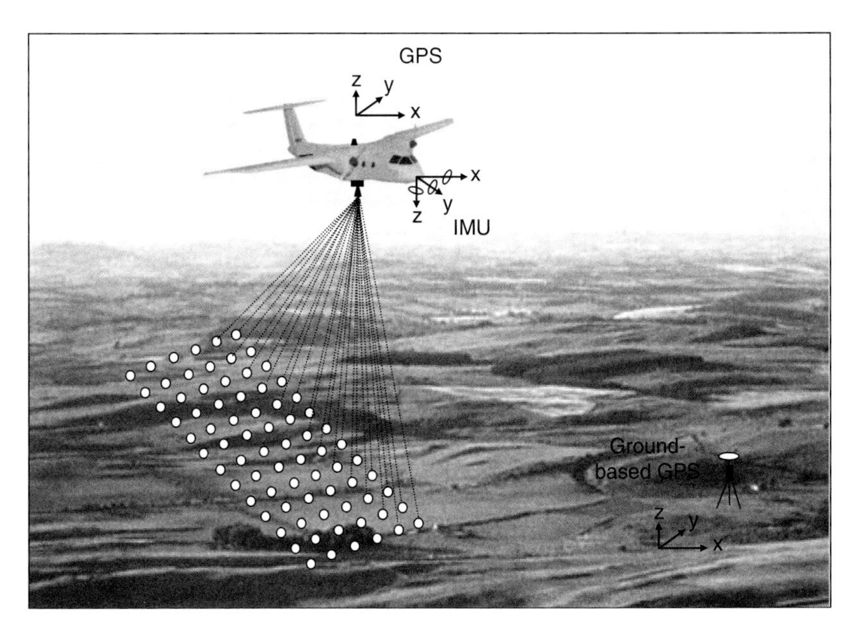

Fig. 1.1 Typical operation of an airborne LiDAR survey. The sensor operates in swaths across the terrain surface (shown by small open circles in this diagram), and is georeferenced by an inertial navigation unit (INU) in the aircraft and a ground-based differential GPS.

which operates at 167 kHz laser pulse repetition frequency. The spacing of these points is determined in two directions. In the in-flight direction, point spacing is determined by aircraft speed and altitude, whereas in the cross-flight direction (normal to the angle of flight direction), point spacing is defined by scan angle and altitude. In terms of what is actually emitted, each pulse has a diameter, or 'footprint' (typically between 0.5 and 1 m) and a length defined by the time between the laser pulse being switched on and off. In essence therefore, each pulse is a cylinder of light. On their own, these reflected pulses are not enough to construct a terrain surface; accurate *x-y-z* position using differential GPS is needed relative to ground-based GPS base stations, the roll, pitch and yaw of the aircraft needs to be measured by an inertial measuring unit (IMU), which in turn allows the angular orientation of each laser pulse to be determined. Finally, the times taken for each laser pulse to reflect off the ground (or whatever surface) and return to the sensor is measured. This is termed the 'return'. In essence then, laser scanning depends on knowing the speed of light, approximately 0.3 m/ns. Using that constant, we can calculate how far a returning light photon has travelled to and from an object:

$$\text{Distance} = (\text{Speed of Light} \times \text{Time of Flight})/2 \tag{1.1}$$

The return operates in two modes (both of which are described by contributors in this volume):

First-pulse: Measures the range to the first object encountered – in many this is vegetation, for example tree foliage.

Last-pulse: Measures the range to the last object – the ground surface under the foliage.

Machines like the Optech ALTM (Airborne Laser Terrain Mapper) system can measure both tree-heights and the topography of the ground beneath in a single pass.

LiDAR sensors are capable of receiving multiple returns. As some are up to five returns per pulse, a 30 kHz sensor has to be able to record up to 150,000 returns per second. Multiple returns are characteristic of 'soft' cover (e.g. vegetation) with a 'first return' indicating, for example, the top of the tree canopy, and other returns being indicative of branches etc. While in theory the last return represents the underlying ground, this is not always the case. Danson *et al.* and Overton *et al.* (this volume) discuss the issues and opportunities associated with LiDAR and vegetation cover in greater detail. LiDAR technology has also proven extremely versatile for studies on the Earth's atmosphere; NASA's Cloud Aerosol and Infrared Pathfinder Satellite Observations (CALIPSO) satellite will use LiDAR to determine atmospheric composition based on scattering of the laser pulse accurately measured to micrometre level (10^{-6} m).

LiDAR accuracy and precision

The integration of airborne LiDAR with GPS facilitates the wider use of high resolution DEMs in a wide range of physical applications. The method of survey is rapid, relatively economic and allows survey of difficult terrain, and large areas. According to Marks and Bates (2000) LiDAR permits rapid gathering of topographic data for areas at rates of up to 90 km^2 per hour. This make it an attractive alternative to ground-based survey methods and, due to the advantages outlined above, airborne-laser scanning has become the primary choice for gathering precise and dense DEMs of large areas for a wide range of applications. While LiDAR is currently the most efficient method for data acquisition, airborne laser-scanning does have reliability issues. It is difficult to determine the level of precision of LiDAR measurements for any one survey and, in complex topography, small lateral offsets associated with the IMU and GPS will be translated into vertical error in the LiDAR surface. This is less of a problem on open, unvegetated surfaces than in areas with tall vegetation cover. If the flight layout can be optimised for GPS (with at least six satellites in view) then precisions of 7–8 cm are theoretically achievable.

Katzenbeisser (2003) has outlined a number of other issues with a primary concern being the fact that users are frequently confronted with artefacts, mis-match of flight strips and distortions in the rendering of data. These can arise for a number of reasons but invariably involve the measurements and calculations of three basic elements: (i) sensor position, (ii) distance to reflecting object and (iii) viewing direction to reflecting object. Sensor position is calculated by post-processing DGPS on-the-fly algorithms (Katzenbeisser, 2003) and requires acquisition at 1 Hz dual frequency and a stable satellite constellation (PDOP <2.5) with no signal disruption. The ground-based GPS reference station should ideally be located within the survey area but be no more than 25 km away. As distance cannot be measured directly, the time from emitting a laser pulse until an echo is received (termed 'time of flight') is measured and converted into distance, typically by counting the number of cycles of an oscillator operating at a certain frequency. According to Katzenbeisser (2003), any offset will cause a shift of the elevation model with a slight distortion and widening of the swath width. Should the offset be negative, the elevation model will be heightened, distortion will dip at the margins and the swath width will narrow. Hetherington (Chapter 6, this volume) also discusses these reliability issues. In terms of beam direction, all LiDAR systems use a GPS for navigation along with an inertial measurement unit (see Figure 1.1). Using the continuous DGPS position and the vectors of movement, direction and speed, a very precise beam attitude can be calculated. Finally, it should be noted that the individual measurements necessary for a precise final result (position, sensor attitude, and distance and beam deflection) are all taken by different parts of the LiDAR sensor system at different times. Therefore, it is necessary to know precisely the time at which each measurement was taken, or the time difference between measurements, to allow correction. Ultimately, these corrections as well as other outlined here need to be applied at an early stage of the data processing (Katzenbeisser, 2003).

Other issues persist. Airborne LiDAR has problems accurately delineating stream channels and shorelines normally visible on the ground or in photographic images. As an example, in gravel-bed rivers deeper sections of the channel show up

Table 1.1 Dissemination of research on airborne laser altimetry 1999–2004 (percentages in brackets).

Topic area	Scientific journal	Conference proceedings	Website/ e-newsletter	Unpublished PhD	Book chapter
Overview papers	4 (44%)	3 (34%)	1 (11%)	1 (11%)	
Filtering, flight adjustment: algorithms and methods	2 (6%)	29 (88%)	2 (6%)	2 (10%)	1 (5%)
DEM generation: terrain and fluvial applications	11 (55%)	6 (30%)			
Calibration, error assessment and quality control	2 (12%)	11 (69%)	3 (19%)		
Commercial LiDAR	1 (33%)		2 (67%)		
Forestry applications	11 (69%)	3 (19%)	2 (12%)		
Urban applications	7 (29%)	11 (46%)	6 (25%)		
Other applications: glaciated landscapes, earthquake hazards, fire, flood modelling, vegetation structure, bird population models, coasts, roads	13 (76%)	3 (18%)		1 (6%)	
Intensity	1 (20%)	2 (40%)	2 (40%)		
Totals	52 (36%)	68 (48%)	18 (13%)	4 (3%)	1 (<1%)

Source: Values adapted from Allen (2004).

as blanks in the dataset (Charlton *et al.*, 2003). In addition, as manual processing is required, contours derived from LiDAR are not normally-hydro-corrected to ensure down-contour flow of water in the digital elevation data – a bane of earlier efforts to model flood scenarios in the UK and elsewhere. For such applications, unedited LiDAR-derived contours may therefore be unacceptable, requiring the incorporation of manual break lines along linear features such as river channels (adding to project costs). Generation of airborne LiDAR-derived surfaces also remains problematic in highly sloping terrain.

While a significant amount of material has been published in relation to airborne laser scanning, a survey of the literature between 1999 and 2004 illustrates a dissemination issue (Table 1.1). Of 143 publications surveyed, only 36% were in (a rather narrow selection of) peer-reviewed international scientific journals, almost half (48%) were in conference proceedings and a further 13% published *via* the Internet. The publications in scientific journals focused more on applications of the technology, whereas those on methodology issues and specifications tended to appear as less widely-circulated conference proceedings.

LiDAR has been rapidly taken up in the natural sciences because digital elevation data created using the technology are less expensive than those created from traditional surveying methods. In addition, the increasing application of laser surveying systems and the ease of data capture that they offer have enabled non-specialist operators from outside traditional surveying disciplines to efficiently generate detailed information in ever-more challenging and complex environments. The result is that the technique is now widely recognised as a leading technology for the extraction of information of physical surfaces (Filin, 2004). At the same time, a new range of issues are arising. Of paramount importance is the question of how data is gathered, collated, processed and managed. This is particularly important in the more recently developed technologies of terrestrial laser scanning (TLS) and the allied field of High Definition Survey (HDS).

TERRESTRIAL LASER SCANNING

Since the development of the first terrestrial laser scanner in 1999 (Bryan, 2006), laser scanning technology has seen a continued phase of product development, growth and expansion into many areas of survey (e.g. Bellian *et al.*, 2005). As Lim *et al.* (Chapter 15, this volume) state, the development of sensors able to rapidly collect 3D surface information, has enabled high-density measurements to be made across landscapes that are unsuited to more conventional approaches due to their inaccessibility, hazardous nature or spatial extent. In addition, the increasing application of TLS systems and the ease of data capture they offer have enabled non-specialist operators from outside traditional surveying disciplines to efficiently generate detailed information in evermore challenging and complex environments. This gives rise to a problem: of all the survey techniques available, TLS has the least standardised control practices and error assessments (Lichti *et al.*, 2005). This is due to both the relative infancy of TLS as a survey tool, the ease of its operation and the apparently complete and satisfactory outputs it provides.

A major advantage is in the rate of data acquisition. Using *Leica* machines as an example, early pulse-scanners laid emphasis on long range, high precision machinery while later generations (from 2004) concentrated on speed of data acquisition and shorter ranges (Table 1.2). From machines in 1998, which collected 100 points per second, the newest generation machines collect approximately 500 times that amount. The ability to accurately position objects so rapidly involves production of a large amount of data. This data, commonly referred to as a 'point cloud', can provide a 3D shape or visualisation of the feature being measured. It should not be forgotten that the 3D terrestrial laser scanner is still providing 3D positional information in a similar way to a total station. The outcome is that the user can either collect denser amounts of data points or significantly reduce survey time (or achieve a combination of both). The maximum range achievable with a laser rangefinder depends strongly on the meteorological visibility; at lower visibility, the maximum range is reduced due to atmospheric attenuation.

The traditional scanning methodology is to use measurements to a number of common targets (reflectors). This allows multiple scans to be related to each other, a process known as 'meshing', or to be related to an existing control network. In essence, the scanner is placed at one location about the survey site and measurements taken to a number of targets as well as to the actual feature of interest (see case study below). The scanner is then moved to a second location and the process repeated, using at least three common targets from the first scanner location.

Table 1.2 A sample of available laser scanning machinery showing evolution of terrestrial laser scanning technology since 1998.

Date	Machine	Emphasis	Data collection rate
Pulse scanning			
1998	Leica Cyrax 2400	Long range, high precision	100 points sec^{-1}
2001	Leica Cyrax 2500	Long range, high precision	1000 points sec^{-1}
2005	Trimble GX	Short range	5000 points sec^{-1}
2004	Leica HDS3000	Long range, high precision	2000 points sec^{-1}
2006	Leica ScanStation	Long range, high precision	4000 points sec^{-1}
2006	Optech ICRIS-3D	Long range (up to 1500 m)	2500 points sec^{-1}
2006	Optech CMS	Short range	100,000 points per survey max.
2007	Leica ScanStation 2	Long range, high precision	50,000 points sec^{-1}
2007	Riegl LMSZ390i	Short range	Up to 11,000 points sec^{-1}
2007	Riegl LMSZ420i	Long-range	11,000 points sec^{-1}
2007	Riegl LMSZ210ii	Short-range	Up to 10,000 points sec^{-1}
Phase scanning			
2004	Leica HDS4500	Short range, high speed	250,000 sec^{-1}
2006	Leica HDS6000	Short range, high speed	500,000 points sec^{-1}

The process continues until full coverage of the site is achieved and any 'shadow areas' fully scanned. Newer machines (e.g. Optech ICRIS-3D) do not require reflector targets to be located in the area being surveyed, but 'memorise' common features to allow meshing of scans to take place. This is especially advantageous in areas where there may be difficulty in access, for example, opposite banks of rivers in spate.

TLS survey data remain subject to many issues that will generate inaccurate, misleading or inappropriate information if not considered. As Lim *et al.* (Chapter 15, this volume) state, error in TLS measurement is spatially variable, given the variation in survey range, spot size and incidence angle onto the target surface. The combination of separate scans, either spatially or through time, has the potential to introduce inconsistencies in the orientation, resolution and positioning of individual surveys. Clear attention has to be given to planning, data collection and processing to minimise these disadvantages. In addition, most currently available laser scanners are not well specified regarding accuracy, resolution and performance (Hetherington: Chapter 6, this volume) and only a minority are checked by independent institutes regarding their performance and whether they actually comply with manufacturer specifications (Boehler *et al.*, 2003).

CASE STUDIES REFLECTING EVOLUTION AND AVAILABILITY OF LASER SURVEYING TECHNOLOGY

A series of case studies on river systems in the north of England provide a useful insight into the evolution of laser scanning technologies and the attendant change in the way landscape models can be rendered and interpreted. The period of research, 1997–2007, coincided with the development of LiDAR and terrestrial laser scanning, and application of these techniques is described here. As such, the case studies described below provide useful context to research on other fluvial systems, sandy beaches, cliffs, archaeological and heritage sites, forest systems, engineered features and geological sites described elsewhere in this book. The section below describes how different scanning and surveying technologies have been used for modelling system function and sediment transport over different scale resolutions on two river systems, the River Coquet and the River South Tyne, both of which flow through Northumberland, northeast England.

The River Coquet, Northumberland, UK: EDM and LiDAR surveys

The River Coquet rises in the Cheviot Hills (776 m) in Northumberland, northern England. Here, the focus of geomorphological research over a ten year period has been a river system characterised by a high degree of lateral instability and channel avulsion (Fuller *et al.*, 2003) due to its position in a piedmont setting at the upland fringe in the catchment, draining an area of *circa* 255 km^2 (Figure 1.2). The valley as a whole is similar to other gravel-bed rivers in northern England, and displays a characteristic 'hourglass' valley morphology with alternating confined and unconfined sections.

Research in the period 1998–2000 was aimed at accurately quantifying annual sediment transfers across a 1 km long reach of the River Coquet at Holystone, Northumberland. As this was prior to widespread use of LiDAR, channel planform and cross-profile surveys were based on theodolite-EDM survey across the active wandering gravel-bed river system, focusing on breaks of slope. This exercise generated a series of *x-y-z* coordinates from which Digital Elevation Models (DEMs) of the reach were constructed – 'detailed' was a term used in resultant publications (e.g. Fuller *et al.*, 2003) to describe the survey resolution based on the technology available at the time (Figure 1.3). Calculating the difference between DEM surfaces provided a measure of volumetric change between surveys. Error analysis, comparing the surveyed cross-profiles with sections abstracted from the DEMs, indicated a mean gross error between surveyed and DEM profiles of around twice the value of the D_{50} of the surface sediment in the reach (51 mm: Fuller *et al.*, 2002, 2005).

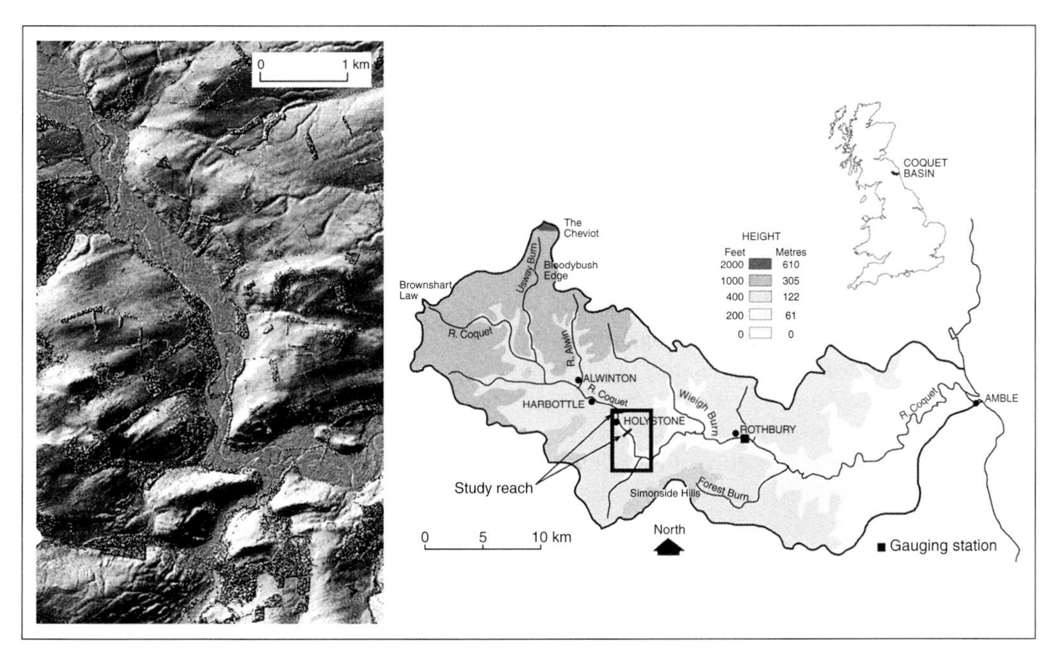

Fig. 1.2 The River Coquet catchment, Northumberland, UK with inset detail showing the valley configuration as depicted by airborne LiDAR flown in March 2006 by the UK Natural Environment Research Council.

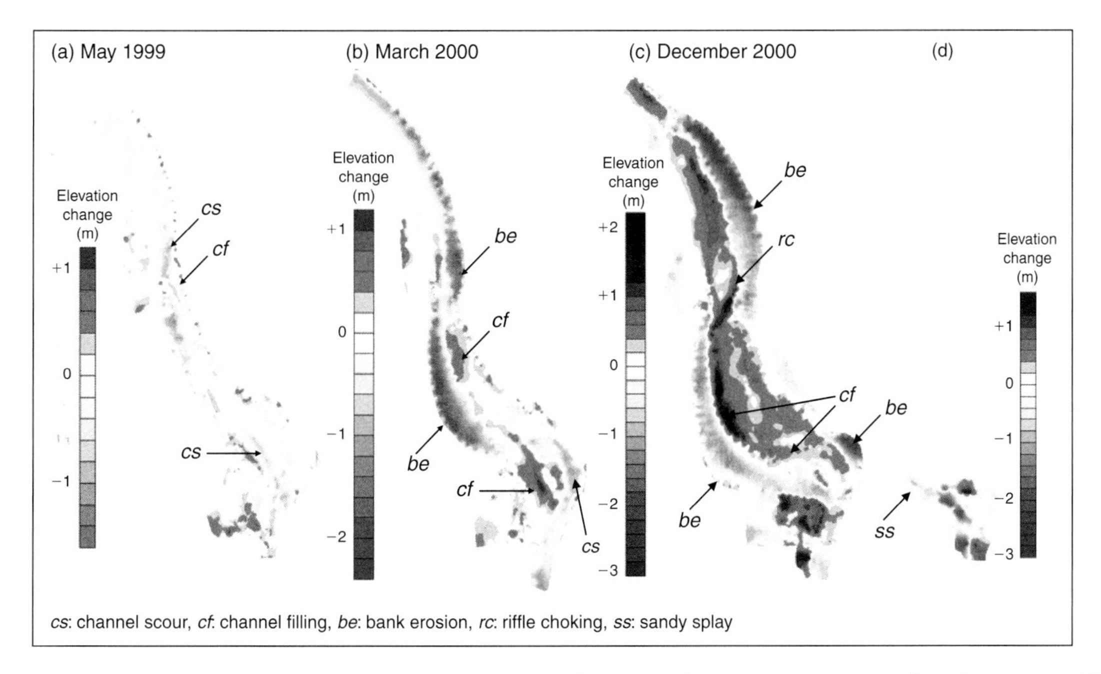

Fig. 1.3 Theodolite repeat surveys of the River Coquet, Northumberland, UK at ~0.05 points m^{-2} resolution. Inset (d) shows a re-survey 7 days after a flood event caused avulsion at the lower end of the reach (after Fuller *et al.*, 2003). See Plate 1.3 for a colour version of this image.

A Sokkia Set 5F Total Station (with a precision of ±5 mm) was used to survey channel planform and monumented cross-profiles in March 1999 and 2000. Sediment budgets were calculated using the morphological budget described in detail by Fuller *et al.* (2002). This approach integrates both planform and cross-profile data, with vertical changes in area along each cross-profile calculated and standardised to net gain/loss values per square metre. These values were then multiplied by the corresponding planform area values to give a net gain/loss value in cubic metres for each morphological unit within the sub-reach. In 1999, 2661 points were surveyed within the active channel area (50,509 m^2) at Sharperton Northumberland, producing a mean point density in the reach of 0.05 points m^{-2}. In 2000, 2985 points were surveyed within the same area (Figure 1.3), producing a mean point density of 0.06 points m^{-2} (Table 1.3). However, these low reach-scale averages mask much higher sampling densities in areas of rapid topographic change (e.g. bar edges, riffles, banks).

Subsequent surveys used DGPS technology to produced georeferenced DEMs allowing greater resolution than previously available using theodolite-EDM. The construction of a DEM for each survey was undertaken within Surfer™ GIS, with data interpolation using kriging at a grid interval of 0.25 m. While mathematical interpolation of elevation for DEM construction assumes that the data are spatially dependant, the data from long linear systems (such as river channels) are usually anisotropic. As such, this grid interval was necessary to ensure a DEM sensitive to more subtle changes in morphology; coarser grid intervals may lead to the loss of potentially important breaks in slope within the channel environment (Brasington *et al.*, 2000; Brunsden: Chapter 5, this volume). Kriging, being a geostatistical gridding method, which can use irregularly spaced data (Dixon *et al.*, 1998), was ideally suited to the morphologically-driven but irregularly-spaced datasets collected for the River Coquet. The accuracy of the DEMs produced using theodolite EDM was quantified using residual analysis, which indicated that more than 96.3% of the interpolated surface is accurate to ±5 cm (equivalent to the surface sediment *D50*) for both the 1999 and 2000 DEM surfaces. DEM methodology based on the morphological unit scale was seen, therefore, to provide a rigorous identification of spatial patterns of erosion and deposition.

Airborne LiDAR survey

In the late 1990s the UK Environment Agency commissioned LiDAR surveys of a number of river and coastal environments as part of a flood prediction exercise. The River Coquet was flown in March 1998. As such, the exercise proved timely and permitted comparison with the theodolite-EDM and DGPS surveying in progress during that period (Charlton *et al.*, 2003). Figure 1.4 depicts a mosaic of two adjacent 1 km^2 tiles for the River Coquet at Sharperton, Northumberland, viewed as an illuminated surface from the south-west. No vertical exaggeration has been applied. As well as showing the undulation of the valley sides, and some faintly visible palaeochannels on the floodplain, vegetation cover is clearly visible. In the study area, white patches indicate where

Table 1.3 Comparison of laser surveying methods used on rivers in Northumberland during the period 1999–2008 and their respective resolutions of survey.

Survey date	Method	Point resolution
1999 (50,509 m^2)	Theodolite EDM (Sokkia Set 5F™ Total Station)	2661 points (0.05 points/m^{-2})
2000	Theodolite EDM (Sokkia Set 5F™ Total Station)	2985 points (0.06 points/m^{-2})
2002–2004	DGPS (Scorpio and Thales Promark 3)	15,000 points (0.30 points m^{-2})
2002 and 2006	Airborne LiDAR (NERC-ARSF)	~64,000 points km^2 (3.9 points m^{-1})
2005–2008	Terrestrial laser scanning (Riegl Instruments LMS-210™ scanning laser)	Up to 26,000,000 points per survey (~515 points m^{-2})

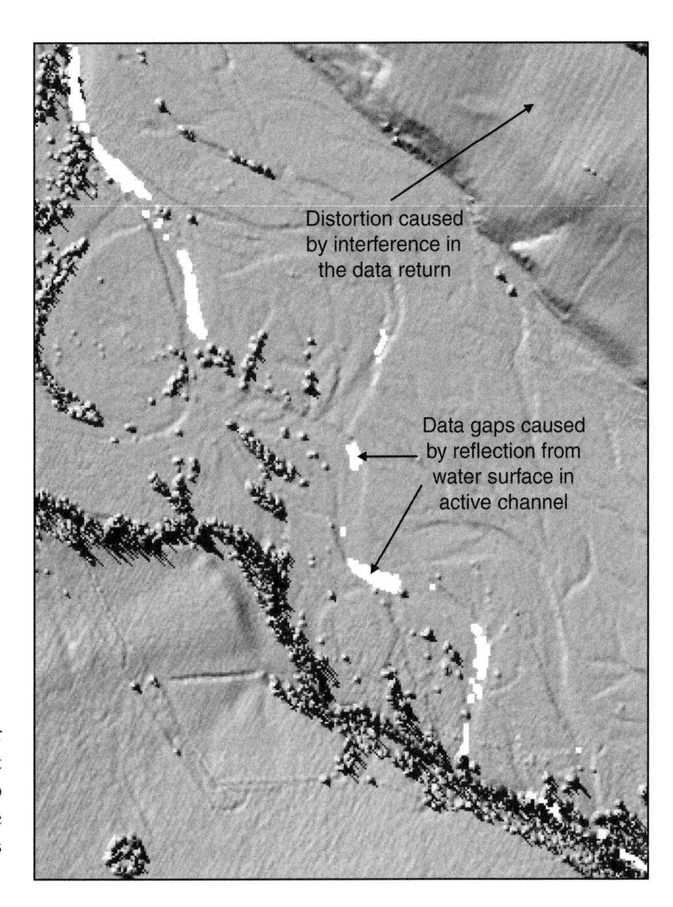

Fig. 1.4 1998 LiDAR survey of the River Coquet valley, flown for the UK Environment Agency for flood mapping purposes. The two juxtaposed 1 km × 1 km LiDAR tiles equate to UK Ordnance Survey national grid squares NT 9502 and NT 9602.

the LiDAR sensor has failed to make a reading. It should be noted that these data gaps are all in the river channel and represent lack of data from the water surface itself (the LiDAR data was first-return only, although last return data also encounters this problem, as still water acts as a specular surface – see Chapter 2). Coincident with the LiDAR flight, channel cross-profiles were surveyed from monumented pegs on the riverbank using theodolite-EDM survey with a Sokkia Set 5F Total Station. This ground survey took one week around the date of the LiDAR acquisition flight, thus ensuring minimal discrepancy between LiDAR and ground survey cross-profiles due to morphological change over time. The ground survey measurements were made at every break of slope across the channel. The locations of the ground survey measurements were georeferenced by comparison with OS LandLine data. Six cross-sections were selected for detailed comparison on the basis that they covered the full range of channel morphology within the reach. Georeferenced height values were calculated from the LiDAR data at 0.25 m intervals along the six cross-sections using the surface profiling facilities available in Arc/INFO GIS. In total, this supplied 9152 estimates of elevation for the cross-sections. By comparison, there were only 551 measurements of elevation available from the ground survey; there is a drawback in that measurements derived from the LiDAR surface were not necessarily taken at precisely the same positions along the cross-profile as those from the ground survey. To resolve this, the surface profiling facilities in Arc/INFO were used to generate interpolated values from the digital elevation model at a set of regularly spaced locations, within the mesh spacing of the model. Cubic splines (Press *et al.*, 1989) were fitted to these measurements (splines have the useful property that they pass through all the observed data points), and values were then interpolated from the spline function at positions corresponding to those along each cross-section at which ground survey elevations were obtained (Charlton *et al.*, 2003). Elevations from the ground survey data were adjusted to the LiDAR elevation at the monumented peg for each cross-section, so that the first data point in all cross-sections had the same *z*-value as the LiDAR elevation for that specific location (thus removing any systematic bias component of the error associated with both the LiDAR measurement and the survey of the monumented peg). Two features were apparent from this exercise:

- Sections of the banks colonised by stands of mature trees were evident on the cross-section LiDAR plots as 'spikes' in the elevation data.
- The 1998 LiDAR data traded off the frequency of measurement by the detector against pulse separation, and the nature of the single return measurement meant deeper water was poorly represented.

From the above exercise, it could be concluded that, in relatively vegetation-free environments, use of single return LiDAR offers a rapid method of acquiring high resolution data. In reality, however, while this survey method operates in three-dimensions, it is often only '2.5 dimensional' in nature due to shadowing, data loss in densely vegetated areas and absence of bathymetric resolution. While not fully addressing the bathymetric issues, terrestrial laser scanning offers the potential of bridging the gap between the need for survey over extended areas and the need for suitable ecological resolution at the micro-scale (incorporating edges and boundary zones in ecosystems). To improve ecological status, there is a need to advance the understanding of geomorphological, hydrological and ecological functional links in ecosystems. Key international legislation now mandates this: the EU Water Framework Directive, for example, has the concept of 'ecological health' or status at its core. In river systems an emerging key question is whether hydromorphology can be characterised at a spatial scale that truly accounts for instream ecological dynamics. However, interactions are highly complex and remain poorly understood. Terrestrial laser scanning systems offer potential for accurately characterising ecosystems at ecologically-relevant scales, bridging scale issues concerning environmental protection legislation. Using the EU Water Framework Directive as our example again, policy is aimed at the catchment ('River Basin') scale with the reach being the unit of measurement for a range of indicative variables. However, biota respond on scale of the morphological unit down to the micro-scale (substrate, vegetation etc.) and therefore ground-based terrestrial laser scanning with its ability to measure surface elevations to sub-centimetre accuracy may offer real potential as a management tool.

Terrestrial laser scanning – an ecologically-applicable surveying technique?

For a section of the River South Tyne in Cumbria, surface roughness was measured using a random field of spatial elevation data collected using a Riegl LMSZ210 scanning laser. The aim was to reliably quantify instream hydraulic habitat defined by water surface characteristics using random field terrestrial laser scanner *x-y-z* data. A range of research has demonstrated that the mosaics of hydraulic habitat types present in a reach (and defined by roughness) are very important in determining biodiversity (e.g. Dodkins *et al.*, 2005). Over the last decade a number of international initiatives have focused on characterisation of instream habitat using hydraulic variables as these are deemed central to the inhabiting biota. Biotopes (Padmore, 1998; Newson & Newson, 2000) provide a standard, descriptive assessment of instream physical structure based on consistent recognition of features. They have their basis in the development of typologies to underpin the 'Habitat Quality Index' developed as a framework for the protection of rivers (Raven *et al.*, 1997), and provide a means of integrating ecological, geomorphological and water resource variables for management purposes. Essentially, the biotope concept allows

Table 1.4 Descriptions of flow types used to field map fluvial biotopes (Newson & Newson, 2000).

Flow type	Description	Associated biotope
Free fall	Water falls vertically and without obstruction from a distinct feature, generally more than 1 m high and often across the full channel width	Water fall
Chute	Fast, smooth boundary turbulent flow over boulders or bedrock. Flow is in contact with the substrate, and exhibits upstream convergence and downstream divergence	Spill – chute flow over areas of exposed bedrock Cascade – chute flow over individual boulders
Broken standing waves	White-water 'tumbling' waves with crest facing in an upstream direction. Associated with surging flow	Cascade – at the downstream side of the boulder flow diverges or 'breaks'. Rapid
Unbroken standing waves	Undular standing waves in which the crest faces upstream without breaking	Riffle
Rippled	Surface turbulence does not produce waves, but symmetrical ripples which move in a general downstream direction	Run
Upwelling	Secondary flow cells visible at the water surface by vertical 'boils' or circular horizontal eddies	Boil
Smooth boundary turbulent	Flow in which relative roughness is sufficiently low that very little surface turbulence occurs. Very small turbulent flow cells are visible, reflections are distorted and surface foam moves in a downstream direction. A stick placed vertically into the flow creates an upstream facing 'V'	Glide
Scarcely perceptible flow	Surface foam appears to be stationary and reflections are not distorted. A stick placed on the water's surface will remain still	Pool – occupy the full channel width. Marginal deadwater – do not occupy the full channel width

for a standard, descriptive assessment of instream physical structure based on consistent recognition of features over a range of spatial and temporal scales (Table 1.4). However, to become an alternative to hydraulic models, biotopes need a robust, empirical and practical channel typology/taxonomy to be developed to allow rapid characterisation of reaches.

Figure 1.5 outlines a potential framework for investigating climate, hydrology and their impacts on physical habitat in UK rivers, emphasising the importance of quantification of biotope types in delimiting characteristic reaches. It can be hypothesised that accurate biotope quantification *via* ground-based LiDAR survey can be key to spatial definition of ecological status and its scaling up towards the reach scale (still the most appropriate scale for monitoring and management of system biodiversity).

With this in mind, efforts were made to use terrestrial laser scanning to accurately quantify biotope distribution for a series of rivers in the north of England. Here, we can outline the approach for the River South Tyne, Cumbria, a river which has been described (Macklin & Lewin, 1986) as divided into five 'sedimentation zones', separated by more stable reaches. As mentioned above, a Riegl LMSZ210™ scanning laser was used to collect water surface roughness data from a bridge. The instrument works on the principle of 'time of flight' measurement using a pulsed eye-safe infrared laser source emitted in precisely defined angular directions controlled by a spinning mirror arrangement. A sensor records the time taken for light to be reflected from the incident surface. Angular measurements are recorded to a precision of 0.036° in the vertical and 0.018° in the horizontal. Range error

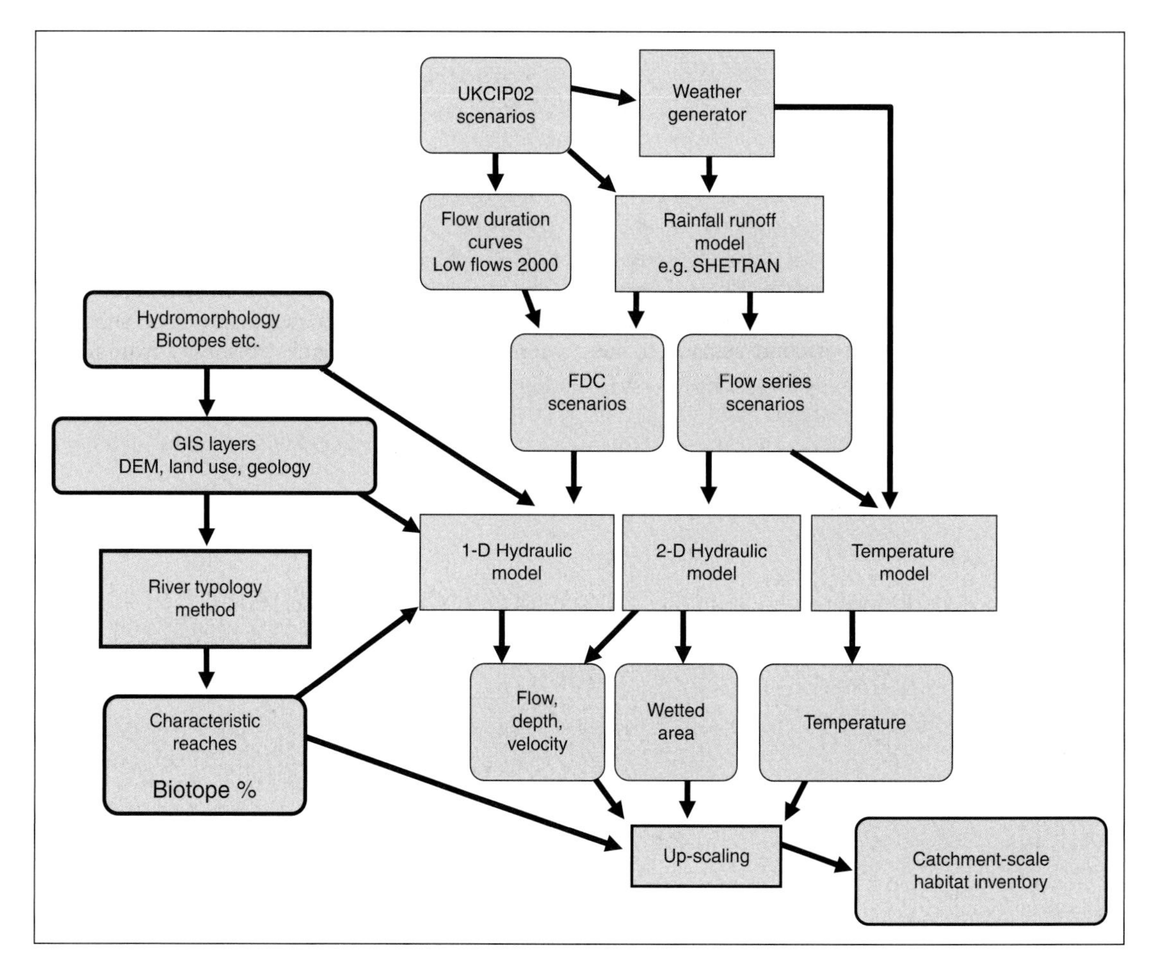

Fig. 1.5 A framework for investigating climate, hydrology and their impacts on physical habitat in UK rivers, emphasising the importance of quantification of biotope types (Chris Kilsby, personal communication).

is 0.025 m to a radial distance of 350 m. Survey control was facilitated by RiScan-Pro™ survey software, capable of visualising point cloud data in the field. Scans were generally restricted to 240° in front of the scanner and repeat scans were collected ensuring that any effects of water surface turbulence on biotope definition were minimised. This also increased the point resolution across the surface and reduced the possibility of unscanned areas due to the shadowing effect of roughness elements along the line of each scan. Before scans were taken, a total of 20 reflectors were placed on and around the study reach. These reflectors were tied into the project co-ordinate system using an EDM theodolite and these were automatically located by the RiScan-Pro™ software and matched to the project coordinates using a common point configuration algorithm. Estimation of laser scan accuracy and calculation of the threshold of accurate detection for the base model is addressed in Heritage and Hetherington (2005). The laser scan data were used to generate a 0.05 m resolution DEM of the point-bar surface. Delauney triangulation with linear interpolation was employed as an exact interpolator.

Characterisation of DEM uncertainty is difficult since validation requires comparison between the derived DEMs and a second, more accurate, surface (Brasington *et al.*, 2000, 2003). Usually acquisition of this second surface is not possible; DEM validation is thus often based on quantifying model uncertainty through diagnostic surface visualisations or field 'ground truthing' (Wechsler, 2000; Heritage & Hetherington, 2005). In this study the use of the 3D terrestrial laser scanning allows the second surface to be obtained up to a vertical accuracy of 2 mm. As the machine can operate over distances of several hundred metres while maintaining comparable levels of accuracy, it potentially offers a valuable tool for quantifying hydraulic habitat based on water surface characteristics. If the technology allows accurate characterisation of water surface roughness (a primary determinant of biotope classification in the field), it in turns offers a tool for such typologies to be derived at the reach scale – thus approaching the scale at which management policies for ecological status operate.

The surface data were initially transformed into a regular grid with a 0.02 m spacing to determine the local standard deviation of sub areas across the surface. Figure 1.6 shows, from top, (a) the data cloud produced from the laser scanning exercise and (b) biotope distribution as defined by visual observation – the method derived by Newson and Newson (2000) and used by the UK Environment Agency, the primary management

Fig. 1.6 Reach of the River South Tyne showing (a) data cluster cloud from laser scanning run, (b) visual classification of biotopes using the methodology of Newson and Newson (2000), and (c) 'edges' unclassified by the standard biotopes methodology. Rif = riffle, Run = run, Cas = cascade, Dw = Deadwater (Large & Heritage, 2007).

authority for river systems in England and Wales. The bottom image (Figure 1.6c) shows 'edges' unclassified by the standard biotopes methodology. These edges constitute what may be termed *critical channel components*, that is they are the areas used by biota for a range of activities including oviposition, providing refugia during spates, feeding sites, sites for emergence, and shelter under shade in vegetated systems. These parts of the instream mosaic of hydraulic units have inherent environmental value as part of the living space or habitat for instream biota, yet are missed under standard monitoring approaches in the UK (Newson & Newson, 2000) and certainly on the scale of the proposed rapid habitat survey methods for streams on mainland Europe (e.g. EAWM, 2004) which emphasise typing of entire reaches under a single biotope denotation.

Figure 1.7 shows the DEM derived from the laser scanning exercise with biotopes as defined by the factored standard deviation surface of the study reach. The results show overlap between the riffle and run habitat types. This is to be expected; these categories lie beside each other in the visual definition field (see Table 1.4) and, as stage rises and fall, biotopes merge and change

Fig. 1.7 Spatial distribution of biotope types for the reach shown in Figure 1.6 defined by laser scan water surface roughness measurement (Large & Heritage, 2007). See Plate 1.7 for colour version of this image.

from one type to another. One implication from the point of view of instream hydraulics may be that these biotopes are providing very similar habitat to each other (Heritage *et al.*, 2009; Milan *et al.*, 2009).

SUMMARY

This series of case studies over the decade 1997–2007 show how evolution of laser scanning technology and its application in morphologically similar systems (here gravel and cobble-bedded rivers) allow significantly higher resolution surveys to be taken with commensurate lowering of error. It also shows how the technology has potential value in feeding into major legislative approaches involving measuring and monitoring of these systems. Issues arise: the amount of data being generated is vastly increased – protocols are required for data archiving. Data mining and data redundancy are also an issue; the earlier theodolite-based surveying exercises focused attention on breaks of slope and more dynamic sections of reaches in order to gain higher resolution. Using LiDAR and TLS, data acquisition is as fine-scale over stable ecosystem components as those more prone to change. How to deal with this data 'overload' has not yet been adequately considered. Other chapters in this volume begin to address these issues.

THIS BOOK – LASER SCANNING FOR THE ENVIRONMENTAL SCIENCES

Both airborne and terrestrial laser scanning (LiDAR) are now well established methods for the acquisition of precise and reliable 3D geoinformation. Beyond the primary tasks of digital terrain model generation, airborne laser scanning has also proven to be a very suitable tool for general 3D modelling tasks and landscape analysis. At the same time, terrestrial laser scanning is successfully used to acquire highly detailed surface models of objects like building facades, statues and industrial installations. Despite large differences in resolution and accuracy between airborne and terrestrial laser scanner data, the research problems such as automatic registration, feature extraction and 3D modelling are very similar. As Lim *et al.* (Chapter 15) point out, as TLS systems mature from state-of-the-art instruments to become standard practice for recording complex topography, a change of emphasis is required; moving from using the system simply as a method of data capture to its use within a systematic surveying framework with well-established data collection and analytical protocols. This will be vital if terrestrial laser scanning is to achieve its undeniable potential in a wide range of environmental systems, each posing their own particular challenges to the accurate collection and interpretation of information.

This volume is divided into three main sections. The first – *new directions in data acquisition* – places particular emphasis on laser technology and reviewing the range of sensors and platforms available. Heritage and Large (Chapter 2) examine general theory and principles of laser scanning; scanning technologies and their advances as related to conventional approaches. Charlton *et al.* (Chapter 3) describe issues of data models and their representation, general data handling, accuracy, protocols, 2D and 3D representations (rasters, vectors, DEMs, TINs, etc.) and computational and visualisation issues. In the second section – *Land surface monitoring and modelling* – Devereaux and Amable (Chapter 4) describe data acquisition, instrumentation, deployment and survey design, as well as data processing while, in Chapter 5, Brunsden discusses principles and procedures of interpolating spatial variation; inverse distance weighting; spline models, kriging, covariates and general spatial modelling. Finally, Hetherington (Chapter 6) discusses data handling issues; there are many issues concerning the new challenges of database integration, dealing with extremely large datasets and grid integration, and new survey protocols are undoubtedly required.

The final section (Chapters 7–16) examines the application of the new technology in a range of environmental systems and for a range of purposes including flood and flow modelling; vegetation cover and landscape change modelling and archaeological surveys. Issues are addressed here with regard to managing data in terms of scale issues, dissemination, databases, software, and methods for DEM construction (2D and 3D triangulation) described for a range of systems. In Chapter 7, Entwistle and Fuller describe a method to reliably quantify the population grain-size distribution of natural gravel surfaces using random field TLS data and generate maps from which large-scale facies assemblages are identifiable. Straub *et al.* (Chapter 8) describe approaches for automatic feature extraction for segmentation and classification of forest areas, road extraction and building extraction. Using spatially distributed datasets, Milan and Heritage (Chapter 9) examine the effect of changing the resolution of topography and height grids on velocity and turbulence flow predictions in a three-dimensional hydraulic model.

The advent of airborne LiDAR makes it possible to quantify three-dimensional change in beach topography at the spatial scales needed to monitor erosion over long segments of coastline quickly, accurately and economically. Starek *et al.* (Chapter 10) present a detailed study of the application of airborne LiDAR to quantify beach dynamics. Using case study examples, Hodgetts (Chapter 11) discusses the LiDAR workflow in geology, covering collection of data, processing, geological interpretation and visualisation. In Chapter 12, Crutchley describes the development and use of LiDAR in archaeological survey, as well as examining practical issues *via* a series of case study examples. Danson *et al.* (Chapter 13) consider the nature of the interactions of laser light with vegetation canopies, emphasising the importance of multiple scatters within a single laser beam. The authors also illustrated this with reference to a range of airborne and ground based laser scanning experiments. In Chapter 14, Overton *et al.* describe the use of LiDAR for flood modelling in large river systems, using the example of the River Murray in Australia. Even in such a large system, small elevation differences create a multitude of different flooding regime habitats, and flows are often driven by water level differences of only a few centimetres. Airborne LiDAR is the only technique able to capture such levels of detail over such large distances and, as such, is ideal for these types of environmental application. Lim *et al.* (Chapter 15) complete the section, emphasising field survey procedures and data protocols using linear features (transportation links and coastlines) as exemplars. Finally, in Chapter 16, we ask: what of the future for laser scanning survey? What should be the priorities as the technologies evolve in terms of data volume and archiving, extraction and handling, modelling and rendering; and what potential does evolution of the technology offer for monitoring and mapping natural environments *via* high definition survey?

REFERENCES

Allen D. 2004. Airborne Laser Altimetry: Annotated Bibliography. http://calm.geo.berkeley.edu/ncalm/pdf/bibliodraft13.pdf. Last updated 29/4/2004.

Ancellet G, Ravetta F. 2003. On the usefulness of an airborne LiDAR for O3 analysis in the free troposphere and the planetary boundary layer. *Journal of Environmental Monitoring* **5**: 47–56.

Baltsavias EP. 1999. Airborne Laser Scanning: basic relations and formulas. *ISPRS Journal of Photogrammetry and Remote Sensing* **54**: 199–214.

Bellian JA, Kerans C, Jennette DC. 2005. Digital outcrop models Applications of terrestrial scanning LiDAR technology in stratigraphic modelling. *Journal of Sedimentary Research* **75**: 166–176.

Bissonnette LR, Kunz G, WeissWrana K. 1997. Comparison of LiDAR and transmissometer measurements. *Optical Engineering* **36**: 131–138.

Boehler W, Bordas-Vincent M, Marbs A. 2004. Investigating laser scanner accuracy. *Proceedings of the 19th CIPA Symposium*, 30th September–4th October 2003, Antalya, Turkey.

Brasington J, Rumsby BT, McVey RA. 2000. Monitoring and modelling morphological change in a braided gravel-bed river using high-resolution GPS-based survey. *Earth Surface Processes and Landforms* **25**: 973–990.

Brasington J, Langham J, Rumsby B. 2003. Methodological sensitivity of morphometric estimates of coarse fluvial sediment transport. *Geomorphology* **53**: 299–316.

Brinkman RF, O'Neill C. 2000. LiDAR and photogrammetric mapping. *The Military Engineer*, May–June 2000: 4p.

Brügelmann R, Bollweg AE. 2004. Laser altimetry for river management. In *International Archives of Photogrammetry and Remote Sensing. 20th ISPRS Congress*, 12–23rd July 2004, Istanbul, Turkey.

Chappell A, Heritage GL, Fuller IC, Large ARG, Milan DJ. 2003. Geostatistical analysis of ground-survey elevation data to elucidate spatial and temporal river channel change. *Earth Surface Processes and Landforms* **28**: 349–370.

Charlton ME, Large ARG, Fuller IC. 2003. Application of airborne LiDAR in river environments: the River Coquet, Northumberland UK. *Earth Surface Processes and Landforms* **28**: 299–306.

Dixon LFJ, Barker R, Bray M, Farres P, Hooke J, Inkpen R, Merel A, Payne D, Shelford A. 1998. Analytical photogrammetry for geomorphological research. In *Landform Monitoring, Modelling and Analysis,* Lane SN, Richards KS, Chandler JH (eds), Chichester: John Wiley & Sons, Chichester, 63–94.

Dodkins I, Rippey B, Harrington TJ, Bradley C, Ni Chathain B, Kelly-Quinn M, McGarrigle M, Hodge S, Trigg D. 2005. Developing an optimal river typology for biological elements within the Water Framework Directive. *Water Research* **39**: 3479–3486.

EAMN (European Aquatic Monitoring Network). 2004. *State-of-the-art in data sampling, modelling analysis and applications of river habitat modelling.* Harby A, Martin-Baptist M, Michael J, Dunbar, MJ, Stefan Schmutz S (eds). COST Action **626** Report.

Filin S. 2004. Surface classification from airborne laser scanning data. *Computers and Geosciences* **30**: 1033–1041.

Fuller IC, Large ARG, Milan DJ. 2003. Quantifying channel development and sediment transfer following chute cutoff in a wandering ravel-bed river. *Geomorphology* **54**: 307–323.

Gauldie RW, Sharma SK, Helsley CE. 1996. LiDAR applications to fisheries monitoring problems. *Canadian Journal of Fisheries and Aquatic Sciences* **53**: 1459–1467.

Geist T, Lutz E, Stötter J, Jackson M, Elvehøy H. 2003. Airborne laser scanning technology as a tool for the quantification of glacier volume changes – results from Engabreen, Svatisen (northern Norway). *Geophysical Research Abstracts* **5**: 12684.

Godin-Beckmann S, Porteneuve J, Garnier A. 2003. Systematic DIAL LiDAR monitoring of the stratospheric ozone vertical distribution at Observatoire de Haute-Province (43.92°N, 5.71°E). *Journal of Environmental Monitoring* **5**: 57–67.

Heritage GL, Hetherington D. 2005. The use of high-resolution field laser scanning for mapping surface topography in fluvial systems. *International Association of Hydrological Scientists Red Book Publication.* IAHS Publication **291:** 278–284.

Heritage GL, Hetherington D, Milan DJ, Large ARG, Entwistle N. 2009. Utilisation of terrestrial laser scanning for hydraulic habital mapping. In *Proceedings of the 7th International Symposium on Ecohydraulic,* Concepción, Chile, January 12–16, 2009.

Hollaus M, Wagner W, Kraus, K. 2005. Airborne laser scanning and usefulness for hydrological models. *Advances in Geosciences* **5**: 57–63.

Innes JL, Koch B. 1998. Forest biodiversity and its assessment by remote sensing. *Global Ecology and Biogeography Letters* **7**: 397–419.

Irish JL, White TE. 1998. Coastal engineering applications of high-resolution LiDAR bathymetry. *Coastal Engineering* **35**: 47–71.

Jackson TJ, Ritchie JC, White J, Leschack L. 1988. Airborne laser profile data for measuring ephemeral gully erosion. *Photogrammetric Engineering and Remote Sensing* **54**: 1181–1185.

Katzenbeisser R. 2003. About the calibration of LiDAR sensors. Proceedings ISPRS workshop on 3-D reconstruction from airborne laser-scanner and InSAR data, 8–10 October 2003, Dresden, Germany.

Krabill W, Thomas R, Jasek K, Kuvinen K, Manizade, S. 1995. Greenland ice thickness changes measured by laser altimetry. *Geophysical Research Letters* **22**: 2341–2344.

Large ARG, Heritage GL. 2007. Terrestrial laser scanner based instream habitat quantification using a random field approach. *Proceedings of the 2007 Annual Conference of the Remote Sensing & Photogrammetry Society*, Mills J, Williams, M (eds), The Remote Sensing and Photogrammetry Society CD-ROM.

Lichti DD, Gordon SJ, Tipdecho T. 2005. Error models and propagation in directly georeferenced terrestrial laser scanner networks. *Journal of Survey Engineering* **131**: 135–142.

Macklin MG, Lewin J. 1986. Sediment transfer and transformation of an alluvial valley floor: the River South Tyne, Northumbria, UK. *Earth Surface Processes and Landforms* **14**: 233–246.

Marks K, Bates P. 2000. Integration of high-resolution topographic data with floodplain flow models. *Hydrological Processes* **14**: 2109–2122.

McHenry JR, Cooper CM, Ritchie JC. 1982. Sedimentation in Wolf Lake, Lower Yazoo river basin, Mississippi. *Journal of Freshwater Ecology* **1**: 547–558.

Milan DJ, Heritage GL, Entwistle N, Hetherington D. 2009. Relationship between turbulent flow structure and biotopes in n upland Trout steam. In *Proceedings of the 7th International Symposium on Ecohydraulic*, Concepción, Chile, January 12–16, 2009.

Newson MD, Newson CL. 2000. Geomorphology, ecology and river channel habitat: mesoscale approaches to basin-scale challenges. *Progress in Physical Geography* **24**: 195–217.

Padmore CL. 1998. The role of physical biotopes in determining the conservation status and flow requirements of British Rivers. *Aquatic Ecosystem Health and Management* **1**: 25–35.

Parson LE, Lillycrop WJ, Klein CJ, Ives RCP, Orlando, SP. 1997. Use of LiDAR technology for collecting shallow water bathymetry of Florida Bay. *Journal of Coastal Research* **13**: 1173–1180.

Pereira LG, Wicherson RJ. 1999. Suitability of laser data for deriving geographical information: a case study in the context of management of fluvial zones. *ISPRS Journal of Photogrammetry and Remote Sensing* **54**: 105–114.

Press WH, Flannery BP, Teukolsky SA, Vetterling WT. 1989. *Numerical Recipes in Pascal*. Cambridge: Cambridge University Press.

Price WF, Uren J. 1989. *Laser Surveying*. London: Van Nostrand Reinhold International.

Raven PJ, Boon PJ, Dawson FH, Ferguson AJD. 1998. Towards an integrated approach to classifying and evaluating rivers in the UK. *Aquatic Conservation: Marine and Freshwater Ecosystems* **12**: 439–455.

Ritchie JC. 1995. Laser altimeter measurements of landscape topography. *Remote Sensing of Environment* **53**: 91–96.

Ritchie W, Wood M, Wright R, Tait D. 1988. *Surveying and Mapping for Field Scientists*. New York: Longman Scientific & Technical.

Ritchie JC, Jackson TJ. 1989. Airborne laser measurements of the surface-topography of simulated concentrated flow gullies. *Transactions of the ASAE* **32**: 645–648.

Ritchie JC, Everitt JH, Escobar DE, Jackson TJ, Davis MR. 1992a. Airborne laser measurements of rangeland canopy cover and distribution. *Journal of Range Management* **45**: 189–193.

Ritchie JC, Jackson TJ, Everitt JH, Escobar DE, Murphey JB, Grissinger EH. 1992b. Airborne laser: a tool to study landscape surface-features. *Journal of Soil and Water Conservation* **47**: 104–107.

Ritchie JC, Grissinger EH, Murphey JB, Garbrecht JD. 1994. Measuring channel and gully cross-sections with an airborne laser altimeter. *Hydrological Processes* **8**: 237–243.

Ritchie JC, Humes KS, Weltz MA. 1995. Laser altimeter measurements at Walnut-Gulch watershed, Arizona. *Journal of Soil and Water Conservation* **50**: 440–442.

Smith-Voysey S. 2006. *Laser scanning (LiDAR); a tool for future data collection?* Ordnance Survey Research Labs Annual Review 2005–06. Southampton: Ordnance Survey.

Staley DM, Wasklewicz TA, Blaszczynski JS. 2006. Surficial patterns of debris flow deposition on alluvial fans in Death Valley, CA using airborne laser swath mapping data. *Geomorphology* **74**: 152–163.

Wadhams P. 1995. Arctic sea-ice extent and thickness. *Philosophical Transactions of the Royal Society of London Series A* **352**: 301–319.

Wealands SR, Grayson B, Walker JP. 2004. Quantitative comparison of spatial fields for hydrological model assessment – some promising approaches. *Advances in Water Resources* **28**: 15–32.

Wechsler SP. 2000. *Effect of DEM uncertainty on topographic parameters, DEM scale and terrain evaluation*. Unpublished PhD thesis, University of New York, Syracuse.

Wehr A, Lohr U. 1999 Airborne Laser Scanning – an introduction and overview. *Journal of Photogrammetry and Remote Sensing* **54**: 68–82.

2 Principles of 3D Laser Scanning

GEORGE L. HERITAGE[1] AND ANDREW R.G. LARGE[2]

[1]JBA Consulting, Greenall's Avenue, Warrington, UK
[2]School of Geography, Politics and Sociology, Newcastle University, Newcastle Upon Tyne, UK

INTRODUCTION AND BASIC PRINCIPLES

Laser scanning has started to revolutionize our ability to characterize surface properties, generating high quality data at scales and resolutions previously unachievable. As a consequence, many conventional approaches to measurement are being challenged and new insights are being gained across a wide range of discipline within the Environmental Sciences. The development of robust highly sensitive LiDAR systems now facilitate the rapid remote collection of 3D surface position and character data offering opportunities to research landform dynamics over spatial and temporal ranges not previously practicable, potentially leading to a revolution of our understanding of earth systems.

This chapter briefly outlines the basic physics of surface reflectivity and its importance in the development of airborne and terrestrial LiDAR (Light Detection And Ranging) systems. We describe the principles behind and development of laser (Light Amplification by the Stimulated Emission of Radiation) technology and its utilization in LiDAR instrumentation. The discussion relates principally to pulsed laser scanners although other instrument types are reviewed. Basic scanner components and function are described and fundamental equations are presented to demonstrate how 3D object position in space is accurately determined. Examples of scanner output are shown before we discuss the important potential errors associated with laser scanner output. We conclude with some brief thoughts on the future for laser scanners in environmental research.

PHYSICS OF SURFACE REFLECTION

Electromagnetic radiation (including light) exhibits the properties of both waves and particles. As such it may be characterized by a wavelength, frequency and intensity. As described in Equation (2.1), light is made up of photons which possess no mass and energy (E) proportional to the wavelength (λ):

$$E = \frac{hc}{\lambda} \qquad (2.1)$$

where h = Planck's constant, c = speed of light.

Figure 2.1 shows the standard classification of electromagnetic radiation based on wavelength.

Laser Scanning for the Environmental Sciences, 1st edition. Edited by G.L. Heritage and A.R.G. Large.

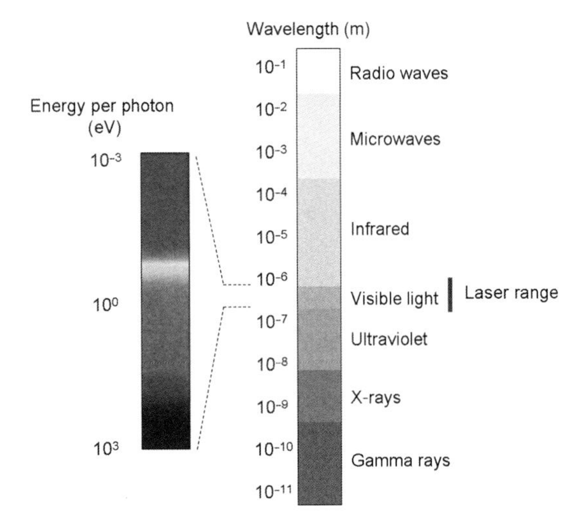

Fig. 2.1 The electromagnetic spectrum and laser wavelengths. See Plate 2.1 for a colour version of this image.

Visible light occurs between wavelengths of 400 nm and 700 nm, while the 'laser range' extends into both the infrared and ultraviolet parts of the spectrum.

Natural light thus consists of a variety of wavelengths. This light is emitted along random trajectories; when this strikes an object a proportion of the electromagnetic radiation reflects from the surface. The ratio of reflected radiation to the amount originally received at the surface is (logically) defined as *reflection*. A surface reflects (R_λ), transmits (T_λ) and/or absorbs (A_λ) received incident (I_λ) radiation energy (E) according to the following equation:

$$E_{I_\lambda} = E_{T_\lambda} + E_{A_\lambda} + E_{R_\lambda} \tag{2.2}$$

where λ = electromagnetic radiation wavelength.

A surface which reflects incident radiation in a single direction away from the source and where the angle of reflection is equal to the angle of incidence (Figure 2.2a), is defined as *specular*. Specular surfaces return no radiation whatsoever in the direction of the source. This type of reflection occurs when the surface roughness is smaller than the wavelength of the incident radiation. In natural environments, this type of reflection is found in water environments under specific conditions of still water and elsewhere where glassy surfaces are found (e.g. ice containing no sediment). A second type of reflection which is extremely common in natural environments is Lambertian, or diffuse, reflection (Figure 2.2b). This is a variation on specular reflection, where reflectivity is dependent on the angle of incidence and the arrangement of roughness on the illuminated surface. Diffuse reflection occurs where the surface roughness is greater than the wavelength of the incident radiation, causing light to be reflected in all directions. In reality most surfaces in natural environments exhibit a combination of reflection types (Figure 2.2c).

For most interfaces between materials, the fraction of the light that is reflected increases with increasing angle of incidence. This effect is particularly pronounced for specular surfaces (e.g. at sunrise and sunset when sunlight strikes a water body surface). It is also apparent from Equation (2.2) that reflectivity is wavelength dependant. Surfaces will absorb certain parts of the electromagnetic spectrum while at the same time reflecting at certain wavelengths (Figure 2.3). The result of this absorption and reflectance are the colours we see in the visible spectrum.

HISTORICAL DEVELOPMENT OF LASER SCANNER TYPES

Einstein (1917) re-derived the Planck law of radiation to set the theoretical framework for the development of laser instrumentation. Progress proceeded through micro-wave amplifiers (MASERS) to infrared lasers thanks to the theoretical calculations of Schawlaw and Townes, published towards the end of the 1950s (Schawlaw & Townes, 1958), and leading to the construction of the first instrument by Maiman (Mossman, 1960). From the earliest light pulse systems in the 1930s through systems based on ruby lasers (e.g. Smullins & Fiocco, 1962) developed in the 1960s, and up to the more recent

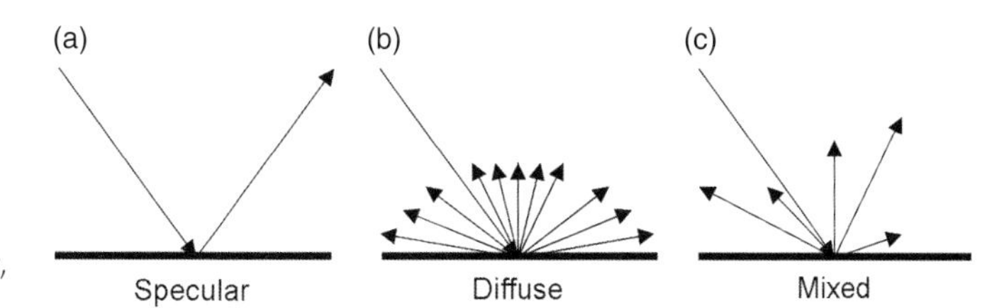

Fig. 2.2 Surface reflection types: (a) specular, (b) diffuse, (c) mixed.

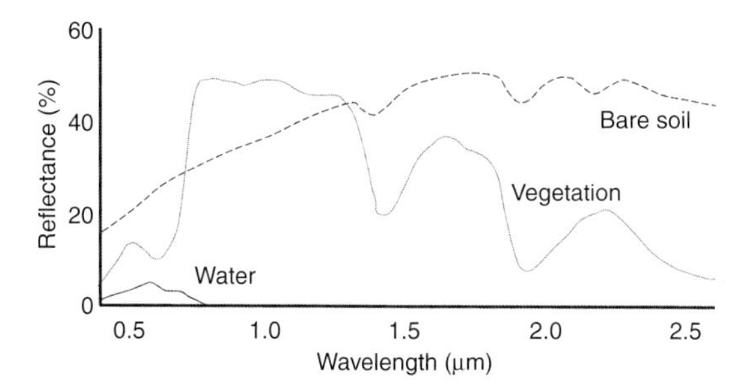

Fig. 2.3 Reflectance characteristics of bare soil, clear water and mixed vegetation.

development of semi-conductor lasers, the way has been paved for more portable and (importantly) eye-safe instruments. Research and instrument development has continued with different wavelength lasers (see Table 2.1), improved pulse width and spectral purity, better optical filters and ever increasing computational power.

BASIC LIDAR PRINCIPLES

LiDAR (Light Detection And Ranging) is a remote sensing technique that utilizes the properties of scattered light to determine certain characteristics of distant objects. Synge (1930) was the first to suggest that, by detecting the scatter of a beam of light into the atmosphere, its density could be determined. Synge used a transmitter (a searchlight) and a receiver (a telescope) separated over a distance of several kilometres, with a photoelectric apparatus to detect the light. As opposed to this, modern equipment is monostatic, that is the transmitter and receiver are sited at the same location. While this type of setup has been used since the 1930s, advances were made in the 1950s and 1960s. Friedland *et al.* (1956) were the first to recognize pulses from the monostatic system, and Smullins and Fiocco (1962) were the first to use laser-generated light in atmospheric research.

We are now observing the extension of the technology into more and more ecosystem types. While the initial use of the technology in the natural environment emphasized airborne remote-sensing applications in a wide range of environments (Large and Heritage, Chapter 1), more recent applications have been ground-based and suitable for distances up to 1500 m from the sensor. These modern instruments use laser light as an emission energy source detecting extremely minute energy returns from distant surfaces. LiDAR technology has been described as extremely versatile (Wehr & Lohr, 1999) and has been used in atmospheric (e.g. Davies *et al.*, 2004) Bathymetric (Guenther *et al.*, 2002) and morphometric studies (e.g. Charlton *et al.*, 2003; Nagihara *et al.*, 2004; Milan *et al.*, 2007). Scanner instruments are able to determine the 3D coordinates of complex object shapes with high precision and point density (Wehr & Lohr, 1999; Heritage & Hetherington, 2007). The production of a final 3D model from raw data is more efficient than conventional methods due to rapid relatively simple data collection and nominal post processing. However, research utilizing the technology remains at a relatively low level due to the cost of the instrument and associated software, a situation that is likely to change as new instruments emerge and production methods become cheaper. Typical entry costs are in the order of £70,000 for hardware and up to £30,000 for the associated data gathering, handling and processing software.

Table 2.1 Principal laser types and their uses.

Category	Type	LASER wavelength	Mode of operation	Applications
Semiconductor diode lasers	Single diodes	Infrared – visible	Continuous and pulsed modes	Optoelectronics
	Multiple diodes	Infrared – visible	Continuous and pulsed modes	Pumping light source for solid state lasers
Solid state lasers	Neodymium YAG laser	1.06 μm	Continuous and pulsed modes	Materials processing, medicine
	Rubin-Laser	Red	Pulsed mode	Pulse holography
Gas lasers	CO_2-Laser	10.6 μm	Continuous and pulsed modes	Materials processing, medicine, isotope separation
	Excimer laser	193 nm, 248 nm, 308 nm	10 ns–100 ns	Micro-machining, laser chemistry, medicine
	Helium Neon laser	632.8 nm	Continuous mode	Holography
	Argon ion laser	515–458 nm	Continuous and pulsed modes	Printing technology, pumping laser for dye laser stimulation, medicine
Dye lasers		Continuous between infrared and ultraviolet (different dyes)	Continuous and pulsed modes	Spectroscopy, medicine

All LiDAR instruments rely on the principles behind laser light (Large & Heritage, Chapter 1). Laser light is generated through the excitation of atoms causing them to emit energy as photons. A light source is used to stimulate the laser source, exciting electrons into a higher state. The new excited atomic structure is inherently unstable and the electron returns to its original state emitting energy as a photon of light. Where the emitted photon strikes an already excited atom, two photons of the same wavelength and direction are emitted. The emitted light radiation is reflected back into the laser source, further stimulating photon emissions, which are allowed through one end of the laser source as highly coherent low divergence Light Amplified by Stimulated Emission of Radiation (or to use its ubiquitous acronym, *LASER*). Lasers thus consist of a lasing material (crystal, gas, semi conductor etc.; see Table 2.1) a light or electrical current pump source and an optical cavity where photons are reflected back into the lasing material to amplify signal. Opaque electro-optical devices control pulse release, becoming transparent when a voltage is applied, allowing picosecond pulsing of emissions.

Whereas natural light is emitted along random trajectories, over many wavelengths and amplitudes, and with no phase correspondence, laser light is coherent, being emitted in a single direction as trillions of individual photons, in phase and with a well-defined energy frequency and wavelength (Figure 2.4). Knowing the emitted character of the laser light allows the characteristics of distant objects to be determined from changes to the return energy from object backscatter. The coherence characteristic of laser light ensures minimal beam divergence as distance increases, enabling relatively small individual objects to be identified especially compared to radar (Figure 2.5) where beam divergence may be three orders of magnitude or more greater.

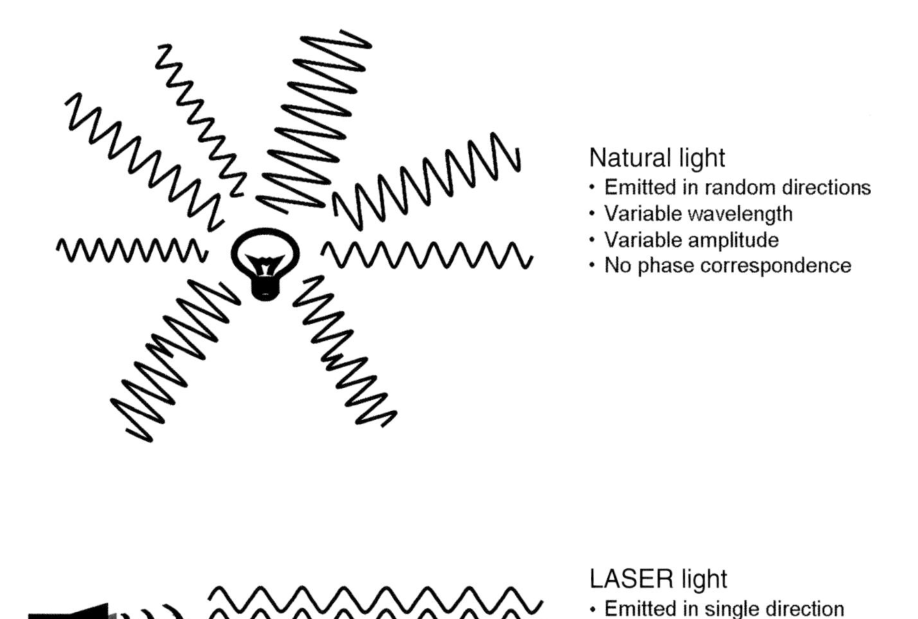

Fig. 2.4 Characteristics of natural light and Light Amplified by Stimulated Emission of Radiation (LASER).

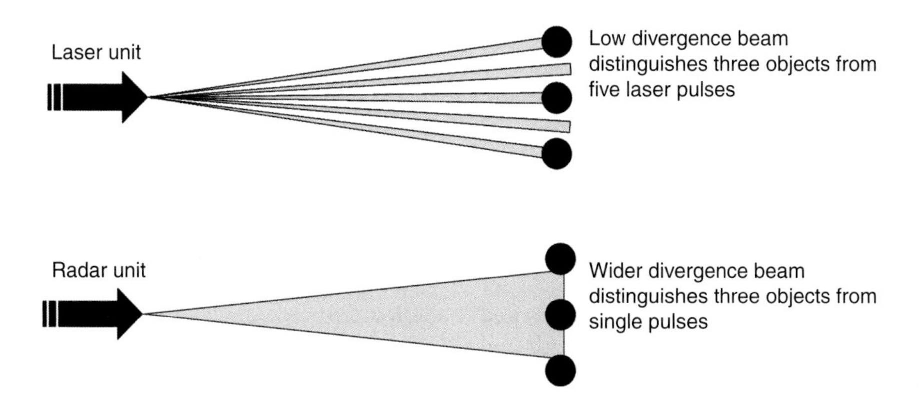

Fig. 2.5 Beam divergence characteristics and its influence on object distinction.

Laser scanners rapidly emit a controlled pulsed waveform of coherent light. The laser pulse is affected by the object or objects it strikes after emission, being absorbed and scattered in all directions. A proportion of the photons emitted may be reflected directly back to the laser instrument as an altered waveform (Figure 2.6). It is the character of this returned energy pulse that is used to determine certain properties of the reflecting object including colour, texture and distance or range away from the scanner. A pulsed laser has a low beam divergence, allows for short

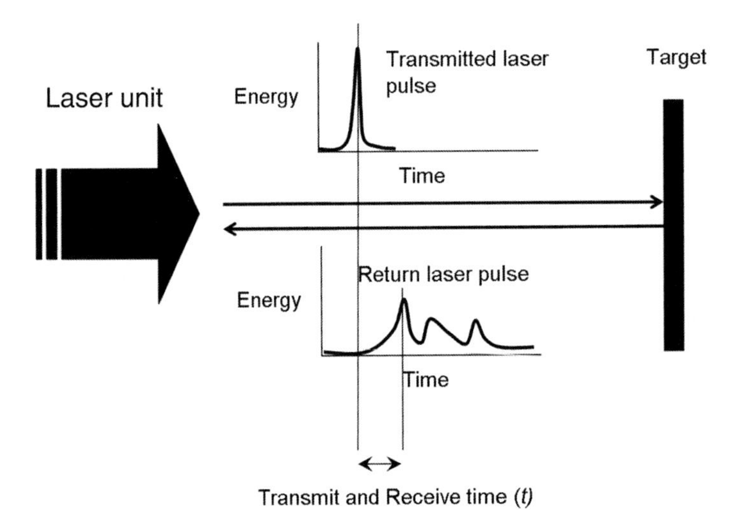

Fig. 2.6 The influence of remote surface character on laser pulse character and time of flight recording.

intense pulses and has an extremely narrow spectral width. The use of a simple telescope arrangement called a beam expander can increase the divergence of the laser depending on the use required. *Large-footprint* systems have a large divergence angle for the beam and can cover a target area of between 10 m and 30 m at the standard flying height (1000 m as used by the UK Natural Environment Research Council and others). Systems that have a smaller divergence angle and typically cover a target area of 6–10 m are termed *small-footprint* systems.

The ability of 3D laser scanners (such as those described in Figure 2.6) to accurately determine the range of distant objects relies on an understanding of the speed of light. Photons travel at a constant velocity of 299,792,458 m s^{-1}. If the time (t) taken for a pulse of laser light to travel to and be scattered back from a distant object is known, the distance (d) between the laser and the object may be determined from Equation (2.3).

$$d = ct/2 \tag{2.3}$$

where c = constant speed of light and t = pulse transmit and return time.

The situation is complicated by the form of the emitted and returned laser pulse and the ability of the scanner recording unit to capture and process the returned waveform information. This has effects on scanner accuracy and often restricts the storage of information from the returned waveform. We elaborate on these issues below.

BASIC PRINCIPLES OF OPERATION

LiDAR instruments consist of three basic components, the laser unit incorporating a transmitter, an opto-mechanical reflector mechanism and a receiver/recorder unit (Figure 2.7). In the case of airborne LiDAR (and with some mobile terrestrial scanners) additional position/navigational systems are necessary. Pulsed laser light emitted from the laser unit via the opto-mechanical reflector mechanism travels from the instrument to a distant object where some of the energy may be reflected back to the recorder unit, where the waveform is analysed to determine object character and location. Airborne LiDAR use wavelengths in the visible light spectrum, varying from infrared to ultraviolet. For vegetation studies, the pulse is usually in the near infrared part of the spectrum as this allows vegetation to be distinguished in distinct shades of red. Danson

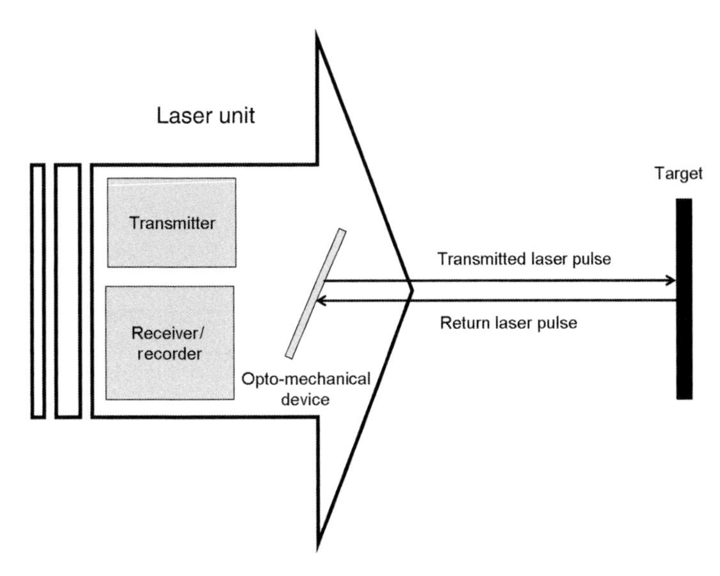

Fig. 2.7 Basic components of a LiDAR instrument.

et al. (Chapter 13) state that the Optech systems use a wavelength of 1064 nm, and the TopoSys Falcon II a laser wavelength of 1560 nm.

Scanner systems may be broadly classified into three types performing different functions:

Differential Absorption LiDAR (DIAL) measures concentrations of chemicals in the atmosphere using two differing wavelengths of laser light, selected such that one of the wavelengths is absorbed by the chemical being investigated. The intensity of return signals may be related to the concentration of the chemical at the measurement location. According to the atmospheric parameter to be measured, LiDAR systems use a variety of interactions such as Rayleigh, Mie and Raman scattering, absorption or fluorescence (Godin-Beekman *et al.*, 2003). The use of pulsed lasers enables range-resolved measurements to be obtained.

Doppler LiDAR machines measure the velocity of a target object based on the slight change to return signal wavelength and frequency from a moving object. These instruments rely on the Doppler Effect, whereby received frequency or wavelength from a source object is modified by the motion of the object, the receiver and the transmitting medium. Assuming both the receiver and the transmitting medium are stationary, an object moving towards the receiver has an elevated frequency above that emitted by the object due to the velocity of the object approaching the receiver. This declines to equal the object transmission frequency when the object and receiver are adjacent, and further reduces as the object moves away from the receiver.

Backscatter refers to the reflection of particles or waves in the direction of emission. A proportion of a laser pulse will often be reflected back to the instrument and time of flight information from the reflected laser pulse may be used together with Equation (2.3) and instrument optical geometric information to determine target object position in 3D space. Much of the text in this chapter relates to backscatter scanners as these are currently the most commonly used in applications in the natural environment.

Backscatter lasers have been widely utilized as positional instruments, with continuous wave systems dominating for close range work (<5 m) and pulsed lasers gaining prominence over increasing scales. Continuous wave systems operate on the principle of recording the phase difference between the emitted and backscattered laser signal. Given the high frequencies at which the scanners operate, the maximum operating range is limited by the time interval between equivalent points on the proceeding wave of the laser signal (Figure 2.8).

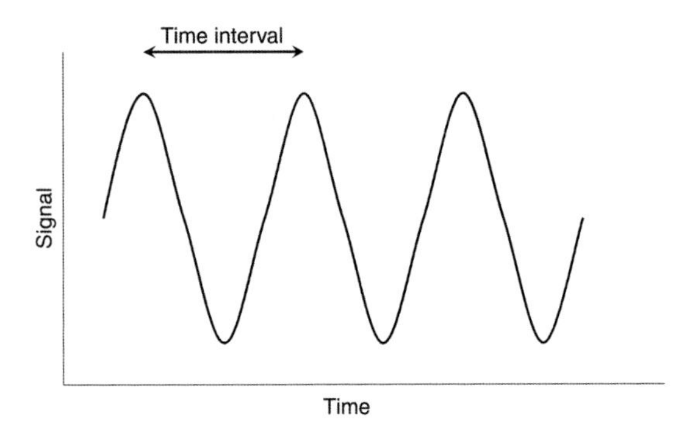

Fig. 2.8 The influence of wavelength on scanner range for continuous wave systems.

Pulsed scanners emit high energy discrete units of laser light with the maximum range limited by the emission to record time (Figure 2.9). Should the laser pulse be returned to the unit after the next pulse is emitted, the target will not be recorded. Effectively this is not an issue as instruments operating at 25 Hz may distinguish returns up to 6 km distant providing the returned energy is sufficient to be picked up by the recorder. As the decline in return energy is inversely proportional to the square of the distance, objects this distance away require a high-energy emission pulse and/or high reflectivity to be detected.

Position finding

Geometric location may then be defined given the optical and mechanical character of the laser scanner mechanism, where precise angular changes to the emitted laser light generates photon pulses along known vectors which can be

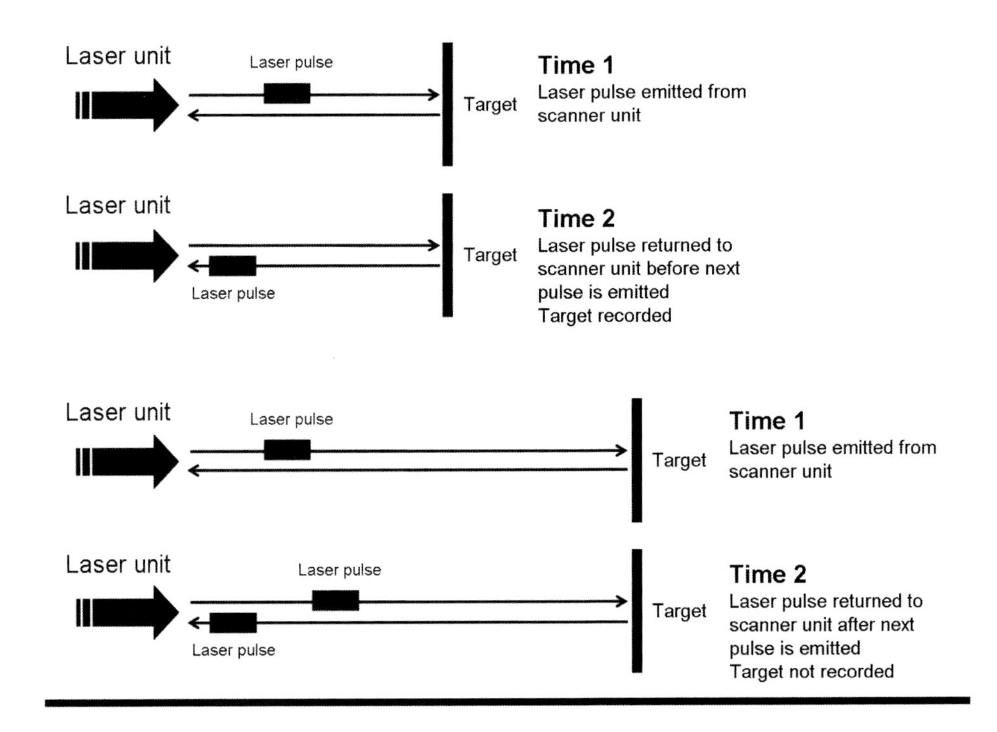

Fig. 2.9 The influence of laser pulse rate on scanner range for pulsed LiDAR systems.

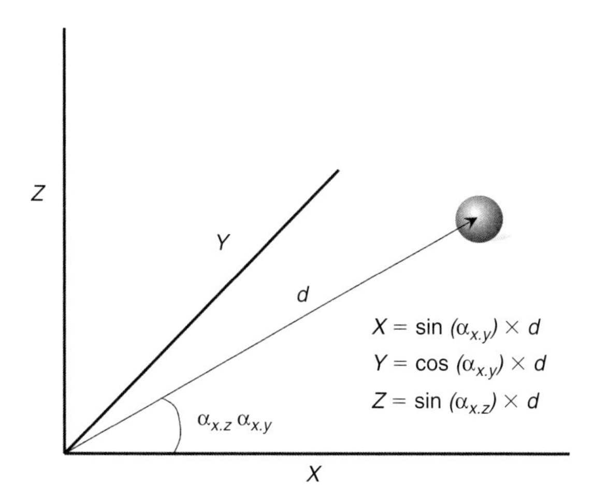

Fig. 2.10 Illustration of the ability of LiDAR systems to position remote objects in 3D space.

simply converted to Cartesian coordinates given the range distance to the object (Figure 2.10). The scanner generates a point cloud of geometric data, defined by a coordinate system initially referenced to the instrument head at the origin (x,y,z = 0,0,0), and object position is computed using the time of flight information combined with the angular data at which the laser pulse was emitted hence:

$$pos\,x = \sin\alpha_{\mathrm{h}}\,d \tag{2.4}$$

$$pos\,x = \cos\alpha_{\mathrm{h}}\,d \tag{2.5}$$

$$pos\,z = \sin\alpha_{\mathrm{v}}\,d \tag{2.6}$$

where, α_{h}, horizontal angle of laser pulse emission; α_{v}, vertical angle of laser pulse emission; d, distance determined from time of flight information (Equation (2.3)).

It should be noted that position accuracy is thus affected by the precision achievable by the opto-mechanical mirror system for terrestrial laser systems and additional navigational positioning issues for airborne and mobile land based systems.

According to Balstavias (1999) LiDAR accuracy is defined from the position of the laser beam, the range of the target point detected and the transformation into the desired coordinate system. Similar to the way theodolite instruments operate, the internal coordinate system of most scanners is not automatically set to a horizontal framework, with the disadvantage that the instrument may often not be levelled with sufficient accuracy in the field to ensure such a horizontal setup. Should this be the case, internal tilt information obtained during the scanning process may be used to level the data. Alternatively a system of known tie-points can be used to re-orientate the point cloud data and to set the project coordinate system for further scans which utilize the same tie-point arrangement. In the case of airborne systems and some mobile terrestrial systems, additional hardware and firmware are required to determine navigational location and position information to allow for real time correction of the returned data into a global coordinate system. Data processing and 3D model production is further described in Chapter 5 and will not be elaborated on further here; however it should be borne in mind that a significant factor affecting the accuracy of pulse-based range measurements is noise (scatter due to dust, wind, rain or temperature; and mixed echoes due to target surface complexity of the target object). Straatsma and Middelkoop (2006), for example, report on the problems of noise in attempting to assess lowland floodplain meadow vegetation using airborne laser scanning.

Scanner output

Most scanners record information on laser pulse time of flight, based on a threshold energy return (see the following section for more detail) and utilize the angular pulse emission values and Equation (2.3) to automatically convert this to a distance away from the scanner. This in turn is recorded as polar coordinate information and converted to Cartesian coordinate output using Equations (2.4–2.6). Additionally, the intensity of the return laser pulse is recorded, together with red, green, blue spectrum information on some scanners. The basic output may be immediately utilized to generate a 3D scan cloud of intensity (Figure 2.11a) which may be coloured using the received spectral information (Figure 2.11b). Rendering the data allows simple digital elevation model construction. Conversion of a set of point clouds from individual scanner locations into a common coordinate system, offers the capability to create a highly detailed model of regional scale terrain variables; a combination that, until recently, had been impossible to achieve as simply, or in as much detail over what are often significant distances across environmental systems.

Factors affecting position accuracy

The principles outlined thus far are complicated by a number of factors that influence the accuracy and reliability of laser scanner data. The following section investigates these factors in detail as they must be considered carefully when collecting or utilizing data from terrestrial and airborne LiDAR systems.

Return signal strength, thresholding and pulse variance

The majority of laser scanners (and at present all terrestrial instruments) do not record the full waveform of the return laser light pulse. Instead they monitor the return pulse recording time of flight information when an energy threshold is crossed (often referred to as first and last return data). Most systems can differentiate between two pulse returns. In vegetated systems, for example, the first pulse return is considered to represent the top of a vegetated element and the last return is taken to be the lowest elevation either in or at the edge of the vegetation. In this way, information can be gathered about the canopy height and dimensions, age and health of the vegetation (see Overton *et al.*, Chapter 14, for further information on this topic).

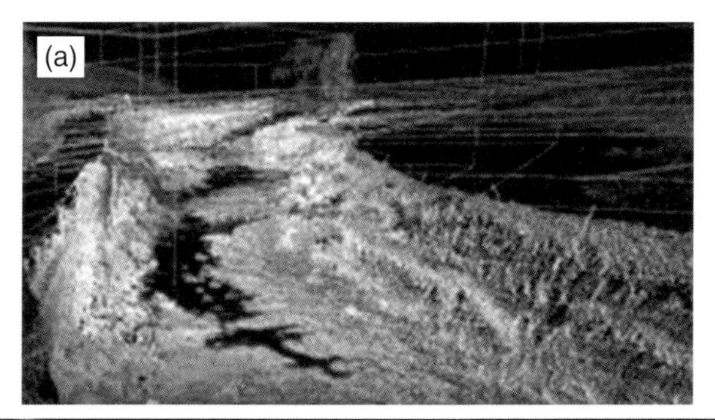

Fig. 2.11 Example laser cloud model and colour-rendered surface for the River Wharfe at High Houses, North Yorkshire, UK. See Plate 2.11 for a colour version of these images.

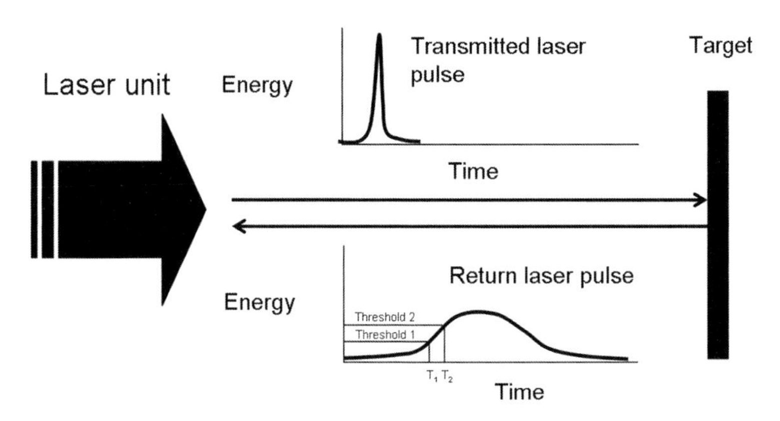

Fig. 2.12 The effect of threshold recording levels on time of flight and distance information recorded using LiDAR instruments.

It is assumed that the emitted laser light is entirely coherent, with phased emission of photons of light at an almost constant known energy state, and this is used to set the threshold for recording of return energy (Figure 2.12). Threshold crossing is thus dependent on the reflected energy (P_r) received by the recording device as defined below:

$$P_r = \rho \frac{M^2 A_r}{\pi R^2} P_t \tag{2.7}$$

where, ρ, object reflectance; P_t, transmitted pulse energy; A_r, illuminated receiver area; M, atmospheric transmission; R, remote object distance

Atmospheric transmission of laser light is best under clear, cool, dry conditions and may be severely affected by elevated water vapour and CO_2 levels (Baltsavias, 1999). It is clear from the form of Equation (2.7) that transmission energy and receiver sensitivity govern the maximum recording distance, especially when lower energy eye-safe laser systems have to be used. Similarly, distance from the scanner becomes a limiting factor in recording object position, especially for low reflectance objects.

Determination of object distance relies on the crossing of an energy threshold in the return signal (first return); this is determined by the instrument manufacturer and is set at a fraction of the emitted energy. It is clear that this effectively ignores the initial low energy return signal even if this came from a closer object (Figure 2.12). This can be a significant drawback, potentially affecting ranging accuracy.

Often a last return threshold is also set, and may be used to record objects impacted further along the laser vector; this has the advantage of effectively 'seeing through' diffuse objects, often to the ground, but is again affected by the arbitrary setting of the signal threshold that triggers data recording. This has the advantage of allowing features like tree cover to be 'removed' during processing, allowing more complete rendering of the ground surface. Many airborne LiDAR systems now record the full return waveform, allowing this signal to be subsequently analysed dependant on the object information required (see Chapters 12 and 13, this volume for examples). In a full waveform system, the system samples and records the energy returned for the full amount of time it takes to reach a surface that cannot be penetrated (e.g. Harding *et al.*, 2001) Datasets from such instruments are, as would be expected, far larger and more complex than first and last return systems; the process is complicated from a computing point of view, and consequently few systems are currently able to extract the full waveform from a laser pulse. Laser pulse emission is assumed to be constant in terms of wavelength and energy level, however variance is inevitable. This variance will have a direct affect on range accuracy as return energy levels will be affected, altering the point at which recording thresholds are crossed. This is a recognized problem, and many instruments have a variable threshold system whereby levels are calculated for each emitted pulse based on a fixed proportion of its initial strength (Baltsavias, 1999).

Beam divergence and scan angle

Despite being orders of magnitude better than radar, laser-based LiDAR systems are affected by beam divergence (Figure 2.13a,b). This issue is accentuated by sloping terrain (Figure 2.13c) and at increasing distances from the scanner (in both the vertical and horizontal plane) adversely affecting airborne scanners, longer range terrestrial systems operating at low angles and instruments with a relatively large beam divergence. The principal effect of increasing divergence is an increased measurement footprint; this alters the return signal waveform and results in range inaccuracies. For pulsed systems, small bandwidth short wavelength (infrared) instruments emitting very rapid pulses at small angular increments are best able to detect small objects. Object differentiation is then possible at the scale of the data point resolution.

Interestingly, greater beam divergence may be of use in distinguishing the ground beneath vegetation, as the chances of some photons penetrating to the land surface increases with a larger footprint, enhancing the probability of the non-penetrative surface being recorded in the waveform profile or as last return data.

Object character and complexity

Perhaps the most difficult aspect to contend with in laser scanning is object character and complexity and its effect on return waveform. Object reflectance and transmissivity affect photon

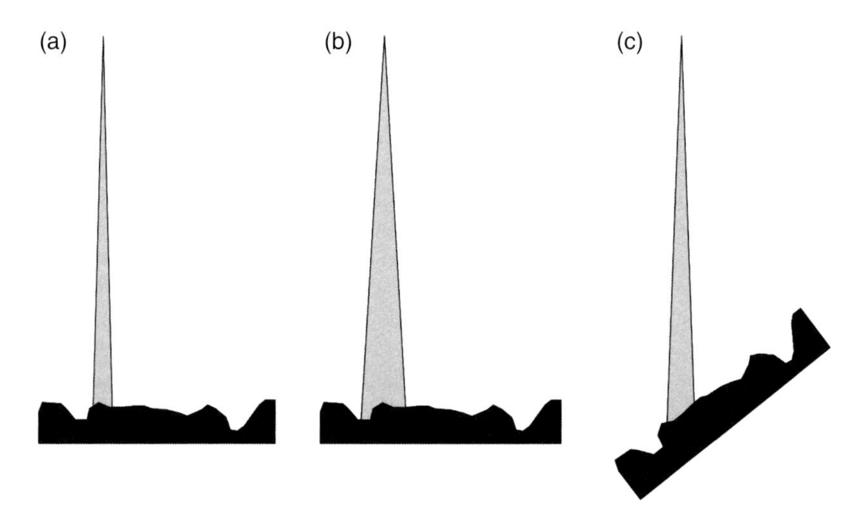

Fig. 2.13 The effect of beam divergence on LiDAR return signal waveform. (a) Narrow beam divergence over horizontal terrain – little ambiguity over surface height, (b) Wide beam divergence over horizontal terrain – much ambiguity over surface height and (c) Narrow beam divergence over sloping surface – much ambiguity over surface height.

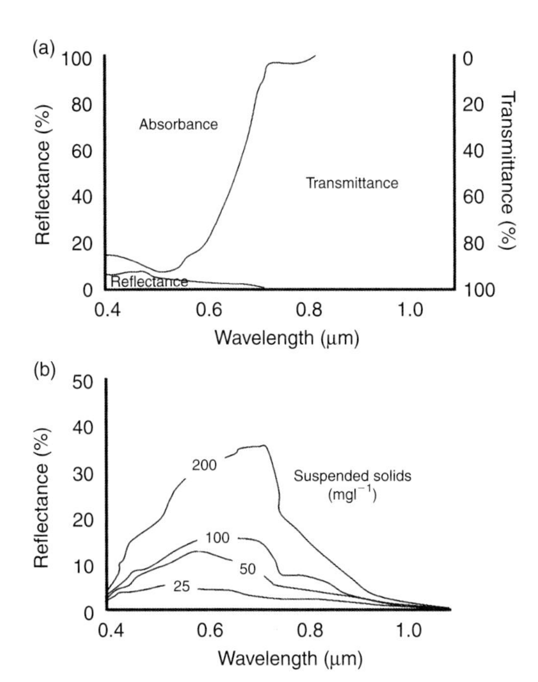

Fig. 2.14 Water character and its reflectance, transmission and absorption of laser light dependent on (a) emission wavelength and (b) water turbidity.

absorption, scattering and passage. Still water is an excellent absorber of infrared light, often resulting in zero return signal from infrared laser sources (Figure 2.14a), however, surface disruption and suspended sediment increase reflectivity significantly (Figure 2.14b). Glass allows the passage of infrared light and is not detected. Many metallic objects are highly reflective, to the extent that they may even be mis-classified as retro-reflective tie-points in scan cloud matching algorithms. Most objects, however, act to scatter the transmitted laser light in all directions, reflecting and re-reflecting some photons back to the laser scanner, this is particularly the case where the surface is incoherent with multiple reflective surfaces. Waveform return is governed by the relative density of the impacted object or objects (Figure 2.15) and may result in erroneous data, particularly for last returns used to detect the ground surface, where thresholds may be crossed over several surfaces before the solid earth is reached. Return waveform character may also be used as a classifier for object recognition tools (see Chapter 8, this volume).

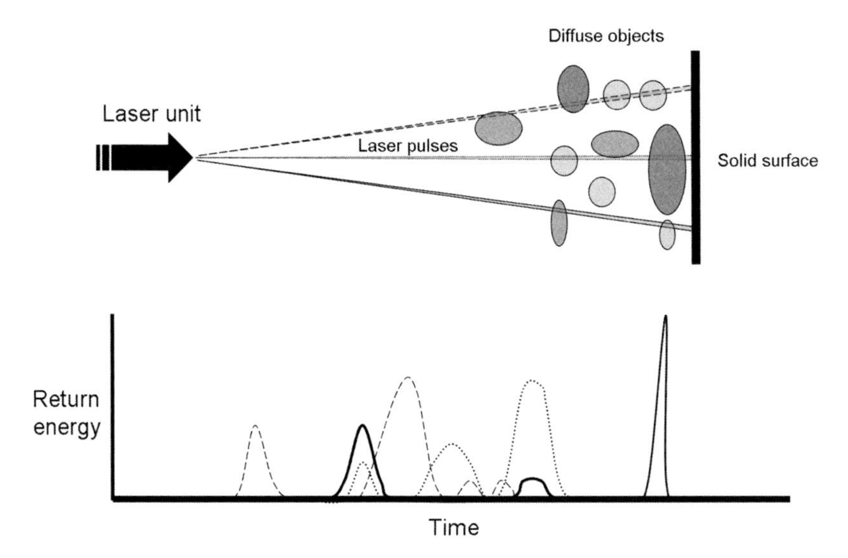

Fig. 2.15 The effect of object density/diffusivity on LiDAR return signal character.

THE OUTLOOK FOR LASER SCANNING

Ever increasing levels of detail and accuracy over increasing spatial scales, captured increasingly quickly as a result of laser scanning technology, is facilitating major advances in the understanding of the nature of the Earth and the processes acting upon it. During the last few decades, airborne LiDAR mapping has gained acceptance as a rapid and accurate method for 3D-surveying of natural environments. The rapid pace of development, coupled with increasing utilization of the technology by the research community, means that this trend can only continue. In this edited volume, we have outlined the tremendous utility of the technology in a wide range of research areas (see, for example, the studies reported in the application section of this volume). We have also highlighted some major issues associated with the output from laser scanners that must be borne in mind when collecting and analysing data using these instruments, and hopefully the volume will play a role in encouraging the innovative yet critical use of LiDAR in new research. With instrument, computing and software costs falling in relative terms, and research consortiums forming to utilize the technology, the future for laser scanner based research looks healthy.

REFERENCES

Baltsavias EP. 1999. Airborne laser scanning: basic relations and formulas. *Photogrammetry and Remote Sensing* **54**: 199–214.

Charlton ME, Large ARG, Fuller IC. 2003. Application of airborne LiDAR in river environments: the River Coquet, Northumberland, UK. *Earth Surface Processes and Landforms* **28**: 299–306.

Davies F, Collier CG, Pearson GN, Bozier KE. 2004. Doppler LiDAR measurements of turbulent structure function over an urban area. *Journal of Atmos. Oceanic Technol.* **21**: 753–761.

Einstein, A. 1917. *Zur Quantentheorie der Strahlung. Physikalische Zeitschrift* **18**: 121–128.

Friedland SS, Katzenstein J, Zatzick MR. 1956. *Improved Instrumentation for Searchlight Probing of the Stratosphere*. University of Connecticut.

Godin-Beekman S, Porteneuve J, Garnier A. 2003. Systematic DIAL LiDAR monitoring of the stratospheric ozone vertical distribution at Observatoire de Haute-Provence (43.92°N, 5.71°E). *Journal of Environmental Monitoring* **5**, 57–67.

Guenther GC, Lillycrop WJ, Banic, JR. 2002. *Future advancements in airborne hydrography. International Hydrographic Review* **3**: 67–90.

Harding DJ, Lefsky MA, Parker GG, Blair JB. 2001. Laser altimetry canopy height profiles: Methods and validation for closed canopy, broadleaf forests *Remote Sensing of Environment* **76**: 283–297.

Heritage GL, Hetherington D. 2007. Towards a protocol for laser scanning in fluvial geomorphology. *Earth Surface Processes and Landforms* **32**: 66–74.

Maiman TH. 1960. Stimulated optical radiation in ruby. *Nature* **187**: 493–494.

Milan D, Heritage GL, Hetherington D. 2007. Application of a 3D laser scanner in the assessment of erosion and deposition volumes and channel change in a pro-glacial river. *Earth Surface Processes and Landforms* **32**: 1657–1674.

Nagihara S, Mulligan KR, Xiong W. 2004. Use of a three-dimensional laser scanner to digitally capture the topography of sand dunes in high spatial resolution. *Earth Surface Processes and Landforms* **29**: 391–398.

Schawlow AL, Townes CH. 1958. Infrared and optical masers. *Physical Review* **112**: 1940–1949.

Smullins LD, Fiocco G. 1962. Optical echoes from the Moon. *Nature* **194**: 1267.

Straatsma MW, Middelkoop H. 2006. Airborne laser scanning as a tool for lowland floodplain vegetation monitoring. *Hydrobiologia* **565**: 87–103.

Synge EH. 1930. A method of investigating the higher atmosphere. *Philosophy Magazine* **52**: 1014–1020.

Wehr A, Lohr U. 1999. Airborne laser scanning – an introduction and overview. *Journal of Photogrammetry and Remote Sensing* **54**: 68–82.

3 Issues in Laser Scanning

MARTIN E. CHARLTON, SEAMUS J. COVENEY AND TIMOTHY McCARTHY

National Centre for Geocomputation, National University of Ireland, Maynooth, Ireland

INTRODUCTION

This chapter is concerned with some underlying general issues in laser scanning. We consider the initial choice of whether to use airborne LiDAR or terrestrial LiDAR, before proceeding to examine some issues which are germane to each, but focusing in the main on terrestrial laser scanning (TLS).

Perhaps the most common form of scanner is based around the time-of-flight principle – that is, a pulse of energy is emitted from a laser and a sensor measures the time between the emission and the return of the pulse through reflection from some distant object. Multiplying the time from emission to sensing by the speed of light gives the distance between the laser and the object. The laser may rotate in one or more planes or, alternatively, mirrors may be used. In airborne sensing, the laser scans from side to side to provide measurements across a path while the forward motion of the aircraft allows measurement along the path. Terrestrial scanners usually have a laser which rotates in the vertical plane, while a mirror above the laser oscillates in the horizontal plane to provide measurements in a defined field of view.

Laser Scanning for the Environmental Sciences, 1st edition. Edited by G.L. Heritage and A.R.G. Large.
 ISBN 978-1-4051-5717-9

The output is generally a set of three measurements of the position of a point relative to the scanner, with perhaps a measurement of the intensity of the reflected pulse, and a red-green-blue colour triple. The scan rate varies between models of scanner, and may be controlled by the user. For the user, the main consideration is that the output is a cloud of geometrically unstructured observations – a large number of individual measurements which form a point cloud in three dimensions. Whilst collection of the data may involve lengthy waits or moderate physical discomfort, the effort required to process the data should also not be underestimated. Unstructured in this context means that the spatial relationships between the individual observations within the data cloud (for example, whether they form an edge, or are part of a metal brace) are not present. Not only are the data unstructured, there is also a large quantity of information. A scanner emitting 8000 pulses per second will generate around 14,000,000 measurements in 30 minutes. Each measurement will contain an x,y,z triple, and may also include intensity information. Some scanners will produce measurements for the first and second returns (that is, where the part of the emitted beam hits and is reflected from a nearer object, followed by a second return from a farther object). If each measurement requires 30 bytes, then 30 minutes could generate around 0.8 gigabytes of data. A file with 28,000,000 records is not easily imported into Excel (which has a limit of 65,536 records, although Excel 2007 has a limit of 2147 million records). The point cloud data are also not always easily visualised in a GIS package. Specialist software is available for editing point clouds – Sycode's Point Cloud software exists as a standalone program or an add-on to AutoCAD. Wand *et al.* (2007) describe a range of point cloud editing software.

Typically, the area of interest may require several overlapping scans. The point cloud generated by a single scan is usually in a Cartesian coordinate system with the origin being at the location of the scanner. The cloud is not usually oriented with a country's national mapping grid system or UTM. Bringing several overlapping scans together requires software that will translate and rotate the point cloud in three dimensions – there are a number of ways in which this is done in practice which will be discussed in some detail below. The resulting single point cloud may then, if necessary, be oriented to some defined grid system, such as UTM.

For many applications in the environmental sciences, reconstruction of the geometric structure of the objects which have been scanned may not be necessary. At the moment, identifying edges of structures such as buildings and bridges, and creating an instance of them in a 3D CAD package, is a highly labour intensive activity. Creating a 3D model of, say, even a simple built structure can involve many hours of painstaking labour at a workstation.

AIRBORNE OR TERRESTRIAL SCANNING?

Long before any data collection begins, consideration must be given to the appropriate mode of data capture. This is related to the size of the actual area of interest, the research question in hand and the equipment which is available. If the area of interest is several square kilometres, then terrestrial scanning is probably out of the question. Whilst scanners differ in scan density, typically we might expect a density of one observation per square metre from an airborne scanner; with a terrestrial scanner the density can be much higher, with one observation per square centimetre being conveniently possible.

There is a trade-off between density of acquisition and accuracy. Whilst data collection from an airborne scanner can yield 8000 measurements per second, the airborne platform itself is mobile – it is moving forward and may be subject to pitch, roll and yaw. The scanner measurements are corrected for these factors using information from GPS and the aeroplane's own inertial navigation system. Even so, a frequently quoted statistic for the vertical accuracy is that the measurements have an accuracy of around ±15 cm.

AIRBORNE SCANNING

Researchers who use LiDAR from airborne survey are unlikely to have the luxury of access to a suitably equipped aircraft. The body sponsoring the research may have access to an appropriate facility, or the researcher may have to approach a commerical organisation. The survey body will at the least require a date and time for the flight, as well as a map on which the area of interest is clearly marked by precise coordinates. Additional information which might be required would include some estimate of scanning density and whether processed or raw information is required.

In the UK the Natural Environment Research Council's Airborne Research and Survey facility (NERC ARSF) uses an Optech ALTM 3033 scanner flown at a height of 1000 m with a scan rate of 33 kHz. The scanner output, which is processed at Cambridge University, includes positional measurements and intensity for both first and second return pulses. The RMS accuracy of the resulting height information is estimated to be ±15 cm.

Application for the NERC facility is a fairly extended process, and applicants are required to provide, *inter alia*, a multi-page justified scientific case for the survey. The application process is competitive, and not all applications are successful. This is perhaps not surprising since NERC is a publicly funded organisation and must be seen to be spending its money wisely.

NERC's ARSF website has a page of useful information concerning mission planning. As well as LiDAR the aircraft carries a range of other sensors, of which several may be used to provide additional information in the survey. Typically a survey will consist of a number of overlapping flight paths along or across the area of interest. The ground speed of the aircraft and its flying height will determine the spatial separation of the laser pulses from the ALTM sensor. The reader is directed to the NERC website (http://arsf.nerc.ac.uk) for an extended discussion of the issues which are important in acquiring airborne LiDAR data.

Area of interest

The area of interest for an airborne survey should be large enough to make the survey economically worthwhile. It will be unlikely that the area can be covered in a single scan by the aircraft, so a survey will consist of several overlapping paths, which are known as swaths. The width of the swath is a function of the side-to-side scan angle of the airborne sensor and the flying altitude of the aircraft – higher altitudes will yield wider swaths, but with potentially greater noise towards the edge of the swath.

Flying conditions

Atmospheric conditions are an important consideration. The aircraft must be able to take-off and land safely from an airfield which may be some distance from the survey site. Typically, survey aircraft are smaller than jetliners and therefore they may be subject to additional operational constraints. They are also subject to air traffic control restrictions as well – these can delay survey flights as well as passenger or freight flights.

Air traffic control is responsible for the organisation of aircraft in controlled airspace and has the final say regarding when and where an aircraft is permitted to fly. Survey sites which are near airfields or military establishments may well require additional permissions before they can be flown. The time of day during which the flight is made may also be subject to restriction. It is sometimes desirable that an entry for the flight is made in the control tower diary, so that the flight does not appear unexpectedly to the controllers who may well appear to exercise caution which, to a researcher waiting for results, seems excessive.

Cloudy or rainy conditions are far from ideal for airborne scanning. The laser pulse can be scattered by water vapour in the atmosphere in the form of mist or fog, and can fail to penetrate low cloud. Should conditions be unsuitable on the days agreed for flying, many weeks may pass between deciding to fly and being able to fly, particularly in the winter months. As an example, a LiDAR survey approved for the lead author and colleagues in January 2004 for a stretch of river in north east England was not actually flown until March 2006. This, however, was an extreme example.

Processing

Processing the raw output is generally an exercise best left to professionals. NERC's data is post-processed at the University of Cambridge. A commercial supplier should post-process the data to your requirements. There are a variety of outputs. Typically unstructured point cloud data is shown in Figure 3.1. The first column is a timestamp. The second to fourth columns are the information for the last return of the pulse, and the fifth to ninth columns are the information for the first return. The fields within each return are: easting, northing, elevation and intensity.

The scan half angle in this example was 17.9 degrees, with an average flying height of 2250 m, so the swath is slightly over 1450 m wide. The 24.9 million last/first pairs were scanned in 35 minutes, although the delivery of the processed point cloud data took several months.

The first and last returns do not equate to 'landscape with vegetation' and 'landscape without vegetation'. To achieve a bare-earth model (that is, ground surface without vegetation and buildings) is a labour intensive process, and

	Last return ---→				First return ---→			
Timestamp	Easting	Northing	Elev'n	Int	Easting	Northing	Elev'n	Int
384781.193014	395857.74	597383.87	162.98	21	395857.73	597383.87	163.03	21
384781.193044	395859.56	597383.96	163.12	20	395859.56	597383.96	163.12	20
384781.193074	395861.45	597384.05	163.04	21	395861.41	597384.05	163.17	21
384781.193104	395863.35	597384.14	162.90	18	395863.34	597384.14	162.93	18
384781.193134	395865.18	597384.23	163.02	22	395865.16	597384.23	163.10	22
384781.193164	395866.99	597384.33	163.20	21	395866.97	597384.33	163.26	21
384781.193194	395868.84	597384.42	163.25	18	395868.81	597384.42	163.33	18
384781.193224	395870.62	597384.51	163.52	21	395870.60	597384.51	163.57	21
384781.193254	395872.38	597384.61	163.87	20	395872.36	597384.61	163.92	20
384781.193284	395874.40	597384.70	163.37	23	395874.37	597384.70	163.46	23

Fig. 3.1 Typical airborne LiDAR output.

may require additional large scale map data or orthophotographs in order to determine what can be removed from the scene. Processing LiDAR for rural areas to create a bare-earth model is generally more rapid than processing LiDAR in urban areas where identifying and removing buildings is time-consuming. It may be difficult, for example, to differentiate in the point cloud between a building and a shipping container on a wharf. The first returns may indeed be *terra firma*, and the last returns may include a considerable proportion of vegetation canopy, as well as fences, walls, buildings, vehicles, animals and any other object usually found on the Earth's surface.

If a point cloud represents one end of a spectrum, then a georeferenced digital terrain model (DTM) perhaps represents the other end. Creating the DTM from the point cloud requires the removal of extraneous measurements – standard GIS software is poorly equipped to deal with this type of editing. Interpolating the regular grid of measurements from the irregularly spaced point cloud is a computer-intensive task. The analyst may decide to mask out observations well away from the real area of interest but which may have been also captured during the flight. Options here include inverse distance weighting methods, splines, radial basis functions, polynomial regression and geostatistical approaches such as kriging. An alternative is to retain the irregular spatial sampling locations and use these to form a triangulated irregular network (TIN). Some of these techniques are discussed by Brunsden (Chapter 5).

Alternative aerial platforms

Alternatives to propeller-driven aircraft as LiDAR platforms exist, and these include helicopters and unmanned aerial vehicles (UAVs). The advantage of the helicopter as a platform is that its ground speed can be reduced to zero, thereby permitted much higher scan densities along the length of a swath. Fischer *et al.* (2008) describe the collection of LiDAR and photogrammetric data using helicopter borne LiDAR in the European Alps. They scanned part of the east face of Monte Rosa – as with aircraft-borne systems the LiDAR data was integrated with positional information from GPS and orientation information from an Inertial Navigation System. The scanning of high altitude steep slopes was facilitated by using an obliquely-mounted LiDAR scanner. The outcome of their data collection was the identification of rock and ice changes, as well as volumetric analyses of past and present slope change.

Unmanned aerial vehicles are remotely controlled small aircraft. Typically these have a much lower payload than either piloted aircraft or a helicopter. Progress in this area has been slow. The challenges for UAV-borne LiDAR are ones of weight – keeping the payload within the limits of the UAV – and power. Sufficient power has to be available on the UAV to drive the scanner and associated position/orientation sensors. Fantoni *et al.* (2004) describe the development of a UAV-based system for large sea surface and territory monitoring; they also describe a LiDAR system based on an underwater remotely operated vehicle (ROV) for sub-surface monitoring in the Ross Sea as part of the XVII Italian Antarctic Expedition. Both of these provide exciting possibilities for aerial and submarine LiDAR collection.

TERRESTRIAL LIDAR

An alternative to airborne LiDAR collection is to use a terrestrial LiDAR scanner to collect the data oneself. The principal differences are one of (a) scale – very much higher scan densities are possible, and (b) immediacy – the scanner can be loaded into a van and driven to the collection site for immediate data collection. However, the researcher becomes primarily responsible for subsequent data processing – this is sometimes a more time consuming and labour intensive process than the basic collection of the data. The entry cost is considerable – the equipment is expensive compared with GPS – and there is a set of operational skills to be developed before rapid and reliable data collection can begin.

Equipment purchase

There are a wide variety of terrestrial laser scanners (TLS) available in an even wider variety of configurations. The entry cost is might be some 5–10 times that of RTK GPS systems. Because of the operating skills which are required, it is not feasible to borrow or hire a LiDAR scanner and expect immediate usable and reliable results. This generally means that, unless the analyst is part of a team that has access to existing equipment and skills, a LiDAR scanner will have to be purchased or the data collection commissioned from a third party. Again, because of the entry cost, it would be difficult to justify the purchase of a scanner for a short term project.

For LiDAR scanning in environmental applications, potential purchasers would find the products of Riegl, Leica, and Optech useful starting points. These are not the only manufacturers of TLS equipment and their mention here does not imply endorsement. Two of the editors of this volume have extensive experience with a Riegl system and the other has experience with Leica equipment. Comparison of the data sheets does not always reveal the advantages of a particular system nor does it always reveal important differences. For a first purchase, demonstration of alternative equipment is a necessity. Important initial questions concern the scanner range and density – short, medium and long range scanners are available. The reflectivity of the material to be scanned affects the range, and information on the scan range is often quoted for different albedo measurements. Long range scanners have a scanning range up to 500–2000 m. More general purpose scanners would be expected to operate up to 200–500 m. In practice, this can vary depending on the light condition when scanning takes place, and the size of the object being scanned relative to the diameter of the light beam. A shorter range scanner (150–400 m measurement range) may in practice provide reliable data up to 150 m. Accuracy information is available on manufacturers data sheets – for a general purpose scanner 6 mm at 100 m representing one standard deviation provides workable and reliable information for many applications. In practice, the acquisition error is often hard to define as the error from RTK GPS is in a similar range.

Demonstration

Any potential suppliers should be asked to provide a full demonstration of their equipment. This should involve some test scans, as well as a brief demonstration of the associated software in use. This will reveal the physical characteristics of the equipment, the complexity and time requirements for its set-up and calibration, the rapidity of data collection, the ease of use of the system in the field and the flexibility of any associated software. The field of view is another factor. Some systems offer 360° rotation around the vertical axis and 270° about the horizontal axis for a vertically-mounted scanner. Some scanners may be tilted for oblique acquisition, other have to remain horizontal. Some systems come with a built-in camera for acquiring image information in the visible spectrum, which can be integrated with the data from the scanner itself. With other systems a calibrated camera may be an accessory. Other accessories may include vehicular mounts, GPS linkage to provide real time georeferencing, and targets which are required for integrating overlapping scans. The output should be at least 3D Cartesian coordinates of the scan points relative to the scanner, and a measure of the intensity of the reflected pulse. If the system permits, there should be a choice of first and/or second return pulse information.

Scanner components

What forms a reasonable working minimum configuration – what's in the box? There will be some variation between individual manufacturer's products. The Leica ScanStation integrates the scanner and camera. Riegl's scanners require the additional purchase of a separate calibrated camera. There will be a standard surveyor's tripod – it might be prudent to purchase a tripod stabiliser as the scanner units are heavy and, although the tripod may have rubber or spiked feet, the stabiliser will prevent the legs from sliding apart. A Leica ScanStation weighs some 19–20 kg and the HDS3000 16 kg – positioning the head on a tripod is easier with two people than one. The tripod should have a suitable tribrach and downsight for positioning over a ground control point.

Figure 3.2 shows the Leica ScanStation in use. The scanner is mounted on a sturdy tripod, with power connection to a battery (located at the base of the leg nearest the operator). There is also an ethernet connection to the laptop. The laptop is placed on the transport case for the scanner. The carrying case for the battery is at the base of the leg furthest from the camera. For scanning on surfaces without a good grip a tripod stabiliser is useful. The ScanStation incorporates a camera. The illustration gives a good idea of the number and size of the components. In the field, two additional batteries are used during scanning to maximise data collection time – at the National Centre for Geocomputation, Maynooth, the authors have five in total. They also possess additional batteries for the laptop, and a second tripod.

Fig. 3.2 Leica Scanstation scanner, tripod, battery pack, laptop and carrying case.

The scanner requires power. Batteries come in a range of shapes and sizes – they may weigh some 7 kg, provide between 4 and 8 hour's power, and are expensive – depending on local taxes €500–750 per battery may be a worst case scenario. Buying too few batteries is a false economy – two are a minimum and, given the charging time, two further ones can be charging while the units are being used in the field. There may also be a mains power unit, and a lead for powering the system from a vehicle battery.

The scanner is usually controlled using the manufacturer's software running on a laptop.

Whilst an ordinary office laptop will suffice, it is a good idea to have a rugged laptop (similar to the Toughbook™ or General Dynamics GoBook XR-1), dedicated for use with the scanner, with a screen that can be viewed in conditions of bright sunlight (particularly if the sun is low in the sky). A ruggedised laptop will usually survive being dropped on the ground, a standard office laptop is unlikely to, and rapid repairs are rarely possible if the survey is being carried out in a remote location. Laptops require power as well – there may be a mains unit, but battery chargers will be required. Again, purchasing several rechargeable batteries may be prudent – batteries have an unfortunate characteristic of expiring in the middle of a survey. Again, the power source may be a vehicle battery with a suitable lead for the cigarette light socket. If the scanning is to take place in a location where the electricity supply is on a different voltage, ask yourself if your chargers will still function correctly.

Unless the survey area is small, most surveys will require several overlapping scans to cover the required surface. Each scan will require a different position for the scanner, so some means of identifying common points is needed. Typically reflective targets are used – these are laid out across the survey area prior to the first scan. Different systems have different ways of locating and identifying the targets – these will involve interaction with the software control program (Cyclone™ in the case of Leica scanners, RiSCAN in the case of Riegl scanners). A coarse scan within the field of view, followed by several fine scans of the targets in the field of view, may be required prior to the scan of the topography, building or structure. The targets themselves vary between manufacturers – they are usually high quality metallic pieces with a matt finish on a tilt-and-turn mount. Again, starting with too few means that targets will need to be repositioned between scans where they are not in the overlap area. However, they are not cheap, and they may require padded carrying cases for transport. They also need to be kept clean and dry.

The scanning process is usually under the control of the manufacturer's own software running on the laptop. The link to the scanner itself is usually via a wired LAN link – this may require initial setup on delivery. The software controls the scanning setup and operation during data capture in the field. As a minimum the software should permit the meshing of adjacent surveys, and registration with an external coordinate system (for example, a national mapping agency's grid system). The software should permit some display and editing of the point cloud (for example to remove noise introduced by passing vehicles or animals during the scan). Meshing is usually carried out by matching the targets between adjacent scans – the software should provide some goodness-of-fit measure as meshing involves the translation and rotation in three dimensions between the local coordinate systems of the scans being meshed. This will allow the identification of those targets where the scanned position is unreliable. With practice, it is possible to mesh scans of buildings using well-defined features on the buildings which have been scanned in both scans. The software may also allow the modelling of the surface topography from the point cloud – this may involve the fitting of a TIN to capture the elevation variation. The software should also permit data export in a variety of formats.

You may decide to purchase additional software at this initial stage. Both RiSCAN and Cyclone are reasonably flexible and powerful tools for controlling the scanner in the field and manipulating point clouds in the office or at the lab desk. Cloudworx™ is available to provide a link into AutoCAD, although unless the desired output is a CAD drawing of the geometry of some regular feature or a building, this need not be in the first purchase. Likewise, one may consider Polyworks™ as an alternative editing and modelling tool.

Transport and storage

The scanner, batteries, laptop, targets and tripod(s) are generally bulky or unwieldy items – suitable transport cases will be needed. These should be strong – will the contents survive rough handling in the field or during transport? If fieldwork is being carried out abroad, will the cases be strong enough to avoid damage during air or sea transport? Some of the carrying cases may be wheeled – a small trolley may also be a sensible purchase as it will enable a single person to transport the equipment over smooth ground. In the field, a small group of willing helpers is invaluable!

A clean and dry storage location is essential. Dust is a hindrance to successful scanning. The equipment needs to be cleaned and kept clean. The targets should be dried and cleaned after use. Lab space is needed for the storage of the equipment, and there needs to be table space which is provided with suitable power supplies for the safe recharging of the batteries (the following day's activity will probably require recharging two batteries for the scanner and two or more for the laptop). It helps if the temperature is kept constant as well – moving the scanner from cold to warm conditions may lead to condensation on the internal mirrors and surfaces.

Training

It is tempting to minimise the amount of time and expenditure spent on training. If you do not have access to any other source of expertise in laser scanning it is certainly worth considering taking a manufacturer's or supplier's training course. If you are in an environment where there are other users who can pass on information on operating the scanner in the field and subsequent point cloud processing, then you will be fortunate. Some manufacturers and suppliers provide training courses at additional cost. These will be general purpose, or may be oriented towards the user of the equipment for building survey (environmental science applications are still novel, hence the justification for this book). They will cover the basics of the use of the scanner, and also the general operation of the control and editing software. They should cover the equipment set up in the field, the use of the control software to obtain scans, editing, meshing, georegistration and simple modelling – in other words what you will require to operate the scanner in field conditions and successfully acquire data which is appropriate for the research task in hand. They may also suggest suitable field operation procedures ('workflow') for reliable scanning, as well as introduce basic survey methods and perhaps more advanced techniques.

Practice in the early stages of gaining scanning experience is invaluable. The Conference Room at the National Centre for Geocomputation at the National University of Ireland Maynooth may be the most scanned location in the Republic of Ireland! The mistakes made and the experience gained during these sessions meant that when data was being acquired in the field the users were familiar with the procedures, and also had a good idea of what to do when problems arose.

WEATHER

It may seem unusual to consider the weather, however, some conditions are more favourable to scanning than others. In general, 'fine bright weather' provides a favourable scanning environment. Direct sunlight can affect reflectance with no return being measurable from the pulse; in this case a hole may appear in the point cloud. This may also be the case if the sun is reflected towards the scanner from a very highly polished surface. Night-time scanning is possible, although it may be difficult to set a suitable field of view without some experimentation, and using the camera to obtain RGB values for the data points will not be available. Other adverse weather conditions include rain, snow, fog and mist – backscatter from the water droplets in the atmosphere may cause increased noise in the point cloud which will be difficult to remove after the scan is completed. However, we have obtained reliable scans in conditions of light rain with the scanner suitably shielded from the raindrops with a golfing umbrella – water may also damage the batteries. However, should any water droplets fall on the scanner glass there are likely to be noticeable measurement errors – the glass needs to be kept dry. Some scanners will stop functioning if the local wind speed becomes too high – scanning resumes when the wind speed drops. One further consideration is the type of laser which is used in the scanner. If there are people or animals in the vicinity of the scanner they may get scanned at the pass – this may include their looking directly at the laser. A class 1 laser is safe under normal operating conditions. A class 3R laser should not be viewed directly or unintentionally if the beam is reflected from a mirror-like or specular surface. This may have implications for scanning near roads or other locations where people may run the risk of unintentional exposure.

Reflectance – what does the scanner see?

Some materials do not provide reliable scan data. Highly reflective surfaces will get reflect the beam in a different direction to the scanner's output pulse – there will be reduced or no measurements if polished metal or high gloss paint is scanned (cf Figure 2.2, Chapter 2). In the field, naturally occuring materials are unlikely to have these characteristics, but should you be asked to scan a building, this is a relevant consideration. In the field, water surfaces – a smooth pond or puddles – will not provide reliable surface measurements. It may not initially be possible to determine the reflectivity of material which is to be scanned, and the dropped measurements only become apparent when the point cloud is examined back in the laboratory. Similarly some surfaces are highly absorbent to the input energy – matt black for example.

Translucent surfaces may not provide a return, or may give a first return; what lies beyond the translucent material may be sufficiently reflective to provide a reliable measurement. Early experiments in scanning the John Hume Building at the National University of Ireland in Maynooth gave excellent returns on the bottles, shelving, cans and food cartons in the ground floor shop, but not the window which forms part of the shopfront.

Vegetation may also provide scanning problems. In general, if the operator can't see it, neither can the scanner. A tree which lies between the scanner and say a rock face will be scanned, and the part of the rock face which lies behind the tree will not be scanned – when the tree is edited out of the point cloud, there will be a vertical strip of missed measurements in the point cloud. Occlusions are a similar problem – several scans of grass tussocks may be required, but the goal of obtaining reliable 'bare earth' measurements may be more difficult to attain than originally envisaged.

In the field

The temptation, however strong, to simply take the scanner out for an afternoon's scanning should be resisted unless there are clear goals. Manufacturer's or supplier's training may be generic, and may suggest workflows which are too complex. Our experience is that the 'one size fits all' approach does not provide efficient data acquisition. The equipment and software is, for many researchers, sufficiently complex to operate that over-complexity is a hindrance. Creating a simple and reliable set of field procedures and writing them down in a checklist will repay the effort required to create them. An added bonus would be to have the field procedures printed on one side of A4 in a legible type and laminated – this will prolong their life both in storage and in adverse field conditions. If you are new to the scanning business, practice scans made under controlled conditions (i.e. indoors, or near to your lab – see above) repay the time spent on them – the object of field survey is not to learn the scanning techniques, but to acquire data and minimise any downtime.

The laptop which is used for the control software should be dedicated solely for use with the scanner. If this is not the case, with the best will in the world, when the scanner is required the laptop will be elsewhere and not easily retrievable.

Many sites will require several scans to survey them completely. If the scanner positions can be worked out in advance, this saves time in data acquisition. However it may only become apparent where sections have been missed once the point clouds have been displayed and perhaps meshed. If there are missed sections, these will require additional surveying. The scanner positions may need to be placed at locations which require reliable positional measurement with GPS – this depends on whether you need to integrate the surface models with other data in a GIS.

Where multiple scans of the survey site are required, the scans need to be tied together into a common coordinate system. The separate scans are usually made in their own local coordinate systems with the scanner positions representing the origins of each coordinate system. Integrating the separate scans into one point cloud which represents a scan of the complete study site is carried out in the control software (often there is a separate module to do this task) – and here we talk of 'meshing' or co-registering the surveys. The outcome may be a point cloud which is registered to one of the local coordinate systems in the component scans, or it may be geo-registered to a national mapping agency's system (such as the Ordnance Survey National Grid system in Great Britain). Using a non-georeferenced coordinate system (e.g. where repeat surveys are tied to a common temporary benchmark or TBM) is adequate for change surveys, **so long as** the TBM is the same for all surveys. More control points are advisable however (see below).

Registration of overlapping scans requires the identification of common positions in each scan. It is usual to use small targets for this purpose. These vary between manufacturers but may be a round or square piece of matt material with a well defined point on one face. They may be mounted in a turn-and-turn mount, or may be fixed to a solid object with a clamp, or they may be placed on a tripod. Within each scan a suitable number of targets should be located and scanned – this is usually carried out prior to the area scan, and the targets are usually scanned at a high density for accurate positioning – any inaccuracy during the acquisition of target positions will propagate itself into the registration process.

What is a suitable number of targets? The transformation of a point cloud in one coordinate system into another coordinate system involves two operations in three dimensions – translation and rotation. Translation merely involves a shift in the x,y,z measurements. Rotation may involve separate rotations around the x, y and z axes. In practice, some rescaling becomes necessary to account for the slight drift in calibration of the scanning device from standard distance measures. GIS users may be familiar with geo-registration in which a digitised map (using the digitising tablet's local coordinate system) is registered to a standard grid system. This usually involves identifying at least four control points whose coordinates are known in both the digitiser and standard grid coordinate systems – there are six unknown coefficients to be estimated using least squares. More control points are desirable. In three dimensions there will be twelve coefficients to identify:

$$x' = \alpha + \alpha x + \chi y + \delta z$$
$$y' = \varepsilon + \phi x + \gamma y + \eta z$$
$$z' = \iota + \varphi x + \kappa y + \lambda z$$

where $\alpha \ldots \lambda$ are the coefficients which are to be estimated. We will need a minimum of five measurements in each coordinate system to provide us with a solution – that is, at least five targets must be visible in the overlap area between each pair of adjacent scans. Estimation is a similar process to linear regression, and the predicted values of x', y' and z' can be compared with their actual values to provide a residual – an RMS error value is often available.

We can see now why there should ideally be many more than five control points available in the overlap between the surveys – should one initial measurement of the position be incorrect the registration itself will be wrong and, if there are not extra control points available, then the time spent in surveying will have been wasted as the results will be unreliable at best and unusable at worst. The residuals should be examined, and any control point which has a high residual be considered for removal in the estimating process. Additionally if there are more control points rather than fewer, the sample estimates of the coefficients themselves will be more reliable. Obtaining control point measurements for each of the targets in each scan is time consuming, but the more accurate the initial measurements, the more useful will be the resulting co-registered point clouds.

The targets themselves should be regularly spaced over the study area to obtain a good solution to the estimation of the coefficients in the equations. If there are local unfavourable configurations of the targets there is a risk that the matrix operations in the registration algorithms will fail – this would be highly undesirable. Whilst it is possible to compute the mathematical transformations, the sheer size of the point clouds that result from a scan means that you may run up against software limits in spreadsheet packages.

The field of view (FOV) represents that part of the study area to be acquired in a single scan. It's usually only part of the maximum FOV of the equipment (for example you probably do not need to scan the sky). This is usually set in the control software – if a camera is present the FOV may be set from the camera image (for the Leica, Cyclone permits this). If an image is not available, the limits may be entered into the control software manually – a clinometer may well be useful in making surface that the correct area has been scanned or that nothing is missed inadvertently.

It seems rather too obvious to mention that battery condition should be checked – both for the scanner and the laptop. There are few research situations which are quite as frustrating as having the equipment fail in the middle of data acquisition because battery power runs low. At best, part of the scan is lost, at worst, depending on how cleanly the equipment shuts down, the data and file structures used in the control software become corrupted.

An important part of the scanning process is to set the scan density correctly. If the density is too coarse then the surfaces being scanned are undersampled and the resulting models created will not represent the topographic variation reliably. This may have implications for change analysis or budgeting attempts. If the scan density is set too finely, then the land surface will be wildly over-sampled. This will have two unfortunate effects – the scan times will be excessive and the resulting scanned files will be enormous. Determining the appropriate scan density is a matter largely of trial and error.

The density is sometimes expressed as a point spacing at a given distance, perhaps 2 cm at 10 m. This means that if a wall was being scanned, the horizontal and vertical spacing of the sampled locations would correspond to the mesh points of a 2 cm grid. However, 2 cm at 10 m corresponds to a spacing of 4 cm at 20 m, 6 cm at 30 m, 8 cm at 40 m and so on. Locations which are further than 10 m from the scanner are therefore sampled less densely than locations nearer the scanner (5 m distance would imply 1 cm spacing) which are over sampled. This has implications for the scanner position relative to the area being scanned – to reduce the potential under-sampling at longer distances, it may be desirable to locate the scanner as high as possible so that the scan angles are not as oblique and the range of distances from the scanner is smaller. The scan time is then a function of the FOV and the desired scan density – the larger the FOV and the finer the scan density, the longer will be the scan time. Again, experience will show what can be achieved realistically in a scanning session. Complex vegetation or river channel configuration may necessitate several scans – the laser cannot see around corners.

Once a scan has been completed and it becomes necessary to move the scanner, it may be possible to simply pick the equipment up and move it. However, in the case of the Leica ScanStation the scanner weighs some 17 kg, so it is advisable to dismantle the equipment and place it in its' packing case before moving it to the new location. The set-up and shut-down times will take longer on early scans until experience is gained – however, they do limit the length of time that can be spent acquiring the topographic measurements. If the equipment has to be transported from a vehicle over some distance, this should be factored into the day's plans. Once the scanner has been moved to its new location and restarted, the targets for the second scan need to be recaptured. Some scanning procedures may be built around a foresight–backsight routine.

WHAT CAN GO WRONG IN THE FIELD?

The goal of scanning is to achieve successful and reliable data capture with the minimum of downtime. Careful planning and well thought out field procedures will repay time spent in their preparation. However, it is worth considering what problems might arise in the field and what workarounds, if any, there might be. To the list below many of the readers of this volume will undoubtedly be able to add their own entries.

With the multiplicity of interacting pieces of equipment that are required for scanning, a checklist of what should be taken is useful. One might arrive at the study site only to discover that one of the cables has been left behind. Perhaps it was the cable that should have been in the scanner transport case but which someone took out to check, or use for another purpose, and forgot to return. Perhaps it's the mains lead for the charger, or an adapter for the electricity supply in another country that momentarily gets forgotten when loading up the vehicle.

Some scanners will not operate in gusty conditions or when the windspeed is too high. If the scanner can be resited in a more sheltered location, some scanning may be possible. Snow, fog and mist can result in heavily reduced data acquisition – puddles on the ground from recent rainfall, or a heavy shower during the scanning session, may result in missed areas. The equipment usually itself needs to be kept dry – not just the scanner, but the batteries and the laptop.

It is worth reiterating that the batteries should be checked before setting out for the day's scanning. Should the batteries run down, is there a second set which can be used to continue the scan? Should the battery run down unexpectedly, did the hardware/software shut down cleanly?

Acquiring the target locations for multiple scans is vital to successful registration of the scans. A target may be moved inadvertently between scans – perhaps as a result of operator carelessness, the curiosity of some passer-by or a gust of wind. If this goes un-noticed there will be problems when attempting to register a series of multiple scans of the survey site.

Other problems encountered by the authors involved the software on the laptop 'hanging' – this usually requires a 'warm boot' (ctrl-alt-del) or a 'cold boot' (power off and on). In both cases you may lose data. Practising 'safe computing' – updating data files, closing unnecessarily open files, only running the control application software during acquisition – will minimise the risk of data loss.

A final problem we have encountered is that under some lighting conditions with low sun angles, the laptop screen becomes difficult to see – this is exacerbated if the operator is wearing sunglasses or photochromic lenses, doubly so if the operator also has presbyopia (this however is not a condition which usually affects young researchers).

PROCESSING

The output from a scan or set of scans is voluminous. With a scan rate of 4000 points per second, the limit of Excel 2002 can be reached in just over quarter of an hour's scanning. Multiple scans of quite modest areas will typically result in hundreds of thousands or millions of measurements. The manufacturer's control software will have little problem in handling and displaying these data on a typical PC or laptop. However, if a large area is to be scanned, or repeat scans are to be made of the same area, it is worth investing in a reasonably powerful workstation with a rapid graphics, and plenty of disk space. Systems with 1 terabyte (1024 Gbytes) of storage are common, and disk storage costs are expected to continue to decrease in the future.

Data backup should also be taken perhaps more seriously than it is – given the effort required to collect the data it would be tragic if that effort was wasted by a disk failure. It's easy to drop a laptop. Data sticks are easily lost. Backup disks are reasonably inexpensive, and at least one copy of important data should be stored in another building. Those who have lost hours of painstakingly collected and processed work will know the inner feelings of bleakness and isolation which follow in the aftermath of discovering that the data cannot be retrieved. We speak from experience here. We have also had laptops stolen.

It is tempting to regard GIS software as having a valuable place in the data processing. Scanned data is 3D – that is, there are explicit *x,y,z* components – there may also be intensity, and RGB components for each observation as well. Manipulating and editing the point clouds is, at the time of writing, perhaps best left to bespoke software. The linkage with GIS appears later in the chain – feature identification or extraction, integration with other data (perhaps geo-referenced aerial imagery) may be handled in a GIS, or it may require an add-on (for example Visual Learning Systems' LiDAR Analyst exists as a plug-in for ArcGIS, ERDAS Imagine, SOCET SET and GeoMedia and provides LiDAR-specific processing functions over and above what is available in these packages).

When a scan is made, the data points are given Cartesian coordinate measurements relative to the location of the scanner itself – the scanner is at the origin of a local coordinate system. The coordinate measurements themselves are generally in millimetres along the orthogonal axes of the coordinate space. Where multiple scans are made, these can be registered to the local coordinate system of the first scan for convenience. A final transformation to some national grid system is perhaps best left until all the scans have been made and registered together – control points for this will require accurate survey – RTK GPS will suffice.

One of the unfortunate side-effects of scanning is the presence of noise in the point cloud. Points or group of points can usually be selected and removed in the manufacturer's software, although the extent to which this is possible depends on theprogram. Such noise is sometimes easy to see, but less easy to remove. Birds landing and moving around in the field of view may be scanned – their size might mean that there are a few extra points acquired slightly above the ground surface. Larger animals may be easier to identify and remove. People walking across the scan appear as slightly ghostly outlines (depending on the scan density). This may not seem to be a problem, but if the desired output is a TIN or DEM, then the noise will have to be removed – if point removal is a problem in the editing software, then this may entail further manipulation in GIS or another program.

There are many reasons for using scanners to acquire detailed information about local topographic variation rapidly. Converting the point clouds to another form so that they may be displayed in a GIS entails the creation of a TIN (a vector representation of the surface variation) or interpolation onto the mesh points of a regular grid – a Chapter 5 describes the use of geostatistical techniques to achieve this.

Most of the control and editing programs allow the export of the point cloud to other software – an added bonus is that the export options will include more than just ASCII output to a comma separated text file. Our experience has been largely confined to acquiring point clouds to represent buildings. We have been using Cloudworx™ with AutoCAD™ to create wire frame models of the scanned buildings, and then to create rendered CAD drawings from the wire frames. These will be exported to a virtual reality environment for visualisation and modelling.

PROBLEMS

As with gaining the requisite skills in using the scanner itself, the control, editing and registration processes all require familiarity and early, regular practice will breed the familiarity that leads to field competence. A major problem occurs when there are insufficient reliably-acquired targets to provide enough control points to register adjacent surveys. One workaround is to register what can be reliably registered and then resurvey what cannot be registered. If the targets cannot be reliably replaced (e.g. they are not at monumented locations or you do not have accurate GPS positions) then there is little that one can do except re-survey the entire area, taking more care.

It may be that the scan density is fine enough that one can identify common locations in adjacent point clouds. We use this approach when scanning buildings where the corners of windows or intersections in brick pointing, or changes of colour, appear clearly in the overlapping scans and have corresponding points in each point cloud that one is registering. Even so, this approach requires some practice before the registration software will accept one's choices of common control points. This approach may not always be possible with topographic survey – if the scan density is too coarse, then the probability of finding corresponding measured locations in either point cloud may be extremely small.

SOME FINAL THOUGHTS...

This chapter might lead the casual reader to think that terrestrial scanning in particular is fraught with problems and drawbacks. This is not the case. Acquisition can be rapid, and with practice processing can be fairly rapid as well. However TLS is not an activity which should be approached either lightly or wantonly. We spent at least six months after acquiring the scanner in rudimentary and practice activity before we started to gain sufficient confidence in scanning to produce reliable scans with our equipment. We have discovered that after the initial efforts in familiarisation, knowledge can be passed on with relative ease. Field operators can be trained fairly quickly – at the National Centre for Geocomputation we have been fortunate to have had valuable scanning assistance from secondary school teachers who were seconded to us. We have also decoupled the scanning and processing activities – one group is responsible for carrying out the scanning, and another group is responsible for the subsequent processing and integration of the data, although one assistant was recruited for his AutoCAD skills. It is useful to find helpful advice such as that produced by English Heritage (2007), even though this is orientated towards heritage applications. The International Society for Photogrammetry and Remote Sensing (ISPRS) run annual laser scanning workshops – at the time of writing the most recent was in September 2007 and the proceedings (Rönnholm *et al.*, 2007) are available on-line. Each workshop provides a forum for the exchange of experience in laser scanning as well as the discussion of new techniques and approaches. Leica also currently run an annual 'Teaching of 3D Laser Scanning' Forum in the UK.

Scanning is one of a range of data acquisition techniques. Airborne and terrestrial scanning operate at different scales. In general, airborne scanning provides moderate detail over a wide area – this may change with the increased use of different aerial platforms. TLS provides, potentially, enormous detail over a smaller area – it may be used to 'fill in the holes' in an airborne survey when and where it is needed, but it is also an effective data acquisition technique in its own right.

In a sense you need to make a leap of faith with scanning. The magnitude of the entry cost dictates that this method of scanning automatically becomes a major data acquisition method in your research. It means that you will probably find that you form interesting alliances with people outside your immediate sphere of interest. The contributors to this book are a reflection of the range of activities in which scanning, whether airborne, marine or terrestrial, is an important component.

REFERENCES

English Heritage. 2007. *3D Laser Scanning for Heritage*. Swindon: English Heritage.

Fantoni R, Barbini R, Colao F, Ferrante D, Fiorani L, Palucci A. 2004. Integration of two LiDAR flouro-sensor payloads in submarine ROV and flying UAV platforms. European Association of Remote Sensing Laboratories. *EARSeL e-Proceedings* **3**(1): 43–52.

Fischer L, Huggel C, Haeberli W, Eisenbeiss H, Zublin M, Vallet J. 2008. Helicopter-borne LIDAR and multi-platform aerial photogrammetry for stability-related terrain analysis of steep high mountain areas: Monte Rosa east face. *Geophysical Research Abstracts* **10**. EGU2008-A-09525.

Leica. 2006. *Leica ScanStation/HDS3000 User Manual*. Leica Geosystems AG: Heerbrugg.

Natural Environment Research Council. 2008. NERC Airborne Research and Survey Facility. URL: http://arsf.nerc.ac.uk/howtoapply/planning.asp

Rönnholm R, Hyyppä H, Hyyppä J. 2007. *Proceedings of the ISPRS Workshop 'Laser Scanning 2007 and SilviLaser 2007'*. ISPRS. URL: http://www.commission3.isprs.org/laser07/ls2007.pdf

Wand M, Berner A, Bokeloh M, Fleck A, Hoffmann M, Jenke P, Maier B, Staneker D, Schilling A. 2007. Interactive Editing of Large Point Clouds. In Bosch M, Pajarola R (eds). *Proceedings of the Eurographics Symposium on Point Based Graphics*, IEEE.

4 Airborne LiDAR: Instrumentation, Data Acquisition and Handling

BERNARD DEVEREUX AND GABRIEL AMABLE

Department of Geography, University of Cambridge, Cambridge, UK

INTRODUCTION

The last decade has seen the rapid emergence and growth of Light Detection and Ranging (LiDAR) technology as a powerful remote sensing tool that is having an impact throughout the environmental science sector. Its spectacular success is primarily related to its ability to enable the creation of 3D landscape models and their surface features from ground-based, airborne and space-borne devices. LiDAR sensors are characterised by an ability to generate very accurate data at high spatial resolution. When integrated with data from the plethora of spectral remote sensing systems that have been deployed over the last few decades, they offer exciting possibilities for 3D analysis and visualisation. Historically, generation of 3D landscape products in the form of digital terrain models (DTMs) and digital elevation models (DEMs) has been a complex, time consuming and expensive process. Frequently, such products have been characterised by lengthy production processes, interpolated data values and significant levels of error (see for example, Hayward *et al.*, 2006). Whilst it would be wrong to suggest that LiDAR surveying is anything other than complex, it now has a proven ability to measure heights directly, accurately and at very high sampling density over significantly large areas.

This ability underpins its attractiveness and success. Both airborne and terrestrial LiDAR systems have become major scientific tools in a wide range of subjects including Archaeology, Agriculture, Botany, Biology, Earth Sciences, Ecology, Forestry, Geography, Environmental Science and Landscape Ecology. Rapid development of systems has been facilitated by research in Applied Mathematics, Computer Science, Electronics and Physics and the entire process has been driven by high levels of demand from survey organisations, environmental management agencies and planning authorities. LiDAR systems thus offer a resource that is important to a large and diverse market sector. Certainly some parts of that sector have a demand backed up by significant financial resources, and this has been crucial in enabling the extremely rapid uptake of the technology over the last five years.

As with any rapidly emerging technology it can be difficult for the user community to keep abreast with developments and potential. For every existing LiDAR user, there are undoubtedly a significant number of potential users who have to make decisions about what they can gain if they make the significant investment in using the data for their applications. These individuals face a multitude of questions and uncertainties.

Laser Scanning for the Environmental Sciences,
1st edition. Edited by G.L. Heritage and A.R.G. Large.
 ISBN 978-1-4051-5717-9

How much will it cost? How does it work? Will it give me the data that I want to make my application a success? Is this the right time to invest or should I wait while the technology improves and becomes more cost-effective? Can I trust vendors, consultants and suppliers to help me progress my application? What are the risks involved? What kind of LiDAR survey device/data will best suit my needs? How will I go about processing the data and what are the pitfalls? How reliable is the technology?

In the last analysis, the only certain way of resolving these questions is to take the plunge and become a user of the technology. However, doing this against a background of knowledge about how the technology works and how it can be applied can make the process considerably easier and safer. Central to this is an understanding of how data is recorded, what types of error it might contain, what assumptions are made in collection and processing, what impact these might have on final data products and how the technology might develop. This chapter will attempt to address these questions by focusing on LiDAR data collection and processing. It will begin with a review of laser technology and laser ranging principles. We then consider how these principles translate into systems that can be deployed for LIDAR surveying and how instrument and platform design impact on the properties of data recorded. Next we describe principles of survey design for airborne LiDARs before presenting an overview of data processing techniques for generation of primary data products. Finally we outline methods for generating the most commonly used derived products in the form of Digital Elevation and Digital Terrain Models, before speculating on where LiDAR processing techniques will develop in the future.

LASER RANGING: THEORY AND PRACTICE

LiDAR is one of many survey techniques based on the use of lasers. As the type of laser used in any particular system has a major influence on its capabilities, it is useful to identify the features of lasers that have a bearing on system capabilities for environmental applications of LiDAR.

The term 'laser' is an acronym for Light Amplification by Stimulated Emission of Radiation. Laser theory dates back to the work of Einstein almost one hundred years ago. Einstein had suggested theoretically that when a photon hits an atom with raised energy levels two photons are emitted with identical properties. Furthermore, the direction of travel followed by the two photons is the same as the original photon. This means that if the electrons in a suitable medium are excited by an external energy source and then the medium is bombarded with photons, a reaction will take place in which the number of photons emitted can be dramatically increased. This is the underlying principle of 'stimulated emission'. In an operational laser (Figure 4.1) the medium is contained within a tube or cavity with a reflective interior and it is 'excited' by some form of 'energy pump'. The photon source is located at one end of the tube

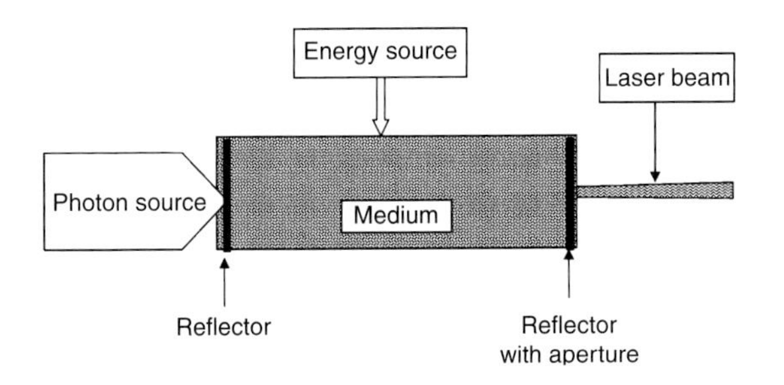

Fig. 4.1 Schematic structure of a laser. For laser surveying key system parameters are laser power, wavelength, pulse repetition rate and beam divergence. Power is a major determinant of the range and sensitivity of the system. Wavelength also influences range and has a major impact on surface reflectance/interaction. Pulse repetition rate has a major impact on resolution while beam divergence influences coverage, target detection and scattering. In any system these parameters interact to determine overall performance.

and photons are reflected back and forth along its length. As they collide with the atoms of the medium the number of photons increases in a chain reaction. A hole at the opposite end of the tube allows some of the photons to escape as a 'laser beam'.

The nature of the photon source, the material of the response medium and the level of energy exciting the atoms determine the properties of the resultant laser light or 'beam'. Laser light has a single wavelength (is monochromatic), is highly focused with minimal divergence (highly collimated) and has a high level of coherence (in phase). As laser technology has developed, a vast number of different mediums have been used to generate laser light in conjunction with a wide range of energy sources. The first operational laser was attributed to Theodore Maiman in 1960 and used a synthetic ruby crystal as its medium (Maiman, 1960). It worked at an optical wavelength (694 nm). Other workers had earlier demonstrated that the principle, which was originally applied to microwave radiation and was hence termed 'maser' (microwave amplification by stimulated emission of radiation), could operate in the visible and near infrared parts of the spectrum (Schawlow & Townes, 1958). Since then there has been an explosion in laser technology. Most lasers operate either in continuous wave mode or in pulse mode in which short bursts of energy are emitted in very rapid succession. Significant developments for LiDAR include greater control over wavelength (tuning), high frequency pulsing lasers, miniaturisation, proliferation of laser media including solid state devices and greater laser energy outputs. Given the strong reflectance of vegetation and other earth surface features at infrared wavelengths, coupled with relatively low levels of noise and scattering, these tend to be used in most environmental LiDAR systems.

The basic principle of LiDAR sensors is thus to fire a laser pulse at a target and use an inbuilt photodiode-based detector (Wehr & Lohr, 1999) to sense its reflection. A highly accurate clock is used to measure the time taken for the laser energy to travel from the laser source to a target and back to the detector – the so called 'time-of-flight'. The distance or range from the source to the target can be calculated from:

$$R = \frac{ct}{2} \tag{4.1}$$

(where, R, the range in metres; c, the speed of light (299,792,458 m s^{-1}); t, the two-way time-of-flight) by virtue of the fact that the speed of light is constant (see Baltsavias, 1999). If the location and precise orientation of the laser source is known, then the precise location of the reflecting target can also be calculated.

Clearly, the nature of the laser is of fundamental importance to LiDAR systems. Because the power of a laser diminishes as the square of distance travelled (the inverse square law), laser power determines the effective operating range of a LiDAR system. It also determines the extent to which it will penetrate vegetation canopies and be reflected by wet or damp and other low reflectivity surfaces such as ice. Similarly, the wavelength of the laser has a major impact on surface reflectance and also on atmospheric scattering. Operationally, the optimum laser for many remote sensing applications is a high power, infrared system. Unfortunately, optimum performance has to be balanced against safety considerations because this type of laser is the most dangerous in terms of radiation exposure. The main risks that arise from operating lasers are damage to eyesight, burning of skin and ignition of flammable targets. Current classification of lasers in terms of safety comprise four main classes, with Class I lasers being the safest, with no exposure of humans to the beam, and Class IV which incorporate powerful lasers being the least safe. Their potential for use in remote sensing applications depends on the laser power, with Class IV lasers being most suitable for airborne operations so long as the necessary safety mechanism can be put in place.

INSTRUMENTATION, PLATFORMS AND LASER OBSERVATIONS

The LiDAR principle is embodied in a wide range of remote sensing systems with a variety of different laser types. With the exception of hand-held laser hypsometers, most involve arranging some form of scanning mechanism to enable area coverage. Usually, this involves the use of a rotating mirror or prism (Burtch, 2002) but there are also designs that rely on pushbroom optics (Schnadt & Katzenbeisser, 2004). Typically scan rates of up to 600 Hz (i.e. 600 scan lines per second) are used across a field of view of up to 75 degrees. In the case of a laser mounted in an aircraft pointed at a flat, horizontal surface below, this means that the laser illuminates a circular area or 'footprint' at the nadir point of the scan with a diameter dependent on the beam divergence. As the laser progresses along each scan line away from the nadir, the footprint becomes an increasingly elongated ellipse. As the platform moves forward the interaction between the scanning pattern and the platform motion results in a 'saw tooth' pattern of laser hits (Figure 4.2). The rate of platform movement, the laser repetition rate, the scan rate and the altitude of the sensor determine the density of points recorded. This is often referred to as the survey resolution. Aerial LiDAR surveys typically have a resolution of up to about 10 points per square metre from fixed-wing aircraft, and up to 200 points per square metre from helicopters. Ground-based systems produce considerably higher measurement densities.

For any laser mounted on a moving platform such as an aircraft, accurate ranging demands that the location of the laser source must be known at all times; so the laser must be integrated with instrumentation for measuring position (x,y,z) and orientation (ω, ϕ, κ). A GPS is usually used to measure the necessary positional information and an inertial measurement unit (IMU) or Inertial

Fig. 4.2 The principle of airborne LiDAR operation. The laser illuminates a circular area (footprint) at the nadir point of the scan with a diameter dependent on the beam divergence. For moving platforms, the laser is integrated with a GPS and an inertial measurement unit (IMU) or inertial navigation system (INS) to provide the necessary orientation data. See Plate 4.2 for a colour version of this image.

Navigation System (INS) is used to measure the necessary orientation data (Figure 4.2).

For a moving platform such as a fixed-wing survey aircraft, typical velocities are in excess of 100 knots (~115 miles per hour) and in modern LiDAR systems laser pulses may be generated at pulse 'repetition rates' of up to 150,000 pulses per second (150 Khz). This means that position and orientation information must be measured at sufficiently high frequency to characterise the change in position and orientation of the sensor. A measurement frequency of 2 Hz or higher would enable the position of the sensor to be determined at these kind of scan rates. Because aircraft attitude varies much more rapidly than position, IMU measurement frequencies of 100 Hz or greater are usually needed for effective recording of attitude. On slower moving platforms such as boats and cars, correspondingly lower frequency of position and attitude measurement may suffice (see for example Barber *et al.*, 2008).

Table 4.1 shows an example of the position and attitude data recorded during a LiDAR survey. These are the basic observations which are essential in geolocating each laser pulse. Every position and attitude record is tied to a GPS time stamp and each record contains an *x,y,z* position and a record of IMU roll, pitch and heading. Because IMU measurements are relative recordings, the roll pitch and heading values at any point in time are the sum of previous incremental changes. This integration process introduces cumulative errors or drift into the measurements which must be modelled and removed by some form of filtering based on the GPS data. The Kalman filter is the most widely used approach (Roy *et al.*, 1997).

In addition to positional data, each LiDAR observation must also contain the scan angle for each laser shot together with some measure of laser reflectance of the target. This reflectance data is processed to compute the range and target surface properties (Table 4.2). Clearly, the form of the outgoing pulse is a major factor in determining the form of the reflection, and most systems record both the nature of the outgoing laser pulse and its reflection when it arrives back at the sensor (Figure 4.3). Figure 4.3a illustrates the basic principle for a solid reflecting surface normal to the laser. The outgoing laser pulse has a Gaussian distribution by virtue of factors including the laser

Table 4.1 Sample records of position and attitude data collected from GPS and IMU systems in a LiDAR survey (angles indicated are in radians and distances are in metres).

WvfTime	Roll	Pitch	Heading	Lat	Long	Elev
139049.706219	−0.009088	0.057801	−0.052701	0.915451	6.279799	789.861206
139049.706249	−0.009090	0.057801	−0.052703	0.915451	6.279799	789.861145
139049.706279	−0.009093	0.057801	−0.052704	0.915451	6.279799	789.861145
139049.706309	−0.009095	0.057801	−0.052706	0.915451	6.279799	789.861084
139049.706339	−0.009098	0.057802	−0.052707	0.915451	6.279799	789.861023
139049.722778	−0.010253	0.057816	−0.053549	0.915451	6.279799	789.837524
139049.722808	−0.010256	0.057816	−0.053551	0.915451	6.279799	789.837463
139049.722838	−0.010258	0.057816	−0.053552	0.915451	6.279799	789.837402
139049.722868	−0.010261	0.057816	−0.053554	0.915451	6.279799	789.837402
139049.722898	−0.010263	0.057816	−0.053555	0.915451	6.279799	789.837341
139049.722928	−0.010265	0.057816	−0.053557	0.915451	6.279799	789.837280
139049.722958	−0.010268	0.057816	−0.053558	0.915451	6.279799	789.837219
139049.722988	−0.010270	0.057816	−0.053560	0.915451	6.279799	789.837219
139049.723018	−0.010273	0.057816	−0.053561	0.915451	6.279799	789.837158

Table 4.2 Sample records of range and associated attribute data collected in a LiDAR survey (angles indicated are in radians and distances are in metres).

GPSTime	ScanAngle	FP_Range	LP_Range	Fpi	Lpi
139049.706219	−0.297930	780.990540	780.994934	191	191
139049.706249	−0.296926	780.840576	780.774963	187	187
139049.706279	−0.295922	780.480591	780.564941	177	177
139049.706309	−0.294751	780.340576	780.354919	192	192
139049.706339	−0.293747	780.030579	780.034912	186	186
139049.722778	0.306670	777.000671	782.144958	0	119
139049.722808	0.307172	776.840698	785.505005	52	87
139049.722838	0.307674	776.550720	785.895020	66	105
139049.722868	0.308176	776.830688	786.045044	46	154
139049.722898	0.308511	776.670715	786.065002	4	220
139049.722928	0.309013	780.440552	786.265015	8	200
139049.722958	0.309347	780.250610	786.275024	32	171
139049.722988	0.309682	779.790588	786.345032	36	125
139049.723018	0.309849	779.700623	784.655029	50	65

generation mechanism and beam divergence. The simple surface results in the reflection of a similar Gaussian shape.

First, the system detects this return pulse (as shown in Figure 4.3a). The calculation of a range for the detected pulse involves measuring the elapsed time between the outgoing pulse leaving the sensor and the receipt of the return pulse. This time-of-flight is then used in Equation (4.1) to calculate the range to the reflecting surface. A decision has to be made about where timing begins for the purpose of range calculations. Once a point is selected for this purpose on the outgoing pulse, a corresponding point must be found on the reflected pulse for calculating the time-of-flight. Usually a location on the leading edge of each pulse is used because the pulse width and amplitude of the return signal are influenced by surface properties, so bias and incorrect results are avoided by looking at the start or leading edge of the reflection. The most commonly used approach is the 'constant fraction detection' (or constant fraction discriminator) method. This involves determining the peak amplitude of the outgoing pulse and accurately locating a point on its leading edge which is a fraction of this amplitude – say 50%. The process is repeated for the return pulse, employing the same (constant) fraction to determine a similar location on the leading edge. The time difference between these two locations is the time-of-flight. This is illustrated in Figure 4.3b. In most LiDAR systems the pulse detection, peak amplitude determination and pulse timing with the constant fraction discriminator are all handled by electronic devices in the hardware.

Pulses of the type shown in Figure 4.3a are relatively rare because most targets have a more complex structure. For example, a laser pulse may clip the edge of a flat roofed building, in which case, a part of the laser energy would be reflected from the roof of the building and a part continues to be reflected from the ground. The result would be two distinct Gaussian pulses in the return signal, one corresponding with the reflection off the roof and one corresponding to the ground. In processing this particular return signal two ranges must be found – corresponding to each of the detected pulses. The time difference between these two pulses corresponds to the difference in elevation, or the height of the building (after the necessary adjustment for scan angle).

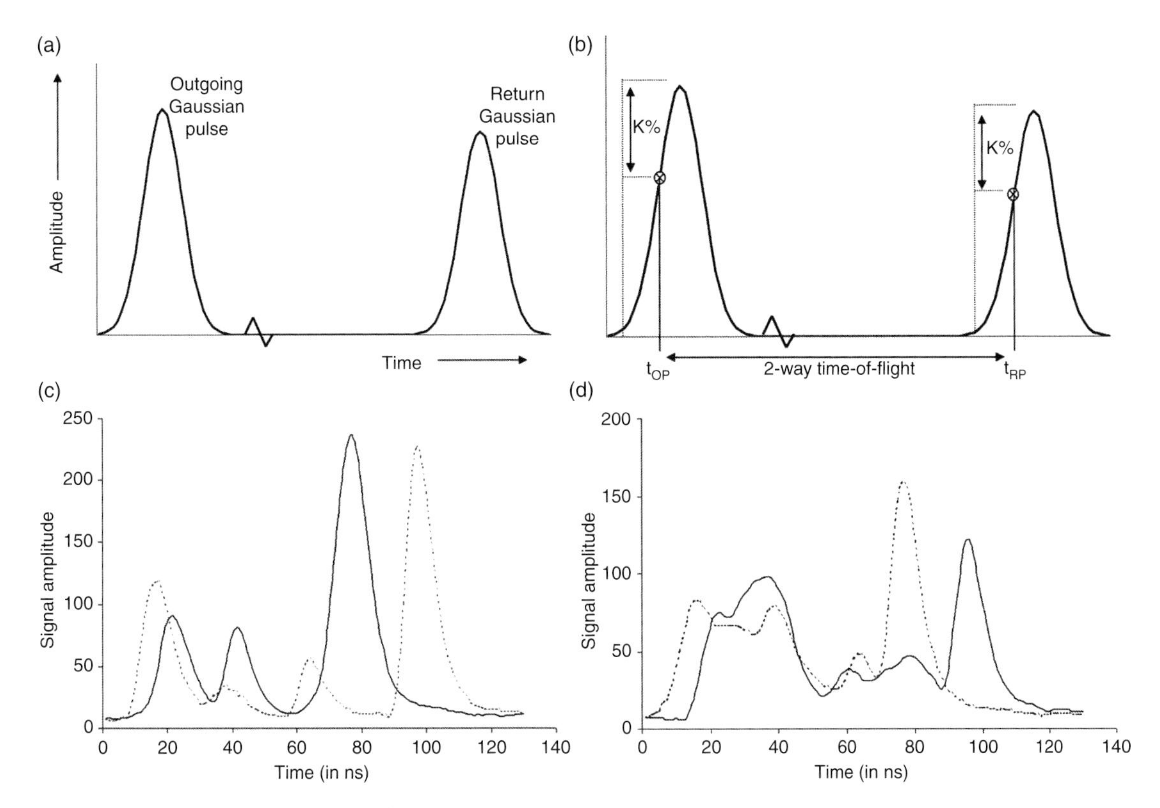

Fig. 4.3 (a) For each laser shot fired the outgoing energy normally has a Gaussian distribution. A reflecting target with a smooth surface normal to the laser beam will also return a Gaussian energy distribution. The time interval between the outgoing pulse and the return pulse is equivalent to the range or distance between the laser system and the target. (b) An illustration of the use of the Constant Fraction Discriminator in the determination of time-of-flight between the outgoing pulse and the return pulse. The 2-way time-of-flight is the time difference between t_{OP} on the outgoing pulse, located at a point which is a constant fraction, K%, of the peak amplitude, and t_{RP} similarly located at the K% point on the return pulse. (c) Examples of return waveforms. These return waveforms present relatively simple structures with distinct, clearly identifiable Gaussian elements, one (solid line) with 3 pulses and the other (dashed line) with 4 pulses. (d) Examples of complex waveforms. These waveforms are composed of several Gaussian pulses some of which are merged or convolved. LiDAR returns from vegetation canopies are generally characterised by complex waveforms. These types of waveforms require advanced Gaussian deconvolution techniques to identify and separate the individual elements. The two examples shown when deconvolved would have 6 or more Gaussian elements.

Figures 4.3c and 4.3d show still more complex possibilities where the laser beam interacts with a tree. Energy may be reflected back from different parts of the tree as each laser pulse clips branches and leaves on its way through the tree column. Each of the reflecting branches will produce a pulse in the returned signal. Ultimately, the energy contained in the outgoing pulse will encounter a solid object such as a large branch or the ground. Thus the return signal may contain more than two detectable pulses. Figure 4.3c shows two examples of return waveforms containing multiple pulses. In principle, a range and a location can be found for every pulse in the waveform. In practice, reflecting objects may be so close to each other that the series of individual Gaussian pulses tend to merge to form complex waveforms (such as those shown in

Figure 4.3d). Extraction of ranges in this case becomes much more difficult and few operational systems deal with this situation. Understanding the relationship between waveform shape and landscape surface properties is a research frontier that requires methods for de-convolving or modelling the component Gaussians (Wagner *et al.*, 2006) and linking them to surface structural properties.

LiDAR systems can be categorised in part by the way they process the waveform reflections for each laser pulse and also by the size of the footprint they record. Systems that record footprints with a diameter of up to 100 cm are often referred to as 'small footprint systems'. They typically fire laser pulses at high frequency (up to around 150 kHz) producing very high measurement density. But due to data storage and bandwidth constraints during the early part of the development of airborne LiDAR technology, most systems only recorded a subset of objects in each return signal. Early small footprint systems only recorded the range to the first reflecting object or the 'first pulse'. In principle, a map of all first pulses results in a digital **elevation** model showing the height of all surface objects; the name digital surface model or DSM has also been used to describe this type of elevation model. For many applications however, there is a need for a **terrain** model (or DTM) with buildings and vegetation removed. This gave rise to demand for a record of the last reflecting object in each return signal if there was more than one reflector – often referred to as the 'last pulse'. As digital elevation and digital terrain models are probably the most widely used LiDAR products at present, it is not surprising that most small footprint systems record just first and last pulse.

It is commonly assumed that a map of all last pulses approximates a terrain model. However, there are distinct constraints on the extent to which these can actually provide an accurate model. Whilst last pulse data clearly has the potential to penetrate vegetation canopies, there can never be any guarantee that the last pulse will have reached the ground and not have been reflected from higher in the canopy. As a consequence these datasets often over-estimate terrain height. This issue is of particular significance in systems with relatively low-powered lasers, flown towards the upper limit of their altitudinal range. Furthermore, where low vegetation is involved, the first and last pulse may be too close together to generate a reliable last pulse range – again leading to over-estimation of terrain height. Extraction of accurate digital terrain models from small footprint data and evaluation of their accuracy is an important area of current LiDAR research (Zhou *et al.*, 2004; Kobler *et al.*, 2007).

By way of contrast, systems that record footprints in the size range 10–100 metres are usually referred to as 'large footprint' systems. These usually have a much lower repetition rate and as a consequence may record just a few (6–12) footprints in each scan line. However, this opens up the possibility of recording the entire return signal for each laser pulse. Usually the waveform is digitised by recording the amplitude of the return signal at fixed time intervals. Most of these systems have been developed by research groups within NASA as precursors to space missions. Examples include SLICER (Scanning LiDAR Imager of Canopies by Echo Recovery: Harding *et al.*, 2001), and LVIS (Laser Vegetation Imaging Sensor: Drake *et al.*, 2002; Blair *et al.*, 1999). These devices have been primarily designed for measuring vegetation properties, and extensive research has shown that waveform shape is related to canopy biophysical parameters including height, basal area, vertical distribution of canopy components, biomass, leaf area index (Lefsky *et al.*, 1999) and light transmittance. A large footprint, full waveform spaceborne system is operational as part of NASA's ICESAT mission. Its primary role is for monitoring the polar regions, but it is also capable of providing data for global vegetation monitoring.

The current state-of-the-art in LiDAR instrumentation are small footprint systems that record the entire waveform. Historically, this has not been possible due to the amount of data involved and the need to record this in real time during each LiDAR survey. For example, a laser repetition rate of 50 Khz will produce over 1.5 GB of data per minute if each pulse is digitised at 1 nanosecond intervals at 8-bits per sample for

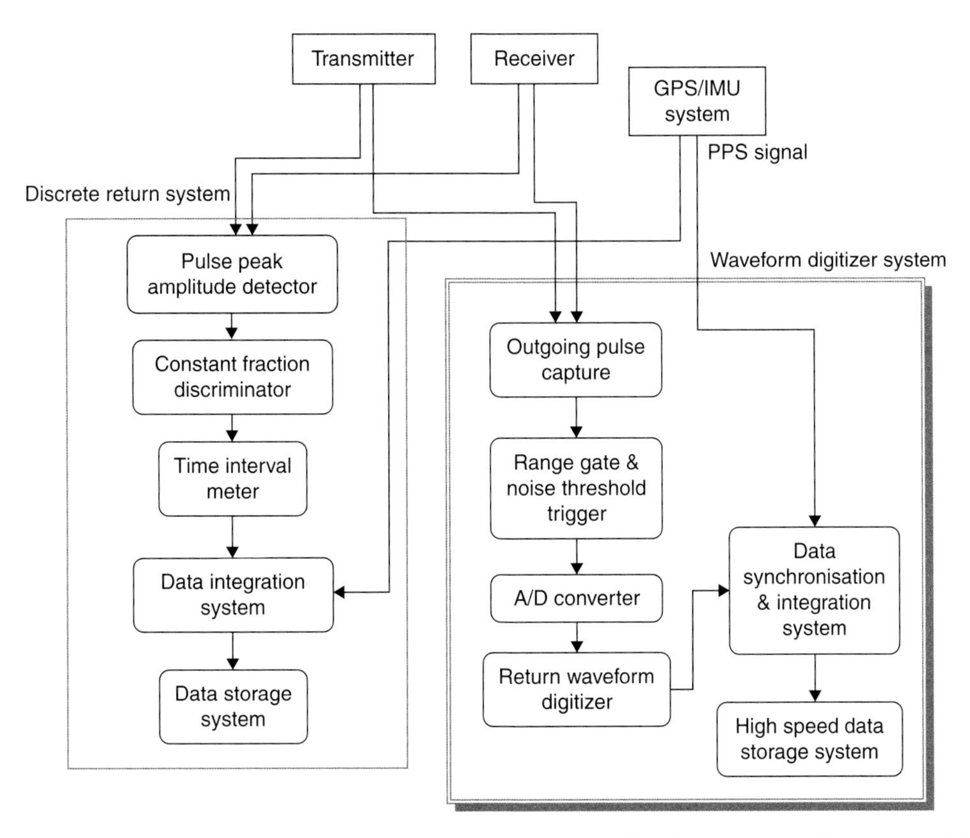

Fig. 4.4 The main components of the Waveform Digitizer System. This design is a product of a collaborative development undertaken by Optech Inc. of Canada and the Unit for Landscape Modelling, University of Cambridge. As shown, the digitiser is integrated with, and operates in parallel with, the existing discrete return system.

a height range of around 75 m (around 512 samples). However, rapid improvements in disk storage technology, bus data transfer speeds and on-the-fly data compaction algorithms have now made small footprint, full waveform LiDAR a practical proposition. Figure 4.4 shows the main elements of such a system developed by Optech Inc. in a joint program with the Cambridge University Unit for Landscape Modelling. (The examples of waveforms shown in Figures 4.3c,d were collected with this system.)

INSTRUMENT DEPLOYMENT AND LIDAR SURVEY DESIGN

Regardless of the type of instrument involved, successful collection of LiDAR data hinges on careful planning and good survey design. This begins with a clear understanding of the data collection objectives. Potential users should ask some basic questions before embarking on any data collection project or programme:

- What is the operational requirement and is it clearly understood? How will LiDAR contribute to achieving the specified requirement? Can it be achieved by less expensive means? Do other types of data need to be collected at the same time as the LiDAR survey – for example aerial photography or line scanner data?
- Is a primary survey necessary? Has a search been conducted to ensure that the site has not been surveyed previously for another user, resulting in existing data that would be much cheaper to acquire?

- Are there time constraints on the survey, such as times of the year when there are no leaves on trees or times of the day when the tide is at its lowest?
- Is there a need for simultaneous collection of field data – vegetation heights, crop state, water/soil state?
- What coordinate system will be used for recording the data and what kind of ground control will be needed for the survey? Is existing ground control provided by national survey organisations likely to be adequate or is some additional ground survey required, bearing in mind that typical levels of accuracy for LiDAR surveys may be higher than those for existing mapping?
- How accurate does the data need to be and what resolution is required? Given that survey cost is highly sensitive to survey accuracy and resolution, are these being specified at the right level to meet the operational requirement?
- What are the risks involved in implementing the survey? Although LiDAR can be deployed both day and night because it is an active sensor, timing constraints such as those referred to above coupled with acceptable weather windows can severely limit the likelihood of successful data collection if they are over-specified.

Too often, all of the above critical questions are overshadowed by one key question that users or potential users of LiDAR understandably have uppermost in their minds – how much will it cost? Regrettably, every survey is different and failure to specify answers to the above questions in a short but clear operational specification can result in massive cost variation. This in turn might result in surveys appearing prohibitively expensive or, possibly worse, users paying more for data collection than they actually need to. The first step in any survey should thus be the preparation of an operational specification dealing with each of the above issues together with a map of the survey area. Once this information is available it is possible to cost the survey.

For most applications the target survey site will be an irregular shape – perhaps the margins of an estuary, an area of woodland or an area of settlement. For the purposes of aerial survey, however, it may be necessary to define an enclosing polygon within which flight lines of survey data can be collected. Generally, though not always, flight lines must be straight and will need to be arranged so that the ground swaths overlap to ensure complete coverage of the survey area. In flat terrain and under stable flight conditions, an overlap of between 10–15% will be adequate, but in mountainous terrain where there are large changes in terrain height, and also under turbulent flight conditions, larger overlaps may be necessary. To minimise cost it is necessary to try and achieve as few flight lines as possible. This minimises the number of aircraft turns required to complete the survey and hence contributes to reducing costs. Because LiDAR survey requires continuous GPS tracking there can be a significant risk that GPS lock will be lost if the aircraft banks steeply. This means that turning movements need to follow wide arcs and can take significant lengths of time (see Figure 4.5). For this reason, long and thin rectangular shaped polygons work out better and are more cost effective to survey.

Once a polygon has been defined for planning the survey it can be used for constructing a flight plan (Figure 4.5). The key parameters for this are the aircraft velocity, the flying height, the swath or FOV of the sensor, the required flight line overlap, the target point density, the laser pulse frequency and the scanning frequency. When combined in a simple geometric model, these parameters enable an estimate of the volume of LiDAR data that will be collected by the survey and also the number of flying hours required for the survey. From these two key variables it is then possible to accurately calculate costs. Simple spreadsheet calculations can approximate this planning process (see for example Figure 4.6). However, for detailed survey planning and generation of a flight plan that can be used by the survey team for data collection, more detailed information is needed. There are numerous proprietary systems available for this and most of these drive an 'on-screen' display which can be mounted in the aircraft cockpit to guide the pilot in data collection. These systems also enable account to be taken of restricted airspace and special features of survey sites that might preclude access, such as danger areas or steep mountain sides.

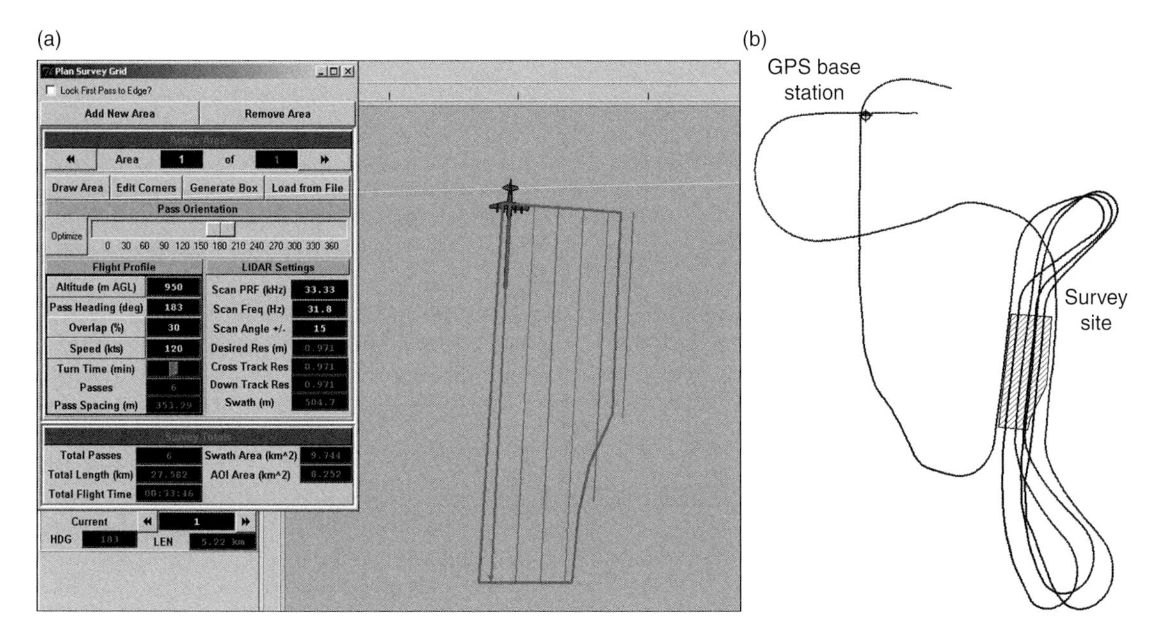

Fig. 4.5 (a) Flight planning for LiDAR survey. The planning involves deciding the correct 'flight profile' (a combination of aircraft/flight parameters) and LiDAR system settings that will generate the required data. (b) A plot of the aircraft trajectory flown during the survey. Note that a substantial part of the flight time is spent to turn the aircraft around as each of the planned flight lines are flown in turn. The aircraft also flies a series of manoeuvres over the GPS base station before and after the survey. This is necessary to ensure correct alignment of the IMU, which is important for the accuracy of the position and attitude measurements.

It is only when a survey plan at this level of detail has been created that accurate costs can be calculated. Additional costs may include the cost of aircraft transit to the survey area and the costs of deploying supporting GPS and other ground-based data collection. Small and apparently insignificant changes in the key planning parameters can result in very significant changes in the average cost per kilometre for a survey so such figures need to be treated with extreme caution, although some survey companies do try to average out the costs of all their flying and quote on this basis. Once actual costs are known, users are in a position to carry out an effective cost benefit evaluation of their planned survey.

DATA PROCESSING AND PRODUCTION

From the preceding description of LiDAR principles and data collection methods, it is evident that significant processing is needed to enable the creation of useable data. As shown earlier in this chapter, typically a survey will generate several data streams including aircraft position data from the on-board GPS; sensor attitude data from the IMU; laser pulse data from the LiDAR instrumentation and a variety of logs used to record aircraft behaviour, instrument behaviour and recording errors. All of these data streams include a highly accurate time stamp using the system's GPS clock. Primary data processing involves integrating these data streams into a database and generation of a so-called 'point cloud' containing locations, elevations and intensities for every laser pulse. Usually, each record in the point cloud contains information for a single laser shot, but it should be noted that each reflecting object (first pulse, last pulse etc.) returned for a laser shot will have its own location, elevation and intensity.

Figure 4.7 presents an overview of the data integration process. This begins by calculation

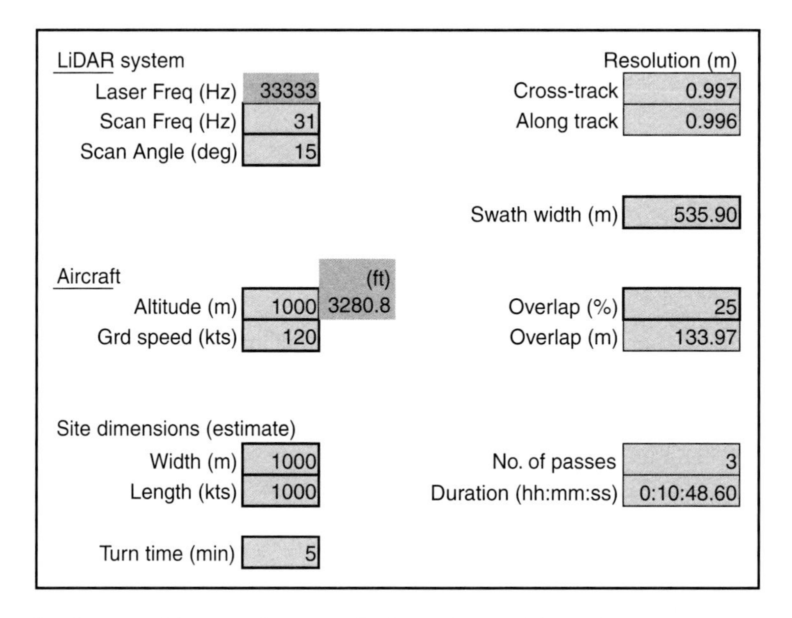

Fig. 4.6 An example of a spreadsheet application which can be used for airborne LiDAR survey planning. The example uses a 33 kHz system. The 1 km by 1 km area will take just over 10 minutes to survey at a point density of approximately one point per square metre. In order to produce an even point spread across the survey site, the survey planner will choose a combination of scanner frequency and scan angle which will ensure that the across- and along-track resolutions are very much the same (A copy of this spreadsheet can be accessed from the website of the Unit For Landscape Modelling at: http://www.uflm.cam.ac.uk/lidar_planning.xls).

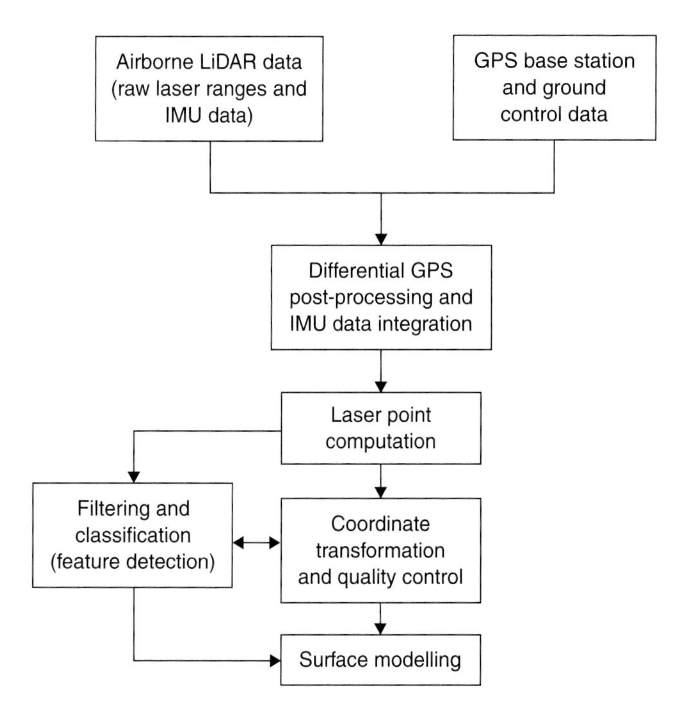

Fig. 4.7 Schematic overview of main data processing stages to integrate and convert airborne sensor and associated ground based recordings into point cloud and derived data products. The entire process requires an efficient data integration system as well as robust database and process management infrastructures.

of the aircraft/sensor trajectory by combining the ground and airborne GPS data and computing a differential solution to yield accurate position coordinates. Next, the attitude data are used to model the sensor geometry at each point along the trajectory. As noted above, filtering is used to integrate the higher frequency attitude information with the lower frequency, fixed points in the aircraft's GPS trajectory. Integration of the GPS and IMU observations with a geometric model of the sensor scanning mechanism enables a vector to be found for each laser pulse. Processing of each laser pulse involves calculating a range and location for each reflection (first pulse, last pulse and any recorded intermediate pulses). Because all calculations are carried out in GPS coordinate space, initial locations and heights are usually generated as measurements on the WGS ellipsoid. The final stage in primary data production usually involves transformation of these values into a local coordinate system such as the British National Grid.

Creation of a point cloud is the end product in a variety of measurement and recording processes involving numerous instruments. Inevitably, all of these processes introduce errors into the data, and an important aspect of generating a laser point cloud is the assessment of and, if possible, removal of errors to achieve a desired level of accuracy consistent with the instrumentation. Production of consistently accurate and reliable point cloud data hinges critically on LiDAR calibration procedures. These involve the use of an accurately surveyed calibration site to provide a 'ground truth' dataset which can be used to evaluate LiDAR performance. The calibration site is surveyed periodically with the LiDAR and the resulting measurements are used to evaluate errors in elevation, roll, pitch and heading. By modelling these errors it is possible to generate correction factors or 'calibration coefficients' that can be used to correct individual surveys. Equipment calibration errors are often manifest in LiDAR data as systematic offsets between adjacent flight lines which can be checked in areas of flight line overlap. Detailed accounts of calibration issues and techniques can be found in Katzenbeisser (2003).

With careful calibration and diligent processing, typical accuracy of LiDAR vertical heights for solid reflecting surfaces is in the region of ±15 cm at ranges of 1200 m. Accuracy evaluation is usually the final stage of data production. Ideally, it involves the use of a sample of independent, ground surveyed, check points to compute RMS height and position errors. Because current LiDAR technology significantly exceeds the resolution and accuracy of most readily available survey data, it is often the case that recourse to primary ground survey is needed to generate check points. It is also often the case that the single largest source of error in LiDAR surveys arises at the transformation stage into local coordinates, so considerable care needs to be taken when interpreting accuracy assessments. Interestingly, in a comparison of LiDAR survey datasets carried out by the present authors for two surveys of the same site by different survey organisations, using different systems for solid, reflecting surfaces, at a time interval of almost five years, mean height differences of less than a couple of centimetres were found. This underlines the potential of the technology for remarkably high levels of recording accuracy.

GENERATING LIDAR-DERIVED PRODUCTS

Point clouds are the natural data structure for storing laser data and have considerable flexibility in this respect. In addition to ranges/heights, other laser pulse attributes can be stored in point cloud records include GPS time, total number of pulses detected and waveform properties such as pulse width and amplitude. However, as a data structure for representing and displaying height observations, they also have certain disadvantages. In particular, they tend to require significant amounts of storage space and can be very slow to process if the density of points is high. As a consequence, many applications begin by conversion of laser point data to a grid format. Grids usually provide a more compact method for data storage and consequently facilitate significantly faster processing as well as better visualisation. Furthermore, they represent the GIS industry standard format for handling terrain height information, so gridding laser points opens up access to a vast and well-established world of processing methodologies and software for applications development. If grids do have disadvantages as a structure for manipulating LiDAR data, it is because they involve some generalisation of the data and in most GIS systems can only be handled as single-valued data sets. This means that a different grid is needed for every laser pulse attribute.

Converting point data to grid formats is a major topic in the GIS world and there is a very well developed literature describing a plethora of approaches – all of which have particular claims in terms of speed, accuracy and quality for particular purposes. Generally, these can be divided into global and local procedures. The former involve fitting a simple or piecewise function to the data which expresses height in terms of x,y location. Examples include trend surface analysis, splines and quadrics. The latter involve use of local estimators such as inverse distance weighted averages, Delaunay triangulation and regionalised variable theory in the form of kriging. Excellent overviews of these techniques can be found in many texts such as Davis (1986). As far as LiDAR data are concerned, almost all of these techniques have found application to a greater or lesser degree. The main issues that must be considered when choosing a procedure are speed, accuracy and preservation of information.

Speed is an issue because most LiDAR data sets include many millions of point observations and gridding procedures, such as inverse distance weighting, involve iterative searches of the input data to identify point sets local to each grid cell that can be used to estimate its value. Whilst algorithms exist for minimising search times (Devereux, 1985) the number of points in a typical point cloud still makes the computational cost very high. Accuracy is also an issue because many of these gridding procedures assume a random distribution of input points. It is evident from the discussion of sensing mechanisms earlier in this chapter that LiDAR point distributions typically follow 'saw tooth' patterns on the ground and are decidedly non-random. A high density of points occurs towards the end of scan lines where they are converging and a lower density occurs where they are diverging. This variation in density can have a distinct influence on triangulation systems and the accuracy of grid point estimation.

It also reinforces the possibility that there may be several point measurements for some grid cells and none for others depending on the relationship between grid resolution and LiDAR point density. As each cell in the grid can only have one value, cells for which there are multiple LiDAR observations become generalised while cells for which there are missing values may require interpolation.

One simple and very fast solution to the problem, which gives good results for visualisation, is to adopt the view that point cloud densities are now so high that the majority of cells in a grid with cell sizes approaching the LiDAR point density can be populated by an actual LiDAR observation. The relatively small number of remaining cells can be estimated by local interpolation between neighbours. This means that one pass through the entire point cloud can be made in which every height value can be written to its correct cell with no searching. At the end of the process, cells for which there are multiple points will contain the value for the last point encountered, and a single pass through the grid enables identification of missing cells and estimation of their value from neighbours.

Beyond format conversion and visualisation there are a range of applications for LiDAR data that do require more complex processing techniques for grid generation. The most important of these is the creation of digital terrain or 'bare earth' models, where surface features such as buildings, trees and even motor vehicles have been 'removed'. There is now a substantial literature dealing with techniques for generating such products, along with their application in fields such as city modelling (Zhou *et al.*, 2004), forestry (Clark *et al.*, 2004) and archaeology (Devereux *et al.*, 2005). In the case of city modelling and forestry there is an implicit proposition that the difference between the digital elevation and digital terrain models will be the height of the surface features. However, such are the problems of estimating bare earth models that this is rarely the case without significant errors and uncertainties. In the case of archaeology, substantial success has been achieved in identifying buried heritage features in digital terrain models after overlying vegetation has been removed. Figure 4.8 shows an example where ancient landscape features are revealed by vegetation removal in a hill-shaded digital terrain model (see also Devereux *et al.*, 2005, *op.cit*).

Two main approaches are used for terrain model construction. The first, which is most suitable for urban areas, involves classification of pulses into 'ground' and 'non-ground'. It is expected that the former is a smooth surface with few or no discontinuities, whilst the latter is characterised by features with sudden changes in elevation defining edges characteristic of surface features such as building outlines (see for example Filin, 2004; Brovelli & Cannata, 2004). Segmentation between edges allows extraction and classification of non-ground features on the basis of other properties such as shape and height. The second approach involves the use of various filtering methods to try and identify the location of the ground. It is generally more relevant to vegetation removal where feature morphology is a less well-defined distinguishing variable.

The simplest approach in this group of techniques is to use a local minimum filter on the last pulse elevations in the point cloud. Whilst this often produces passable results, it suffers from a range of problems – most notably that the results are highly sensitive to the size of the filter and they are also scene dependent. This is because dense vegetation results in fewer last pulse values reaching the ground. Therefore there is an

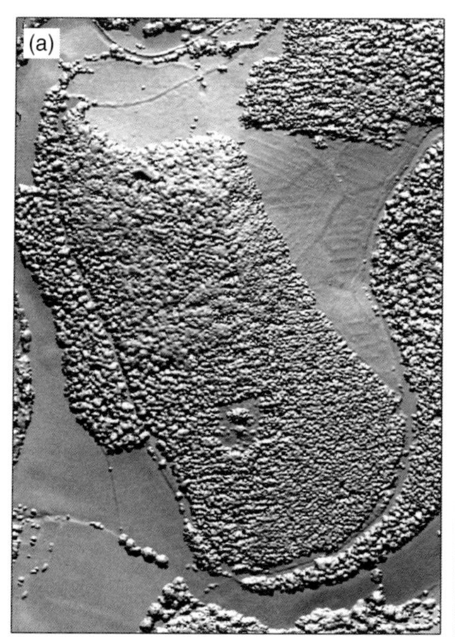

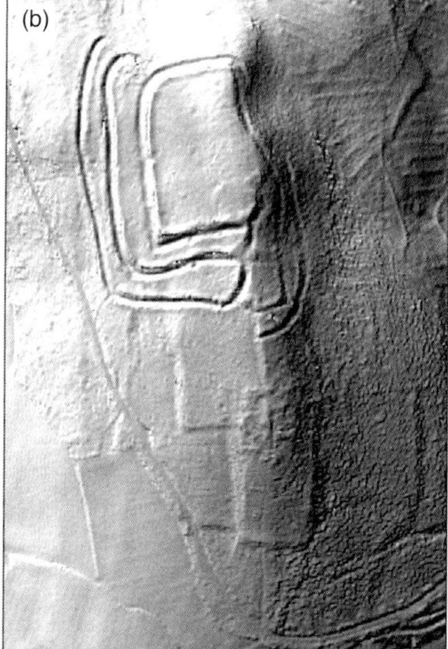

Fig. 4.8 Hill-shaded elevation models created from airborne LiDAR survey of Welshbury hill-fort in Gloucester, England. The hill is covered with dense forests comprising deciduous, coniferous and mixed woodlands. (a) Digital Surface Model (DSM) and (b) Digital Terrain Model (DTM) created after vegetation removal. No other form of aerial survey has been able to reveal the details of the remnant archaeological features of the site. Even field survey techniques have been less successful due to difficulties associated with surveying tree covered landscapes and the concealing effect of the layers of leaf-litter and under-storey vegetation.

increased probability that not all of those pulses recorded as ground points will have reached the ground. Nevertheless, with incorporation of scene knowledge such as expected feature heights and maximum ground slope angles into iterative filtering schemes, it is possible to derive terrain height models with relatively low levels of error. Kobler *et al.* (2007) report RMS errors of between 16 and 37 cm for a terrain model of steep, heavily forested slopes. Zhang *et al.* (2003) describe a similar 'morphological' filtering scheme for general-purpose feature removal based on successively larger filter sizes and a system of size thresholds based on expected feature heights.

Finally, in terms of LiDAR-derived products it is important to mention approaches to visualisation of LiDAR datasets. At first sight, it is tempting to conclude that they represent standard 3D data products which are amenable to display using conventional techniques such as contouring, 3D surface rendering with hidden line/object removal, hill shading and image draping. Current, widely available GIS functionality thus supports all of these approaches to display. However, there are still important practical and research issues affecting the visualisation of LiDAR data. Two are particularly worthy of note.

First, almost all of the GIS techniques described above support surface visualisation. However, LiDAR has the capability of penetrating terrain vegetation and hence of creating accurately measured, volumetric models of vegetation assemblages. Techniques for visualisation of such models are still relatively undeveloped and tend to revolve around rendering of 3D point scatter graphs according to the value of the point. In this respect, immersive reality is an interesting and exciting approach to visualisation in which point observations are projected into a room creating a real space, 3D rendition of the dataset that allows a data interpreter to 'walk in' to the dataset and visualise patterns.

Secondly, conventional techniques such as hill shading need to be used with considerable caution. Selection of inappropriate grid sizes for hill shading products can result in very substantial loss of quality and incorrect interpretation of patterns. Furthermore, as a basis for feature detection based on visual interpretation, hill shading has significant drawbacks because features may be entirely obscured if they are badly aligned in relation to the azimuth and elevation of the light source. Devereux *et al.* (2008) illustrate this problem in relation to archaeological sites and suggest a solution based on principal components analysis of grids illuminated from multiple directions. The results are relevant to a wide range of applications in environmental science and engineering.

CONCLUSIONS AND FUTURE PROSPECTS

Over the last decade, LiDAR has emerged as one of the most rapidly developing and powerful fields of environmental remote sensing. The basic principles have been embodied in a wide range of sensors that can be deployed on ground-based, airborne and space deployed platforms. The power, flexibility, safety and size of lasers has improved dramatically during this period, resulting in rapid improvement in the functionality of LiDAR sensing devices. This has been particularly the case in terms of pulse repetition rates and has also resulted in sensors that operate at very high levels of accuracy and resolution.

This rapid development has resulted in devices which generate vast amounts of data. This not only means that there are challenges in the development of hardware and software for processing this data, but also that major challenges remain in understanding what the data means and what kinds of useful information can be derived from it. This chapter has reviewed basic principles of LiDAR remote sensing and examined some of the more frequently employed processing techniques used for the production of primary data that the broad, environmental user community can use to support their applications development. It has revealed close synergies between these techniques and existing GIS and Image Analyses methods. The most developed areas of LiDAR processing for applications development are in digital elevation and digital terrain model development, reflecting the fact that the strengths of current systems lie in collection of reliably accurate height information. Not surprisingly, these are the areas where access to proprietary software for processing LIDAR such as Terrascan and LasTools are best developed (Hug *et al.*, 2008).

However, the field is developing rapidly both in terms of data collection possibilities and the range of applications that can be supported. Nowhere is this more true than in the processing of LiDAR waveform data. As of yet there are few comprehensive processing systems for dealing with this type of data. This is not because of technological barriers to their production, but more likely because the user community does not yet fully understand how lasers interact with different surface types and features. Without this understanding it is difficult to specify appropriate processing functionality, and more research is needed to extend progress made with large footprint systems in NASA (Harding *et al.*, 2001) to the much higher resolution small footprint devices which are now emerging.

Given the speed of development and the huge applications potential of this technology there is no doubt that this understanding will soon arrive along with appropriate processing functionality. It is the view of the present authors that the next five years will see one fundamental change in LiDAR. It will develop from being a technology for creating DEMs and models of terrain morphology, to a fully fledged remote sensing technique capable of delivering understanding of many Earth surface processes including carbon cycling, vegetation structure and dynamics, canopy energy utilisation and atmosphere biosphere interactions/energy exchanges.

REFERENCES

Baltsavias E. 1999. Airborne laser scanning: basic relations and formulas. *ISPRS Journal of Photogrammetry and Remote Sensing* **54**: 199–214.

Barber D, Mills J, Smith-Voysey S. 2008. Geometric validation of a ground-based mobile laser scanning system. *ISPRS Journal of Photogrammetry and Remote Sensing* **63**: 128–141.

Blair J, Rabine D, Hofton M. 1999. The Laser Vegetation Imaging Sensor: a medium altitude, digitisation only, airborne laser altimeter for mapping vegetation and topography. *ISPRS Journal of Photogrammetry and Remote Sensing* **54**: 155–122.

Brovelli MA, Cannata M. 2004. Digital terrain model reconstruction in urban areas from airborne laser scanning data: the method and an example for Pavia (northern Italy). *Computers and Geosciences* **30**: 325–331.

Burtch RL. 2002. LiDAR Principles and Applications. Paper presented at the 2002 IMAGIN Conference, Traverse City, MI.

Clark M, Clark D, Roberts D. 2004. Small footprint LiDAR estimation of sub-canopy elevation and tree height in a tropical rain forest landscape. *Remote Sensing of Environment* **91**: 68–89.

Davis JC. 1986. *Statistics and Data Analysis in Geology*. Chichester: John Wiley & Sons Ltd. 646p.

Drake J, Dubayah R, Clark D, Knox R, Blair J, Hofton M, Chazdon R, Weishampel J, Prince S. 2002 Estimation of tropical forest structural characteristics using large-footprint LiDAR. *Remote Sensing of Environment* **79**: 305–319.

Devereux B. 1985. The construction of digital terrain models on small computers. *Computers and Geosciences* **11**: 713–724.

Devereux B, Amable G, Crow P, Cliff A. 2005. The potential of airborne LiDAR for detection of archaeological features under woodland canopies. *Antiquity* **79**, 305: 648–660.

Devereux B, Amable G, Crow P. 2008. Visualisation of digital terrain models for archaeological feature detection. *Antiquity* **82**, 316: 470–479.

Filin S. 2004. Surface classification from airborne laser scanning data. *Computers and Geosciences* **30**: 1033–1041.

Harding D, Lefsky M, Parker G, Blair J. 2001. Laser altimeter canopy height profiles: Methods and validation for closed-canopy, broadleaf forests. *Remote Sensing of Environment* **76**: 283–297.

Hayward I, Cornelius S, Carver S. 2006. *An Introduction to Geographical Information Systems*. New York: Prentice Hall. 464p.

Hug C, Krzystek P, Fuchs W. 2008. *Advanced LiDAR data processing with LasTools*. Unpublished Manuscript.

Katzenbeisser R. 2003. About the calibration of LiDAR sensors, *Proceedings of ISPRS Workshop 3D Reconstruction from Airborne Laser Scanner and InSAR data*, 8–10.

Kobler A, Pfeifer N, Ogrinc P, Todorovski L, Oštir K, Džeroski S. 2007. Repetitive interpolation: A robust algorithm for DTM generation from Aerial Laser Scanner Data in forested terrain. *Remote Sensing of Environment* **108**: 9–23.

Lefsky M, Harding D, Cohen W, Parker G, Shugart H. 1999. Surface LiDAR remote sensing of basal area and biomass in deciduous forests of Eastern Maryland USA. *Remote Sensing of Environment* **67**: 83–98.

Maiman T. 1960. Stimulated optical radiation in ruby. *Nature* **187**: 493.

Roy D, Devereux B, Grainger B, White S. 1997. Parametric geometric correction of airborne thematic mapper imagery. *International Journal of Remote Sensing* **9**: 1865–1888.

Schawlow A, Townes C. 1958. Infrared and optical masers. *Physical Review* **112**: 1940.

Schnadt K, Katzenbeisser R. 2004. Unique airborne fibre scanner technique for application oriented LiDAR products, *International Archives of Photogrammetry and Remote Sensing*, XXXVI – 8/W2: 19–23.

Wagner W, Ullrich A, Vesna D, Melzer T, Studnicka N. 2006. Gaussian decomposition and calibration of a novel small footprint full waveform digitising airborne LiDAR. *ISPRS Journal of Photogrammetry and Remote Sensing* **60**: 100–112.

Wehr A, Lohr U. 1999. Airborne laser scanning – an introduction and overview. *ISPRS Journal of Photogrammetry and Remote Sensing* **54**: 68–82.

Zhou G, Song C, Simmers J, Cheng P. 2004. Urban 3D GIS from LiDAR and digital aerial images. *Computers and Geosciences* **30**: 345–353.

Zhang K, Chen S-C, Whitman D, Shyu ML, Yan J, Zhang C. 2003. A progressive morphological filter for removing non-ground measurements from airborne LiDAR data. *IEEE Transactions on Geoscience and Remote Sensing* **41**: 872– 882.

5 Geostatistical Analysis of LiDAR Data

CHRIS BRUNSDON

Department of Geography, University of Leicester, Leicester, UK

INTRODUCTION

Geostatistical analysis, and in particular the technique of ordinary kriging, plays an important role in the modelling of terrains from 'raw' LiDAR data. In many instances – for example when analysing slope or direction of terrain, or modelling flooding, estimates of altitude are required at points placed on a rectangular grid. Although at times data from airborne LiDAR may come close to this, it rarely meets this demand perfectly. There are a number of reasons for this – firstly the aircraft supporting the LiDAR sensor will not travel at constant velocity, so that the time interval – and hence the distance – between LiDAR returns will not be constant. Secondly, the plane will not travel along 'perfect' straight lines – although it may come reasonably close to this. These phenomena will lead to perturbations from a perfect grid. A third issue is that of buildings and vegetation. If the intention is to record the terrain elevation, to give 'bare earth' representations of terrain, then LiDAR returns from anything above ground level are misleading – although very useful in other applications here they are essentially 'nuisance' information.

Thus, algorithms (or manual editing approaches) are employed to remove information about such returns from the raw data. This leads to another kind of deviation from a perfect grid – where holes appear due to the removal of data points deemed to be associated with either buildings or vegetation or some other feature. An example of this is shown in Figure 5.1 – this shows the locations of raw LiDAR elevation points provided by the US Army Corps of Engineers that have been made publicly available[1]. Here, the 'holes' corresponding to buildings, vegetation and other features are apparent, as are the deviations from a regular grid. Also of note, perhaps, is that the lines of observation are not perfectly aligned in a north/south or east/west direction, and that orientation changes at around 3374350 m north.

Since all of these factors imply that the arrangement of raw LiDAR data is not a regular, rectangular grid, there is a need to employ some form of interpolation to estimate height values on such a grid, given the original information. A number of interpolation techniques exist – for example, inverse distance weighting (IDW), nearest neighbour modelling or kriging. In this chapter, one of the most commonly used of these, kriging, will be outlined. The chapter will begin by considering the underlying theoretical ideas of this technique, and then go on to discuss practical implications and how kriging may be carried out using the 'gstat' library from the R statistical

Laser Scanning for the Environmental Sciences,
1st edition. Edited by G.L. Heritage and A.R.G. Large.

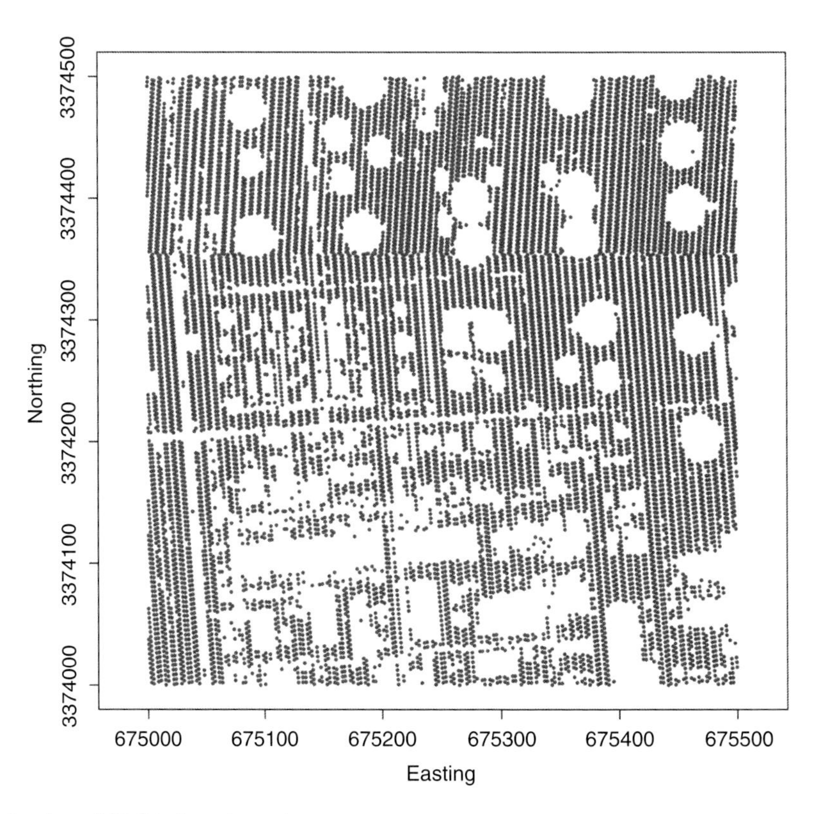

Fig. 5.1 Example of raw LiDAR elevation points.

software package. Throughout the chapter, the data from Figure 5.1 will be used as an example.

ELEMENTARY METHODS OF INTERPOLATION

Perhaps the key underlying principle to spatial interpolation is Waldo Tobler's dictum that 'Everything is related to everything else, but near things are more related than distant things' (Tobler, 1970). With a few exceptions, landscapes are reasonably smooth, so that measured elevation at some point **p** is likely to be relatively close to the measured elevation at another point **q**, provided **p** and **q** are nearby to each other. Thus, if the elevation at **p** is known and that at **q** is not, then the elevation at **p** is a reasonable estimate of that at **q**. Taking this idea further, if there are a number of points {**p1, p2, ...**} with known elevations that are nearby to **q** then some kind of average of these elevations will also make a good estimate of the elevation at **q**. The issue here is in the detail – what exactly are meant by the terms 'nearby' and 'reasonable estimate'. Sometimes these two questions are answered by educated guesses, but kriging attempts a more rigorous and objective approach.

However, before moving on to kriging, two simpler ideas will be outlined. The first of these is referred to as naive, or nearest neighbour, interpolation. Rather than using an average, to estimate the elevation at **q** one simply searches the set of known elevation points for the nearest one to **q** and uses the elevation at that point.

An example of naive interpolation, based on a subset of the points in Figure 5.1, is shown in Figure 5.2. Here, 'raw' LiDAR data in the

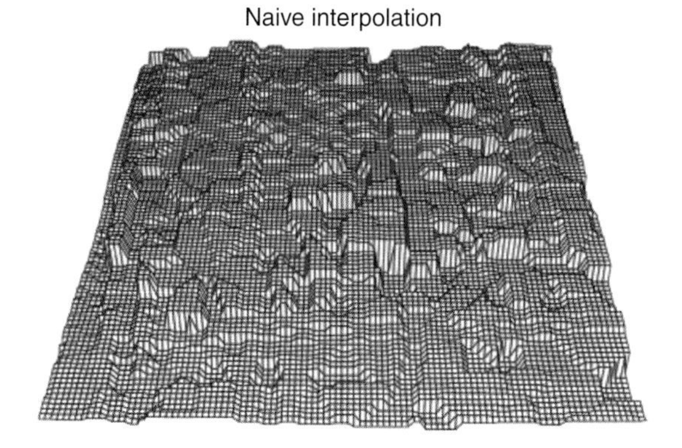

Fig. 5.2 An example of naive interpolation.

rectangle bounded by 675000W to 675200W and 3374150N to 3374350N (units in metres, projection as stated above) is interpolated at points on a regular 1 metre grid. A notable characteristic is the 'stepped' appearance of the landscape. This is due to the fact that in certain places the nearest observation point to the point on the regular grid jumps suddenly. In general, the nearest points to the observation points are determined by computing the Theissen polygons associated with the observation points. The 'steps' generally occur on interpolation grid points that are close to the edges of these polygons.

Unfortunately, although this approach is computationally quite simple, the 'stepped' appearance is an artifact of the computation method rather than of the landscape itself. In terms of faithfully reproducing ground altitude, this is highly undesirable. For example, if this model were used to model groundwater flow or to investigate slope, the results would be far removed from reality!

For this reason, a more appropriate interpolation strategy should be attempted to smooth the input data in some way. As suggested earlier, this involves estimating the elevation at point **q** by taking some kind of average of the known elevations of the points {**p1, p2, ...**}, where these points are close to **q**. Typically, the closer a point **p** is to **q**, the more useful information is contained in its elevation. For this reason, a **weighted average** is a useful approach. If e_i is the elevation at point $\mathbf{p}_i$ then we have the interpolation formula

$$e_q = \frac{\sum e_i d_i^{-k}}{\sum d_i^{-k}} \qquad (5.1)$$

where d_i is the distance between **q** and $\mathbf{p}_i$ and k is a positive constant. Typically $k = 1$ or $k = 2$. In this case, the weights place more emphasis on the elevations of points close to **q**, If $k = 2$, the fall-off in weighting with distance is more rapid than it is for $k = 1$. Two important properties of this approach are that:

1 If **q** is the same point as one of the points with known elevation then the interpolated elevation will be the same as the known elevation. This can be verified by checking the limiting behaviour if one of the distances d_i tends to zero.

2 The weighting function depends only on distance, and not direction.

At times, either one or both of these properties may not be a good reflection of reality. For example, surfaces may show more variability in the north–south direction than they do in east–west, so that weighting fall-off in one direction should perhaps differ from that in another. Second, the property that the interpolated value at a measurement point agrees with the measured value is perhaps inappropriate if there is some degree of error in the measurement process. Thus, although IDW is an improvement on naive interpolation, there is clearly some room for further improvement. The results of applying IDW to the points in Figure 5.1 are shown in Figure 5.3; on the left hand side $k = 1$ and on the right hand side $k = 2$.

It is worth noting that the two surfaces, although smoother than the naive estimate, differ notably in appearance. In each case, distinct 'bumps' are apparent, although they are sharper for $k = 1$ than for $k = 2$. This is another difficulty with the IDW approach: although the two interpolated surfaces differ, there is no clear way to determine which of the two values of k should be used – or indeed if a value other than either of these is more appropriate. This brings us on

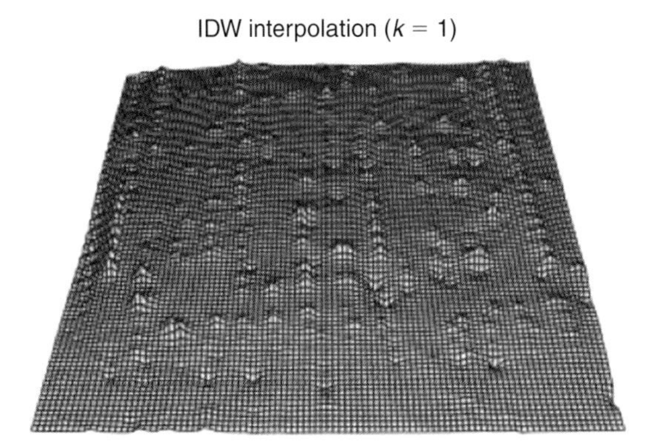

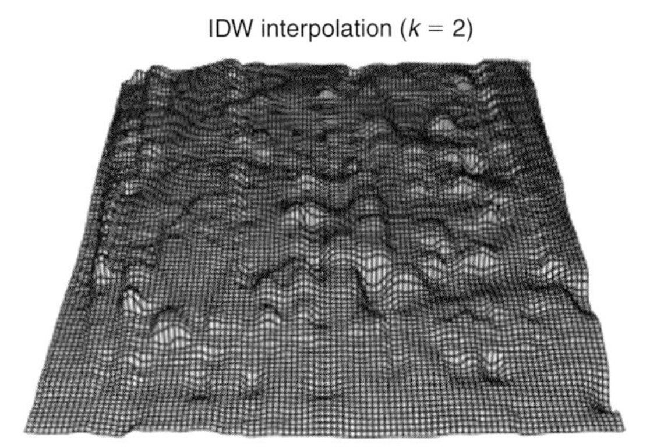

Fig. 5.3 Results of inverse distance weighting interpolation.

to the geostatistical approach. This approach also interpolates elevation values at locations using a weighted average of nearby known elevation values. The IDW approach uses a fairly arbitrary approach to selecting the weighting scheme: although distance-based, the weight values depend on the user providing a choice of k without any real guidance. In contrast, the geostatistical, or **kriging** approach gives a much more 'data-driven' approach to specifying a weighting scheme. This generally gives better results, but the price paid for this is that the geostatistical approach is more complex mathematically, and more resource-intensive computationally. In the next sections, the concepts underpinning the mathematical approach are set out. Following this, a practical example is presented.

GEOSTATISTICAL INTERPOLATION

In order to understand this approach to interpolation, the idea of a **random function** is a concept which should be explained. This idea can itself be thought of as extending the idea of a random number. Random numbers are numbers chosen by chance from some probability distribution – the values of numbers we might expect to see will depend on the mathematical expression of the probability distribution. Random numbers are sometimes **correlated** so that when considering them in pairs, some particular pairings of values are more likely than others. For example, if we regard the amount of rainfall on a given day as random, and similarly the barometric pressure, then these two numbers are likely to be correlated, since lower barometric pressure is often associated with rainfall. Correlation can be positive (i.e. high values of one variable tend to occur with high values of the other) or negative (low values of one variable occur with high values of the other). In the rainfall example, correlation is negative.

A geographical **random function** is a function of location mapping every location in space $\mathbf{q}$ to a random number z – typically one writes $z = f(\mathbf{q})$ where f is the random function. For geostatistics applied to LiDAR data, z refers to the elevation at $\mathbf{q}$, so we could subsitute e_i for z. The concept may seem rather abstract, so perhaps an example may shed further light. If we were to choose a square kilometre of land in the UK, at random, and look up the elevation at each point in that zone, the mapping from the grid reference of a point $\mathbf{q}$, relative to the south east corner of this quadrat of land to elevation, would be a random function. Correlation is an important characteristic of random functions: if we take two points $\mathbf{p}$ and $\mathbf{q}$ then the values of $f(\mathbf{p})$ and $f(\mathbf{q})$ are often correlated if $\mathbf{p}$ and $\mathbf{q}$ are nearby. This is certainly the case in LiDAR data – typically geographical points that are near to one another have similar elevation values. For example, two points near the crest of a hill will both have high elevation values, while two points near the bottom of a valley will both have low values. There are occasional exceptions to this – for example two points on either side of a cliff edge – but generally this is the case.

The degree of correlation between a pair of points typically depends on the distance between them. For example, if **p** is on the crest of a hill, a point **q** several kilometres away could reasonably be on the peak of another hill, in a valley or at any elevation in between. However, a point **q** only a few metres away from **p** is likely to have a similar elevation. Usually, the correlation between elevations of **p** and **q** is positive for small distances between **p** and **q**, but falls towards zero as the distance increases. For this reason, associated with a random function we have a **correlation function** which relates the correlation between elevations at locations **p** and **q** with the distance between them. Typically a graph of this function will be similar to Figure 5.4 in appearance.

Note that correlation falls off from 1 (maximum positive correlation) to 0 (no correlation – i.e. values of elevation between point pairs are unrelated). As can be seen, correlation gradually falls off with distance. The shape of this function, and in particular the scale of distance such that the correlation becomes small, are important characteristics of the random function. Among other things, they contain information about the typical size of hills and valleys, and the 'texture' of the surface (i.e. the frequency with which ridges tend to occur). The fact that the

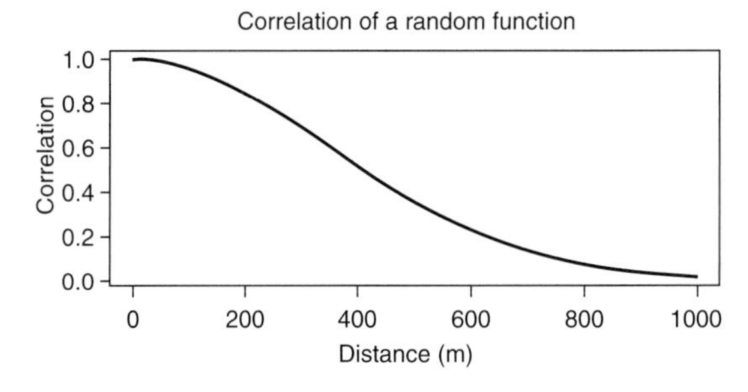

Fig. 5.4 Correlation vs distance for a random function.

correlation at distance zero is one reflects the fact that an observed elevation is perfectly correlated with itself – essentially this is because observations separated by zero distance are just repeated measurements of the same elevation. Assuming the measurement process has negligible error, this amounts to repeated recordings of the same quantity. One can experimentally estimate correlation for measured elevations separated by a variety of distances, and thus estimate correlation function from the observed elevation data.

Some examples of random surfaces are shown in Figure 5.5. In surfaces 1–3 the same correlation function is used; a second one is used for surfaces 4–6. This means that the sizes of hills within each of the two groups are more or less the same between surfaces, although the locations may differ. In surfaces 4–6, there is a more rapid fall-off of correlation with distance, so that hills cover smaller areas and occur more frequently. Surfaces 7 and 8 are different, in that hills tend to be longer in the east–west direction than north–south. In this situation some care must be used, as correlation falls off at different rates in different directions. For now, we will concentrate on the situation where fall-off is the same in all directions (which is referred to as **isotropy**) leaving the analysis of surfaces such as 7 and 8 to be discussed later.

Finally, surface 9 shows the situation where the correlation between any pair of points is zero. The intention here is to illustrate the fact that some kind of correlation pattern is necessary to model landscape realistically – intuitively, surface 9 does not provide a convincing model of terrain.

The idea of correlation is related to the idea of interpolation. If the elevation at a number of points {**p1, p2, ...**} is known then, to interpolate the elevation at some point **q**, a weighted mean can be used – as in the IDW approach – but weighting should depend on the degree of correlation between **q** and each of the points {**p1, p2, ...**}. This is approach is intuitively appealing – essentially one estimates the elevation at a given point by considering points whose **known** elevations are

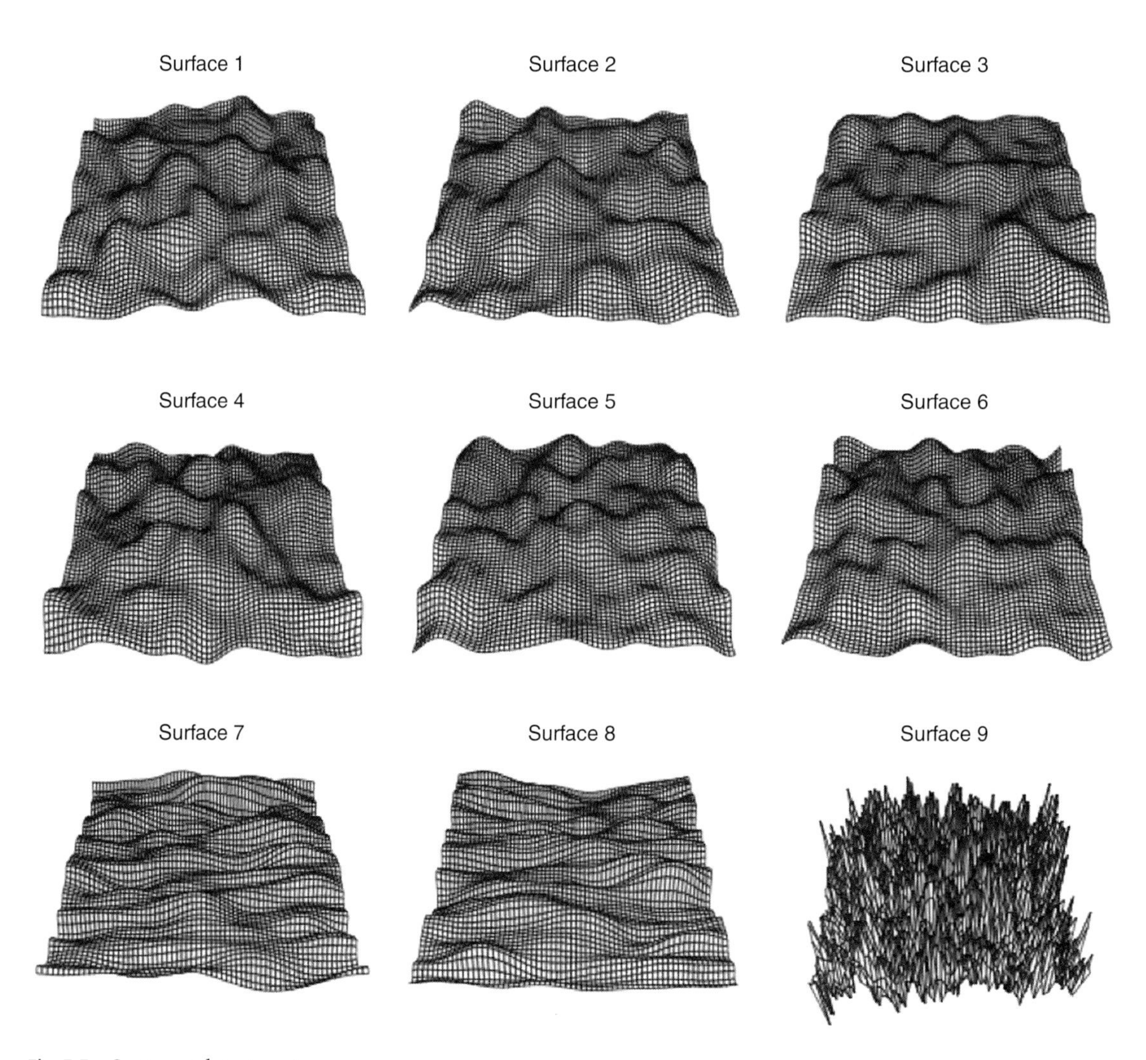

Fig. 5.5 Some random terrains.

strongly correlated with the elevation at **q**. This still implies a dependency on distance – recall that correlation is dependent on distance, and so the weighting, in turn, is also influenced by distance. Recall also that the correlation function can be estimated from observed elevation data, so that the weighting scheme can be derived from observed elevation data, as stated at the end of the previous section. Having outlined the underlying ideas of geostatistical interpolation, some more specific details are outlined in the next section.

SPECIFIC DETAILS OF KRIGING

Geostatistical interpolation – (also called **kriging**) can be broken down into two stages – determining the correlation function, and then deriving the associated weighting scheme for the interpolation. Once these stages are complete it is simply a matter of computing weighted averages of the known elevations.

The Variogram

In the text above, the correlation function has been used to describe the fall-off between correlation of a pair of observed elevations and their separation. However, an equivalent approach is to use an alternative function, the **variogram** to describe the same information. The variogram is related to the correlation function – essentially the variogram is another function of distance – in fact it is the average value of the squared difference in elevation between two points separated by distance d in the study area. When elevations at nearby point pairs are correlated, this quantity will be smaller, since on average higher correlation suggests a smaller difference in observed elevations. As distance increases, to the point where correlation is negligible, it can be shown (see, for example, Cressie 1993) that this squared distance is equal to the variance of all of the elevations in the study area. In fact, the correlation function and the variogram are related by the formula

$$S(d) = 2V(1 - C(d)) \qquad (5.2)$$

where d is the distance between a pair of points, V is the variance of all elevations, and S and C are respectively the correlation and variogram functions. Typically (mainly by convention) it is the variogram rather than the correlation function that is estimated for kriging. Noting that the variogram is a function of d the first stage of estimation is to construct an **empirical** variogram, using the following procedure:

1 Work out the distance between every point pair in the data.

2 Specify a set of 'bins' for distance – i.e. 0 to 10 m, 10 to 20 m, and so on. Assign each distance into one of these bins. The size and number of the bins will depend on the scale of the study.

3 For each bin, compute the average squared difference for each of the point pairs.

4 Create a list of pairs of the form (mean bin distance, mean squared difference).

In the final step, mean bin distance refers to the mean distance between the actual pairs of points assigned to the bin. If, for example, many of the distances between pairs were at the higher end of the bin interval, this mean would reflect this, taking a value above the mid-point of the bin. Plotting these pairs of values gives a scatter plot offering some clues to the mathematical shape of the variogram.

The procedure outlined above was applied to the LiDAR data associated with Figure 5.1, with the result shown in Figure 5.6. As suggested in the text, this quantity is smaller at lower distances, and levels off to the overall variance of the data V as distance increases. In this case the y axis is labelled **semivariance**: this is because the factor 2 has been dropped in equation 2 (this convention is often used).

The graph above gives some indication of the relationship between variogram value and distance, but as it is just a set of discrete points, it does not allow the value to be determined for an arbitrary distance. However, when the interpolation is carried out, we will need to estimate such values since we may wish to interpolate the elevation at an arbitrary location. To determine a more general variogram relationship, a curve must be fitted to the points.

Unfortunately, for the curve to be valid, it has to satisfy a mathematical property called **positive definiteness**. The exact nature of this property will not be discussed here, but the implication of this is that only certain mathematical forms of curve may be fitted. A non-exhaustive list of such functions is given in Table 5.1.

Here, s is referred to as the **sill** and r the **range** of the variogram. The sill is effectively the long-distance variance (i.e. the value that the variogram 'flattens out' to at long distance) and the range is the distance at which the 'flattening out' occurs. The sill and range for a typical variogram are shown in Figure 5.7.

The parameters r and s have to be chosen, typically on the basis of non-linear curve fitting to a set of observed variogram points such as those

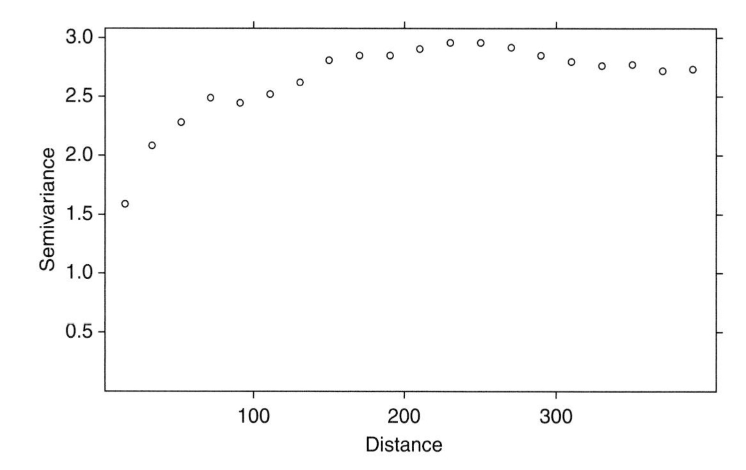

Fig. 5.6 Variogram for LiDAR data.

Table 5.1 Some suitable variogram functions.

Formula	Name
$S(1 - \exp(d/r))$	Exponential
$\begin{cases} s\left(\frac{3d}{2r} - \frac{d^3}{2r^3}\right) & \text{if } d < r \\ s & \text{otherwise} \end{cases}$	Spherical
$S(1 - \exp(d^2/r^2))$	Gaussian

in Figure 5.6. Note that the variogram gets close to zero as the distance gets close to zero. This suggests that measuring the elevation at a given point is without error, so repeated elevation estimates at the same point always give the same elevation. One way of extending the geostatistical model is to consider the situation in which elevation estimates do involve some degree of error. This is important in LiDAR data analysis – random effects such as photon scattering on foliage, for example, lead to measurement error. In this case, repeated elevation measures at the same location will not be identical, so that as distance approaches zero, the variogram function doesn't approach zero but instead tends towards some positive number. This number is the variance of the random error in the elevation estimation.

In graphical terms, this implies that the variogram curve meets the y-axis not at the point (0,0), but at a higher point (0,n). This is referred to as the **nugget effect** and can be incorporated into theoretical variogram functions by the addition of a constant n. For example, the spherical variogram with a nugget effect is:

$$\begin{cases} n + s\left(\frac{3d}{2r} - \frac{d^3}{2r^3}\right) & \text{if } d < r \\ n + s & \text{otherwise} \end{cases}$$

and the graph of this function is shown in Figure 5.8.

It is also possible to estimate the three parameters n, r and s from the empirical variogram estimations. Also, nugget effects can be added to the other variogram functions such as those listed in Table 5.1. For example, using the data from Figure 5.6 and fitting an exponential variogram with a sill gives the result in Figure 5.9, with $n = 1.2$, $s = 1.6$, and $r = 44.1$.

This example shows the typical process of calibrating a variogram. Although the selection of sill, range and nugget is automated[2] the choice of the variogram function often relies on human judgement – in this case all of the three curve types listed in Table 5.1 were fitted and visually this exponential appeared to give the best fit. Having arrived at an estimate of the variogram function, the interpolation process may

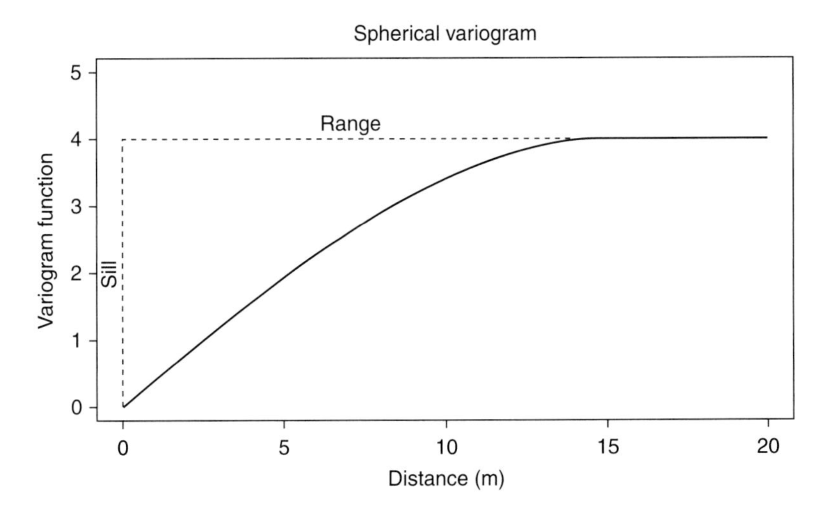

Fig. 5.7 The spherical variogram function.

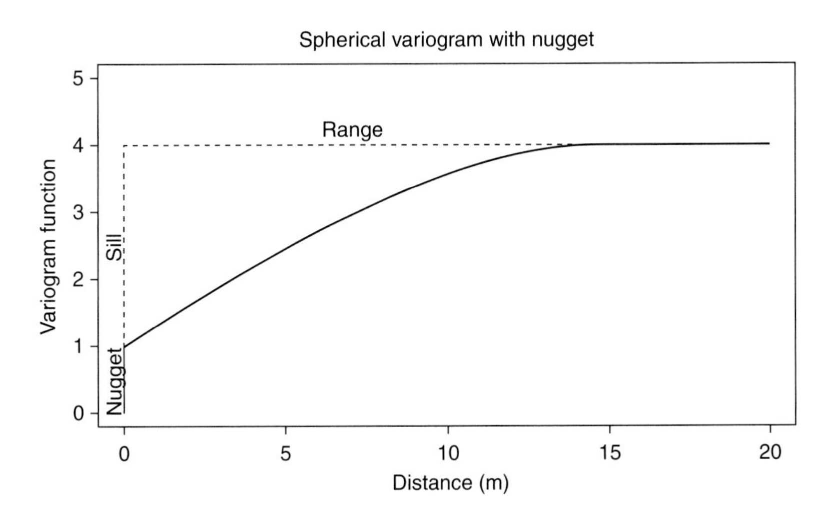

Fig. 5.8 The nugget effect.

take place. This is discussed in the following section.

From variogram to interpolation

Recall the earlier statement, that one would expect elevation measurements near to the interpolation point to be more useful in interpolation than those further away. The variogram is a tool that gives us some idea how near a measured point must be to an interpolation point for it to have a useful input to elevation prediction. As with IDW, the interpolation is achieved by computing a weighted average of the elevations at measured points, and assigning greater weighting to closer locations. However, unlike IDW the

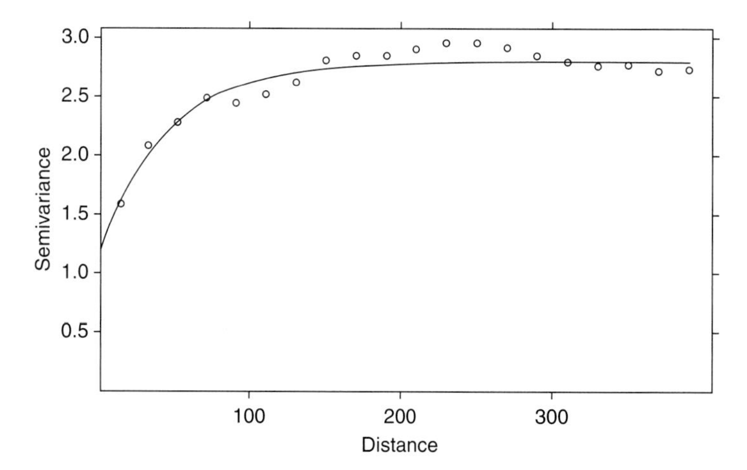

Fig. 5.9 A fitted variogram.

weights are computed from the variogram – and since this is estimated from the observations, the whole weighting process is more 'data driven'.

The connection between the variogram and the weights is a mathematical formula that is derived by minimising the expression for the expected variance of the interpolated value minus the true value:

$$\mathrm{Var}\left(e_* - \sum \omega_i e_i\right) \text{ subject to } \sum \omega_i = 1 \qquad (5.3)$$

This perhaps needs further explanation. Since we have modelled the surface as a random function, the true value of the elevation at the interpolation point is a random variable. Therefore the difference between the weighted mean of the measured elevation points used to estimate elevation, and the true elevation is also a random variable. Since the latter variable is an indicator of the interpolation error, its variance is a measure of the quality of the estimate – the lower the variance, the more reliable the estimator. Therefore, by finding the weights that minimise this quantity, we thus find the weighted mean that provides the most reliable estimate out of all possible weighted means. Finally, the constraint that all weights sum to 1 (the 'subject to' part of equation 3) ensures that the estimate is unbiased – that is, that the mean of the random variable used to interpolate elevation is the mean of the elevation in the study area.

Here, a full derivation of the expression linking the variagram to the weights will not be discussed, but the formula itself is given below, in matrix notation:

$$\begin{pmatrix} w_1 \\ \vdots \\ w_n \\ m \end{pmatrix} = \begin{pmatrix} V(d_{11}) & \cdots & V(d_{1n}) & 1 \\ \vdots & \ddots & \vdots & \vdots \\ V(d_{n1}) & \cdots & V(d_{nn}) & 1 \\ 1 & \cdots & 1 & 0 \end{pmatrix}^{-1} \begin{pmatrix} V(d_{1*}) \\ \vdots \\ V(d_{n*}) \\ 1 \end{pmatrix} \qquad (5.4)$$

where d_{ij} refers to the distance between observed elevation points i and j, where i and j run from 1 to n, where n is the number of observed points, and the asterisk denotes the point at which interpolation is to be carried out – so that d_{i*} refers to the distance between the interpolation point and observed point i. Also, m is the overall mean elevation for the study area and V(.) is the variogram function. Note that the computational overheads here are high – for every interpolation point it is necessary to invert an $(n+1) \times (n+1)$ matrix. If it is desired to interpolate over a fine grid of points, this may take a good deal of time in practice.

The result of applying this weighted interpolation to the example data set is shown in Figure 5.10. It can be seen from here that the result is notably smoother than either a naive interpolation or an IDW approach. The latter tends to show 'bumps' around the observation points, which is clearly a problem – the modelled elevation should ideally be independent of the points at which elevation has been observed.

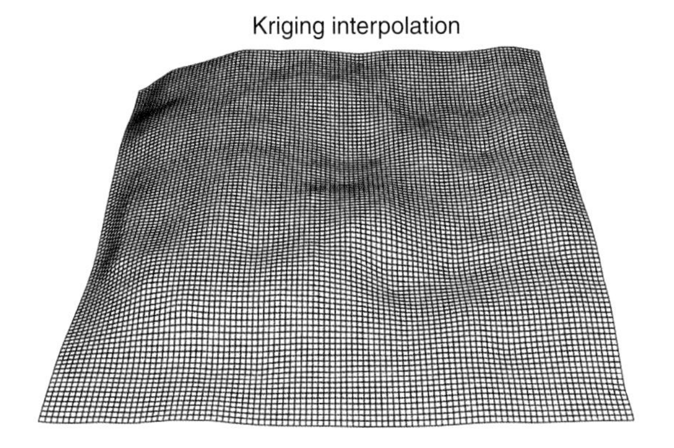

Fig. 5.10 Result of applying kriging interpolation.

The weights applied to the observations are illustrated in Figure 5.11: the circle shows an interpolation point and the shading of the measured points shows their weighting – darker points having greater weight. It can be seen that most of the weighting is given to four or five points that are relatively close to the interpolation point. It is also worth noting that, unlike IDW, it is possible to give some observed elevations a negative weight. Although this may seem strange, this can occur if the structure of the terrain is such that when a given point has a high elevation, then points separated by a certain distance always tend to have low elevations, or vice versa.

A final useful feature of kriging is that as well as providing an interpolation estimate for elevation at unmeasured locations, the random nature

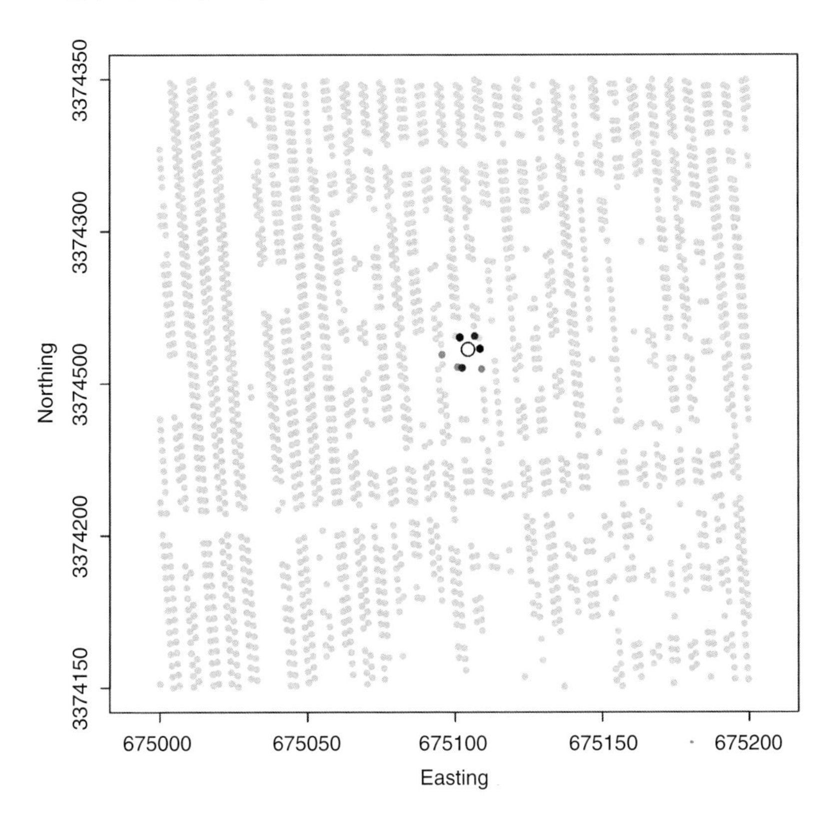

Fig. 5.11 Weighting in kriging.

of the model also makes it possible to compute standard errors of the estimates[3]. This enables the user to judge the reliability of the interpolations and is a feature not easily obtained for naive and IDW estimates. The standard error for the kriging interpolation here is shown in Figure 5.12.

Computational issues

One thing hinted at in the previous section was that computation was an important factor in kriging. This is generally because of the relatively complex nature of the mathematics underlying kriging. Fortunately, the end-user need not be too concerned about this, provided that they are at least aware of some of the issues surrounding it. This is because kriging has been implemented in a number of software packages. This saves many users from having to memorise various formulae, but perhaps more importantly, also exploits expertise in numerical computation – so that, for example, large linear equations are solved in such a way that computation is efficient, and rounding errors are sufficiently low that the answers obtained are reliable.

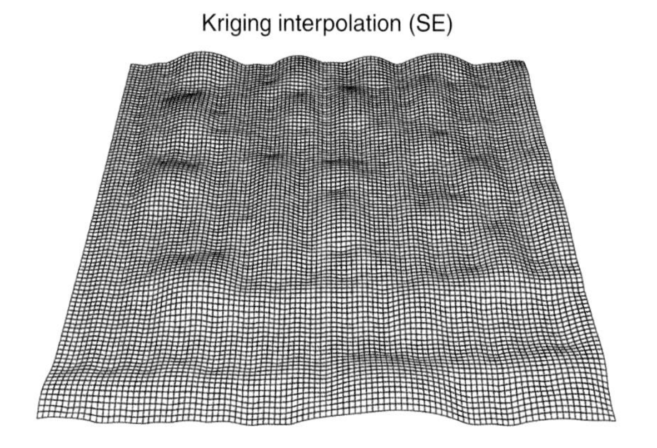

Fig. 5.12 Kriging standard error surface.

The examples in this chapter were computed using the gstat package (Pebesma, 2004) from the R statistical programming language (R Development Core Team, 2008). Although the aim here is not to provide a tutorial in R, the code used to carry out the kriging (together with some comments) is shown below in Figure 5.13.

The R code may seem daunting to the beginner, but provides a powerful tool as it is possible to link the results of geostatistical analysis with other available R packages – there are over 1000 of these. For example, the rgl package connects R to

```
v.emp = variogram(z~1,~x+y,obs,cutoff=400,width=20)
# Compute the observed variogram -  'obs' contains the observed elevations and
their locations
v.mod = fit.variogram(v.emp,vgm(1.5,"Exp",100,1.5))
# Calibrate the mathematical variogram function
surf.est = krige(z~1,~x+y,obs,gr,v.mod)
# Carry out the kriging
sp = array(surf.est$var1.pred,c(101,101))
se = sqrt(array(surf.est$var1.var,c(101,101)))
sx = unique(surf.est$x)
sy = unique(surf.est$y)
# Extract the surface elevation estimates and their SE's
persp(sx,sy,sp,scale=F,box=F,axes=F,border=rgb(0.1,0.1,0.1),phi=45,expand=7)
title('\nKriging Interpolation')
# Plot the surface as a wireframe
persp(sx,sy,se,scale=F,box=F,axes=F,border=rgb(0.1,0.1,0.1),phi=45,expand=7)
title('\nKriging Interpolation (SE)')
# Plot the surface for standard error
```

Fig. 5.13 R Code to compute variogram and kriging interpolation.

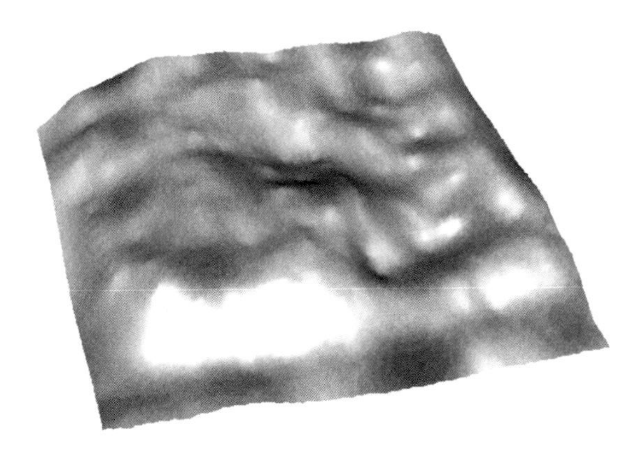

Fig. 5.14 rgl rendering of interpolated surface, with overlaid standard error.

the OpenGL 3D interactive graphics (SGI, 2006). Using this in companion with gstat it is possible to produce interactive terrain models in three dimensions – an example is given in Figure 5.14 – here the surface shown in Figure 5.10 is overlaid (using colour shading) with the standard error of interpolation. One feature that this rendering highlights is the fact that the standard error has large values where there are 'holes' in the measured elevations; this is intuitively reasonable, as it suggests that estimates are less reliable where there are fewer measurements nearby.

Although this chapter makes use of R, a number of other possibilities could be used – for example the Spatial Analyst ArcGIS 9.2 (ESRI, 2008) offers a GUI-based approach to the calibration of variograms and then carries out the kriging operation within a standard desktop GIS environment.

FURTHER ISSUES

The above gives an outline of a basic form of kriging, known as **ordinary kriging**. There are a number of assumptions made in this approach. In particular, the variogram model adopted above assumes that the correlation between measurements for all pairs of points separated by a given distance will be the same. This assumes two key things: first, that correlation depends only on distance and not on direction of separation of points. This property is referred to as isotropy. The second assumption is that it is only the relative separation of points that affects correlation – so that a pair of points separated by 100 m in the east of a study region will exhibit the same correlation between elevations as a pair in the west that are also separated by 100 m. This property is referred to as **stationarity**. There are a number of ways of extending the basic kriging methodology, but here we will focus on two, associated in turn with relaxing the assumptions of isotropy and stationarity.

Relaxing the assumption of isotropy leads to the idea of an anisotropic variogram (and corresponding correlation function) – the correlation (and variogram) functions still depend only on the relative locations of point pairs, but both distance and direction (or equivalently a pair of euclidean coordinates) are taken into consideration – see Eriksson and Siska (2000) for example. This phenomenon is often encountered when studying rivers, for example. The argument to the variogram function V(.) is now written as a vector – such as $V(\mathbf{p}-\mathbf{q})$ where the two vectors $\mathbf{p}$ and $\mathbf{q}$ are a pair of points. In this case, the variogram function is a surface in 3D space rather than a curve in 2D space. Such an approach may be appropriate, for example, for surfaces 7 and 8 in Figure 5.5. Anisotropy can be explored by fitting standard isotropic variograms to elevation measurement pairs that have been partitioned into subsets of the vector $p - q$ as cited above according to the direction of this vector – for example in Figure 5.15, using the example dataset, sectors with mid-angles of 0, 30, 60, 90, 120 and 150 degrees are analysed separately, and the isotropic variogram function is plotted against each of these.

From these it may be seen that the variogram model fits well at angles of 0, 30 and 90 degrees, but under-predicts variability at angles of 120 and 150, and over-predicts variability at 60 degrees. This can provide clues to fitting an anisotropic variance function, which may then be used to provide a set of kriging weights that take this directional variation into account.

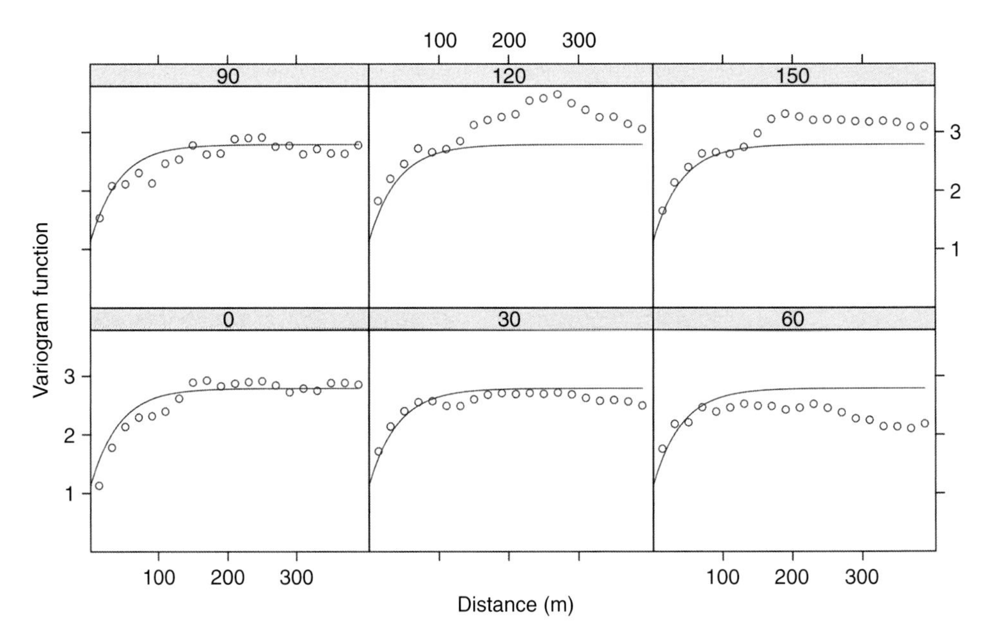

Fig. 5.15 Exploring anisotropy in the variogram.

The second extension of ordinary kriging is non-stationarity. In this case the variogram function depends not only on the relative location of points, but also on their **absolute** location. A typical way of dealing with this is to use a moving-window approach (see for example Atkinson and Lloyd (2007) for an application using this approach). To estimate elevation at some point **p**, consider all of the points within some radius of **p** (a circular window) and fit a variogram on the basis of just these points – from this kriging weights may be found. To estimate elevation at another point, say **q**, then consider only those points within the circular window centred around **q**. As the interpolation point moves, so the window moves, and local 'drift' in the variogram function may be taken into account.

In this chapter, an overview of ordinary kriging has been given, together with a brief discussion of some extensions to the method. However, geostatistics is a very large field of study, and many other extensions to the basic technique exist. A number of specialised texts provide insight into a great many of these techniques – the reader wishing to find out more could benefit greatly from consulting Cressie (1993), Chiles and Delfiner (1999), Goovaerts (1997) or Webster and Oliver (2001).

ENDNOTES

1 Edited LiDAR Elevation Data, NE quadrant of Baton Rouge West quadrangle, Louisiana, UTM 15 NAD83, U. S. Army Corps of Engineers, Saint Louis District, (Obtained March 1999, published 2001) *A subset of the data is used here.*

2 In this case least squares was used, although Cressie (1993) suggests an alternative approach.

3 It is however worth noting that the computed SE's assume that the variogram is known. Allowing for the fact that the variogram is itself an sample estimate would be mathematically more complex – see for example Diggle *et al.* (1998).

REFERENCES

Atkinson PM, Lloyd CD. 2007. Non-stationary Variogram Models for Geostatistical Sampling

Optimisation: An Empirical Investigation using Elevation Data. *Computers and Geosciences* **33**: 1285–1300.

Chiles J-P, Delfiner P. 1999. *Geostatistics: Modeling Spatial Uncertainty.* New York: John Wiley & Sons Ltd.

Cressie N. 1993. *Statistics for Spatial Data.* New York: John Wiley & Sons Ltd.

Diggle PJ, Tawn RA, Moyeed RA. 1998. Model-based geostatistics. *Applied Statistics* **47**: 299–350.

Eriksson M, Siska PP. 2000. Understanding Anisotropy Computations. *Mathematical Geology* **32**: 683–700.

ESRI. 2008. *ArcGIS Desktop Help 9.2 – Kriging*, Redlands: Califonia. http://webhelp.esri.com/arcgisdesktop/9.2/index.cfm?TopicName=Kriging

Goovaerts P. 1997. *Geostatistics for Natural Resources Evaluation.* New York: Oxford University Press.

Pebesma EJ. 2004. Multivariable geostatistics in S: the gstat package. *Computers and Geosciences* **30**: 683–691.

R Development Core Team. 2008. *R: A Language and Environment for Statistical Computing.* R Foundation for Statistical Computing, Vienna, Austria. http://www.R-project.org

Tobler WR. 1970. A computer model simulation of urban growth in the Detroit region. *Economic Geography* **46**: 234–240.

Webster R, Oliver MA. 2001, *Geostatistics for Environmental Scientists.* Chichester: John Wiley & Sons Ltd.

6 Laser Scanning: Data Quality, Protocols and General Issues

DAVID HETHERINGTON

Ove Arup and Partners, Newcastle Upon Tyne, UK

INTRODUCTION

Measurement techniques based upon the principles of travelling sound and radio waves (sonar and radar) have been employed globally over the last century for numerous purposes. However, it is only over the last couple of decades or so that measurement techniques have been widely applied that gauge distance using the time-of-flight of a laser pulse. Increasingly, scanning and single point LiDAR (Light detection and ranging) measurement techniques are being used as standard for a huge and varied range of applications in science and industry.

The required quality of data that can be acquired using LiDAR scanning techniques should be driven by the purpose for measurement, and by what level of representation is required over a given scene or object. Measuring and representing physical surfaces is possible to varying degrees of quality, over various spatial scales and ranges, using LiDAR equipment. The type of measurement platform (e.g. fixed or mobile terrestrial, or low or high level aerial) plays a large role in determining the general characteristics of the resultant data. The type of platform dictates the potential range of perspectives that a subject scene or object can be measured from. Often, variations in the mechanical approach to scanning mean that resolution/point-spacing indices are not directly comparable between different platforms and models in their rawest form – points per specific area (Lichti, 2004). Issues such as this result in data being described in general terms where mean statistics are provided that represent entire datasets. This often leads to the inherent heterogeneity of data character within individual datasets being poorly represented in associated descriptive statistics.Local data should therefore be interrogated manually if a location-specific appraisal is required within a larger dataset. Variation in expected resolution and point-spacing for any given platform is largely determined by the perspective of a platform to the subject scene, the structure and complexity of a subject of scene, and the degree of non-returns in the dataset. Non-returns being when a surface is not measured, either because it is outside of the measurement range of the laser scanning equipment or return signals have been reflected away from the LiDAR sensors.

A common problem with LiDAR scanning techniques is that the resultant data file sizes can be overwhelming and often have to be reduced in size in order for them to be manageable. This contradicts the benefits that are gained from the measurement technique in terms of resolution and point accuracy. If the usefulness and representivity of the data is to be maximised then

Laser Scanning for the Environmental Sciences,
1st edition. Edited by G.L. Heritage and A.R.G. Large.
 ISBN 978-1-4051-5717-9

computational power should not be a restricting factor during post-processing and analysis. As time passes and computational power improves such issues will become less salient.

This chapter describes some of the concepts relating to LiDAR data quality, whilst outlining some of the approaches that can be taken when dealing with LiDAR data. An attempt will also be made to explain some of the key issues that should be taken into account when considering LiDAR measurement options, acquiring and post-processing data, and using the results. This will be done both in general terms, and more specifically for airborne and terrestrial methods.

Data quality and measurement protocols

Measurement point error is only an issue if it has a substantial effect on the parameter that is ultimately required from the work, regardless of the type of approach to LiDAR measurement that is selected. The combined accuracy of measurement is cumulative, and is affected by error from two broad sources. These sources are equipment-based error – as affected by the technical capability of the equipment, calibration levels or unexpected movement during measurement – and human error, due to inappropriate setting-out or improper use of equipment. An approach to measurement can be justified if its resultant data fall within an acceptable level of quality for any specific project. This train of thought is not relevant if the selected technique covers the required area, whilst measuring at such a resolution that all relevant scales of topographic and/or structural variation are captured and represented in three dimensions. Data quality should be considered based upon how well point cloud data, and potentially the resultant DEM, represents the feature or environment as it is in reality. Cooper (1998) explains how it is inappropriate to consider DEMs, and thus the datasets that form them, in terms of their 'accuracy', and that it is better to use surface quality as a measure of appropriateness and representivity. Cooper (1998) continues by describing how terrain surface quality is controlled by data precision, data reliability and the accuracy of individual points. This quality/representivity is dictated by the nature of the data that is acquired, based upon the selected measurement method, and how any acquired data is processed afterwards. In short, the level of data quality is entirely dictated by 'what is good enough' for any given project and it is the best-case results (resolution, point accuracy, coverage) for a specific type of laser scanning measurement which will dictate its potential range of appropriate applications. A key concept to appreciate when dealing with LiDAR measurement data is the difference between measurement error and surface misrepresentation. For example, data can be highly-accurate but, when acquired over a structurally complex scene, unable to digitally-represent the intricate detail of the situation in reality. This situation could arise either due to a lack of spatial resolution of measurements, or due to a multi-layered scene that is too complex to measure from any general perspective and/or platform (Table 6.1).

Testing of data quality

It is explained by Lane (1998) that assessing terrain surface quality is difficult because the 'real' surface is rarely known and has to be based upon exploring the effects of sampling from an existing surface on either surface parameters or surface derivatives. Lane (1998) continues to state that three issues surrounding surface quality must be explored (i) the quality of the surface fitted to the data (ii) the effects of point density of DTM quality; and (iii) the effects of point distribution on DTM quality. The generic affects of point distribution and point density on data quality are shown to a certain extent in Table 6.1, however the affects of the choice of interpolation technique and fitted surface on surface quality are more complex. This complexity stems from the fact that different interpolation techniques work better than others, depending on the nature of the measured surface and the spatial data that are obtained on that surface. A selection of commonly-used interpolation techniques are shown in Table 6.2,

Table 6.1 A schematic representation showing potential scenarios of measurement quality on a complex surface (the open circles represent measurement points). The data quality descriptions assume that representation of the most intricate details of the surface were pre-requisites of measurement.

Surface measurement scenario	Description	Data quality
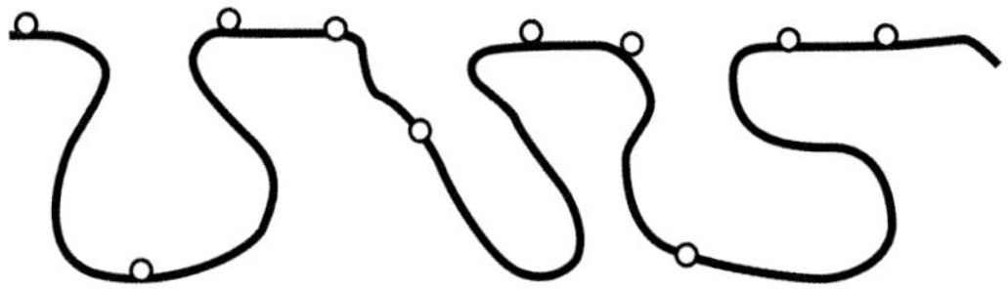	Accurate	Poor
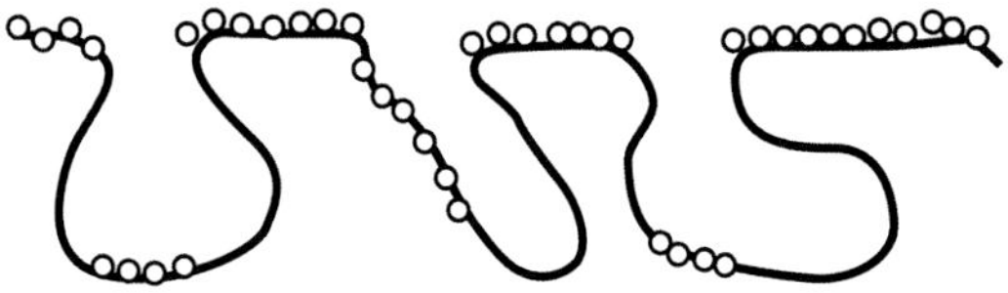	Accurate and precise (allowing high resolution)	Moderate (normal)
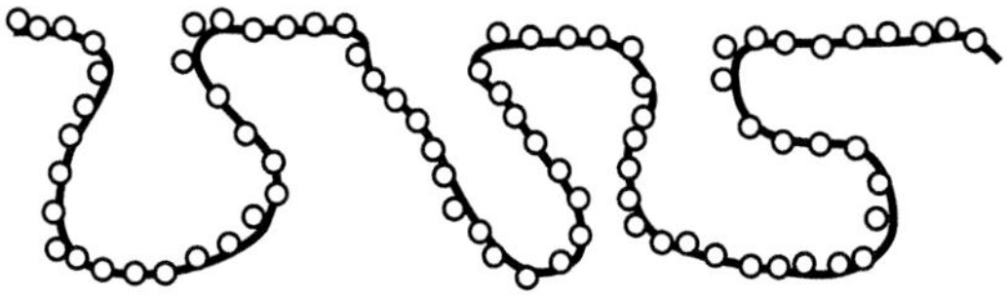	Accurate, precise and representative	Good

along with descriptions of their characteristics and recommendations relating to when they should be used.

The testing of the quality (and accuracy) of individual points can be achieved by comparing their positioning against a network of control measurements that are of a known accuracy. This can be achieved by calculating Root Mean Squared (RMS) 'errors' (the nearest point in 3D space), or by considering elevation as a spatially-distributed random variable (as in geostatistics) and comparing elevations of the nearest point in lateral space. The latter of these techniques is more acceptable if the data that are being tested are of sufficiently-high resolution to capture small-scale detail, and confidence is high in terms of lateral positioning of the global dataset.

Airborne acquisition

Airborne laser scanning (LiDAR altimetry) is an important data source for many environmental applications, being able to map topographic variations, and the height of surface objects, to a high degree of vertical and horizontal accuracy over large areas (Cobby *et al.*, 2001). Filin (2004) explains how the technique is now widely recognised as a leading technology for the extraction of information of physical surfaces. The ever increasing point density that can be acquired by airborne systems enables the achievement of a detailed description of a surveyed surface and provides a wealth of information on physical objects and on terrain (Filin, 2004). The technique has been shown to be effective and efficient in terms of measurement of the fluvial zone

Table 6.2 Some of the various interpolation methods that are commonly used to create DEMs of the natural environment (Source: Keckler, 1995).

Interpolation method	Summary
Inverse distance	• Weighted average interpolator; weight of one data point diminishes with distance from the grid node. • Normally an exact interpolator; honours the data points when they coincide with a grid node; can add a smoothing parameter if desired. • Tends to create concentric rings around data points. • Fast for medium sized datasets.
Kriging	• Geostatistical gridding method; has proven to be popular. • Generates visually appealing contours that attempt to depict trends in the data. • Uses irregularly spaced data; good for random point photogrammetric method. • Can be an exact or smoothing interpolator. • User can define specific variogram model. • Can be slow for large datasets, but results are worth the wait.
Minimum curvature	• Popular in Earth science. • Generates the smoothest possible surface whilst trying to honour each point. • In is not an exact interpolator, so not all points are honoured exactly. • Fast for most datasets.
Nearest neighbour	• Assigns the value of the nearest datum point to each grid node. • Useful when data are already on a grid, or for filling holes when data values are missing for a grid.
Polynomial regression	• Used to define large-scale trends and patterns within the data. • Not really an interpolator as does not attempt to predict unknown z values. • Very fast but detail lost within the resultant grid.
Radial basis functions	• Consist of a range of exact interpolators that attempt to honour the data points. • Multi-quadratic is probably the best for a smooth surface that fits the data. • The results are similar to those produced by kriging.
Shepards' method	• Uses inverse distance weighted least-squares method. • Similar to inverse distance but local fit reduces concentric circle effect. • Can be an exact or smoothing interpolator.
Triangulation with linear interpolation	• Uses optimal Delauney triangulation. • Original points are connected by lines to create triangles, produces a patchwork of triangle faces over the extent of the grid. • Is an exact interpolator. • Good for evenly distributed points; with large enough datasets breaks in slope can be preserved.

(Cobby *et al.*, 2001; Charlton *et al.*, 2003), and as a tool for monitoring geomorphological change in the fluvial environment (e.g. Thoma *et al.*, 2005). Morphology and morphological change has also been measured using this technique over similar spatial scales in coastal zones (e.g. Gomes Pereira & Wicherson, 1999; White & Yang, 2003) and in complex landslide terrain in mountainous areas (e.g. McKean & Roering, 2004).

In standard two-dimensional airborne laser scanning the rotating mirror drives the lateral (y-dimension) adjustment of the signal trajectory, and thus the swath (and resolution) of coverage as dictated by the surface topography and altitude. The mean point spacing between individual swaths in the x-dimension is controlled by the forward speed of the airborne platform. Figure 6.1 shows how the spatial distribution of

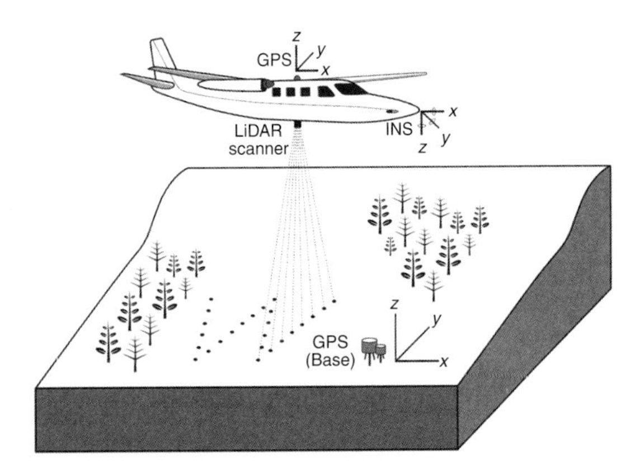

Fig. 6.1 A schematic representation of airborne LiDAR data acquisition (using an inertial navigation system (INS)) and dimensional connectivity to GPS. Note how the swath (in the *y*-dimension) is affected by forward movement of the platform in the *x*-dimension.

measurement points (a key contributor to overall data quality) is controlled by forward movements and height of the plane. The height of the plane is also the key control of the swath width (assuming that the angular range of the scanning equipment remains constant). Sources of measurement point error (and thus point inaccuracy) in 3D space can result from a combination of inherent laser measurement error, GPS-sourced error, and any errors associated with the inertial navigation system (INS).

In general, airborne laser scanning can be categorised into two types; low level and high level. Low level laser scanning is usually done from low-flying platforms such as light aircraft or helicopters and tends to be conducted at altitudes of below 1000 m. High-level laser scanning tends to be conducted from larger aircraft, at relatively high altitudes using more powerful laser sensors. The selection of an appropriate method involves a compromise between coverage/range and resolution/accuracy, which is closely related to the altitude of the scanning platform. The characteristics of the raw datasets that can result from airborne LiDAR surveys can vary a great deal, depending on the platform of measurement and the scanning equipment employed. Gomes Pereira and Wicherson (1999) were able to utilise helicopter-mounted laser scanning equipment to measure the fluvial zone, for the purposes of providing information for two-dimensional hydraulic models. This approach was justified as the immediate floodplain was less wide than the typical 61 m swath that was allowed by the measurement methodology. Here, a vertical accuracy of 0.07 m was quoted in this study which measured over a 20 km^2 area. French (2003) conducted an investigation into the suitability of airborne LiDAR data for supporting geomorphological and hydraulic modelling on an estuarine zone. Here it was found that high-level airborne LiDAR data could be used to measure an area capable of representation over the plain scale (12 km × 4 km) to within a vertical accuracy of between 0.1 m and 0.15 m. French (2003) makes the point that often these vertical accuracies are problematic when using the data for flood modelling due to many flood inundation depths being less than 0.05 m within the used study site. Mark and Bates (2000) explain how high-level (800 m) LiDAR data have been flown within the UK over a significant number of coastal strips and floodplain reaches, especially in eastern and south-western areas. Such data acquisition is typically associated with a swath width of 600 m. French (2003) explains how this data has a typical measurement interval of between 2 m and 4 m. At this resolution it is possible that scales of topographic variation at sub-meso scale would be well represented. Cobby *et al.* (2001) state that, after processing, their UK Environment Agency (EA)-acquired data set was typically accurate to within 0.1 m of the surface in reality. Lane *et al.* (2003) utilised synoptic remote sensing to estimate erosion and depositional change over a large braided river. In this study a 3.3 km × 1 km length of river was measured at a pixel resolution of 1 m, to typical precisions of between 0.1 m and 0.26 m in dry areas. To achieve this; data were acquired using both airborne LiDAR and airborne photogrammetric surveys. This was to attempt to counteract the constant trade off between the increase in coverage that comes from the reduction in image scale (Lane, 2000). Lane *et al.* (2003) add to this by stating that this is particularly relevant with respect to large braided river beds because of the low vertical relief (in the order of metres) compared to the channel width (in the order of kilometres).

Geist *et al.* (2004) describe the results of a two-year glacier measurement and monitoring project in which airborne laser scanning was used to acquire surface elevation data. Here, data were acquired from an average altitude of 2 km, which allowed coverage of 62 km^2 after 28 small flight paths. The data that were acquired over the glacier surface were accurate to within an average vertical distance of 0.25 m, at a resolution of 0.66 points per m^2. Geist *et al.* (2004) comment that these statistics compare well with those that may be acquired using airborne photogrammetric techniques, whilst being far more time and cost efficient. Geist *et al.* (2004) continue by predicting that data quality might improve by accompanying LiDAR with InSAR (Interferometric Synthetic Aperture Radar) technologies in a simultaneous data acquisition flight. White and Wang (2003) utilised airborne LiDAR data to study morphological change along a 70 km stretch of coastline. The authors do not provide statistics relating to the raw data; however data that were acquired were used to produce a DEM with a point every 1.5 m, with a vertical accuracy of ±0.15 m. Surface representation at this level would not be able to represent information on many unit-scale features and sub meso-scale features within the natural environment. Airborne LiDAR was also utilised by McKean and Roering (2004) in order to analyse surface morphology in order to detect landslides. Here, data were captured at a resolution of 0.38 points per m^2 with a typical vertical data accuracy of between ±0.2 m and 0.3 m. McKean and Roering (2004) reiterate the suggestions of Latypov (2002) relating to data accuracy in that the source of inaccuracy is uncertain but has generally been ascribed to errors in data from the aircraft GPS and inertial navigation systems or misalignments in the laser optics. Latypov (2002) quantifies the magnitude of the potential uncertainty of error by explaining that in a LiDAR system that uses a 1 Hz GPS and a 20 kHz laser, a single random GPS error will cause a systematic shift in 20,000 LiDAR coordinate data points. This further confirms the importance of ground control in validating airborne LiDAR survey. Hodgson *et al.* (2003) report on their evaluation of the commonly-used USGS (United States Geological Survey) LiDAR data in conjunction with IFSAR data that was acquired during leaf-on conditions (i.e. when seasonal/deciduous trees bear leaves). They explain that the level one USGS data is provided at a resolution of 0.033 point per m^2 (one point per 30 m), with a point accuracy that varies depending on when, and with which equipment, the flight was undertaken. The USGS quote associated vertical errors of between 3.0 m to 7.0 m with their level-one data, however Hodgson *et al.* (2003) explain that vertical surface errors during leaf-on conditions were in the range of 4.8 m to 10.0 m. This degree of measurement and representation, whether representing leafed or unleafed conditions, would only be suitable for representing very large terrestrial features (large deltas etc.) to any degree of acceptable relative representivity.

Hopkinson *et al.* (2004) attempted to evaluate the respective errors associated with vegetation height and ground height when measuring with airborne LiDAR data. They state that many studies have investigated the relationship between airborne LiDAR field measurements and tree-level vegetation height, with r^2 values typically ranging between 0.85 and 0.95, with the best relationships being shown by Nasset (2002) Nasset and Oakland (2002) and Popescu *et al.* (2002). Both Davenport *et al.* (2000) and Cobby *et al.* (2001) explain how short vegetation (<2 m) height can be predicted using the standard deviation of LiDAR data over large areas; however specific errors were not associated with their measurements. Hopkinson *et al.* (2004) used data covering a 260 km^2 area of wetland, formed by 20 flight lines. The final dataset had a resolution of between 0.6 and 1.3 points per m^2, after 8% of the total laser pulses had been lost due to areas of open-water. Hopkinson *et al.* (2004) presented z difference figures between field data and LiDAR data of 0.84 m for tall shrubs (2–5 m) 0.52 m for low shrubs (<2 m), 0.26 m for grass and herbs and

0.21 m for exposed aquatic vegetation. Rango *et al.* (2000) used airborne LiDAR to study the morphological characteristics of shrub coppice dunes in a large area of desert. They explain how they acquired multiple return data that represented the surface with a point every metre, having a vertical resolution of 0.05 m, whilst quoting the manufacturers stated vertical error of 0.15 m. Rango *et al.* explain that using the system they employed, 1000 m could be covered in 12 hours to a similar level of detail.

Full waveform LiDAR

More recently, developments have been made in terms of two-dimensional LiDAR signal processing in order to allow full-waveform analysis. Hug *et al.* (2004) explain how it is now possible to digitally sample and store the entire echo waveform of reflected laser pulses. This technique analyses the echo signal of a laser pulse in order to elucidate information of multi-layer surfaces such as below trees and low level vegetation. This new development allows the complex 3D texture of overlying surfaces to be understood, which makes this a potentially useful tool for appreciating vegetation height, density and roughness. This technique, although an exciting prospect, is still in development as precise waveform response to different surfaces remains relatively unknown. Hug *et al.* (2004) state that waveform registration in topographic LiDAR systems is not a new approach and that early experimental setups date back to the 1970s (e.g. Marmon *et al.*, 1978). Only recently, however, have advances in electronics and hard disk size and computer performance been advanced enough to construct LiDAR systems that are self-contained and rugged enough for operational use (Blair *et al.*, 1999). The advent of commercial waveform-digitising LiDAR mapping systems like the LiteMapper-5600 (see www.riegl.com) provides the opportunity for much more detailed analyses of distributed vertical surfaces. In terms of published research, little attention has been paid to the use of full-waveform LiDAR in mapping surface structure and topography. Hyde *et al.* (2005) have reported on the use of waveform LiDAR in mapping forest structure for wildlife habitat analysis. Using waveform LiDAR they were able to map vegetation structure using the full footprint of returns over an area of 175 km^2, to within a 0.3 m vertical resolution. This allowed them to quantify and represent canopy height, canopy cover and above-ground biomass using the single dataset. Hyde *et al.* (2005) report various error statistics relating to different layers within different types of vegetation, however overall results were fit-for purpose. The utility of full waveform LiDAR over an unvegetated, non-complex area would be questionable, especially on single layer surface zones such as exposed gravels or fine sediments. However, its potential is undoubted in terms of being able to better represent the internal structure of vegetation and vegetative layers on floodplains, which could contribute to the improved understanding of the vegetative component of floodplain roughness.

Hug *et al.* (2004) explain how, instead of singular return locations as generated by conventional LiDAR systems, the digitised echo waveform reveals all the information the laser pulse collected during its trip to the surface, such as the detailed distribution of targets in the beam path, their reflectance (or relative surface area) and their vertical extent. Hug *et al.* (2004) go on to explain that using waveform LiDAR ahead of conventional single return LiDAR techniques can increase data volumes by between 50 and 200 times. This is an important issue to consider, especially if the user already finds conventional LiDAR data processing to be cumbersome using their available computer equipment.

Measurement below water

Recent advancements in digital photogrammetric data acquisition, storage and processing have meant that it is now possible to couple photogrammetry with high-resolution, multi-spectral remote sensing in order to measure low-relief topography and water depth from airborne platforms. Lane (2001) explains how water depth can be acquired using image analysis and semi-empirical calibration equations in order to obtain maps of water depth (e.g. Gilvear *et al.*, 1995; Winterbottom & Gilvear, 1997; Gilvear *et al.*, 1998; Bryant & Gilvear; 1999; and Legleiter *et al.*, 2004). Lane (2001) continues by describing the theoretical basis of this approach, which is that the level of light reflected from a submerged bed is some function of water depth as a result of absorption process (see Lane 2001 for a more detailed schematic flow-path of this process).

Airborne laser bathymetry is a new technique that relies on the differential timing of laser pulses reflected from the water surface and the underwater surface to determine the water depth at the point where the laser pulses strike the water surface. This approach is commonly referred to airborne blue–green LiDAR scanning. The measurement system generates short pulses of light at two wavelengths; infrared and blue–green. The infrared pulse is reflected by the water surface and the blue–green pulse penetrates the water and, if it penetrates far enough, is reflected from beneath water-surface. The automated laser system scans beneath the aircraft, and sends out pulses in a swath over which the position of the water surface and sea floor is measured. To correct for the aircraft's movements, the motions of the aircraft are measured by an inertial reference system (IRS). These data are then sent for real-time pitch and roll compensation and then recorded for post-processing, while a GPS receiver records the aircraft's position. Meanwhile, a georeferenced digital camera records the area being surveyed. The laser pulse, waveforms, scan angle, GPS data and IRS data are combined to produce accurate real-time and post-processed sounding positions or terrain elevations. This technique was first described by Weinrebe and Greinert (2002) after they utilised it to successfully map the sub-marine morphology off the coast off Norway to a depth of around 30 m. Vertical sea floor measurements acquired using this technique currently have an achievable accuracy of within ±0.25 m and have a range of around 70 m below the water surface (Simmons Aerofilms website, 2007). This makes this technique appropriate for near-shore sea areas or where the permanently submerged sea floor has a low gradient. Swath widths of this technique are typically in the region of 300 m. A major limitation of this technique is that the depth of achievable measurement is dictated by the state of the water surface and levels of turbidity and suspended sedimentation in ocean waters. Because of this, surveys may only be conducted on the calmest of days, when ocean movements and turbulence are at their least.

Terrestrial acquisition

Bornaz and Rinaudo (2004) explain how terrestrial laser scanners can be considered as highly-automated motorised total stations. Unlike total stations, where the operator directly chooses the points to be surveyed, terrestrial laser scanners randomly (within the systematic confines of an angular pseudo-grid) acquire a dense set of coordinate points. The various practical limitations that exist when developing terrestrial laser-scanning systems means that there are unavoidable trade-offs between laser power, range and footprints, and precision, resolution and accuracy. Schultz and Ingensand (2004) explain how this is due to the physical laws relating to laser behaviour which define the functionality between accuracy and intensity, as described by Gerthsen and Vogel (1993). This situation has led to the development of numerous different laser-scanning systems with attributes that vary in many ways. This variation means that no single terrestrial laser-scanning system exists that can be used for all conceivable applications (Schultz & Ingensand, 2004). The availability and variety of terrestrial laser scanners has increased markedly since their widespread employment in the mid-1990s. Lemmens (2007) reports on the results of an in-depth review of all commercially available terrestrial laser scanners, in which all available laser scanners were categorised and defined in terms of various attributes. He describes how, since 2001, the number of available laser scanning systems (of which there are often various models for each) has more than doubled from nine to nineteen. Fröhlich and Mettenleiter (2004) state that classifications by technical properties and specifications are useful as they indicate possibilities

based upon the performance of an individual system. These technical specifications are:
• Scanning speed and sampling rate of laser measurement system.
• Field of view (camera view, profiling, imaging).
• Spatial resolution, i.e. number of points scanned in field of view.
• Point accuracy (for range measurement systems, deflection systems and overall for the systems).
• The potential for combination with other devices that can be mounted on thc laser scanner (e.g. Camera, GPS).
Lemmens (2007) continues to suggest that the most important defining feature of terrestrial laser scanners are their respective ranges, as it is this attribute which largely controls the potential application of a scanner. This is a notion that is echoed by Shultz and Ingensand (2004). With this in mind, Lemmens (2007) suggests that terrestrial laser scanners can be categorised into one of three application groupings:
• High-precision measurement and detailed three-dimensional reconstruction of objects.
• The measurement of outdoor scenes featuring objects of complicated shape (construction, architecture, civil engineering).
• Land survey.
Lemmens (2007) goes onto quantify the ranges that should be associated with each of the application classifications giving a range for short-range scanners of <25 m, a range for medium-range scanners of 25 m–250 m and a range for long-range scanners of >250 m. Regardless of the potential range of measurement, terrestrial laser scanners can also be categorised based upon the way in which they function technically. Fröhlich and Mettenleiter (2004) provide a good description of how terrestrial laser scanners can be broadly classified based upon their technical specifications and data acquisition methodology. It is suggested that terrestrial laser scanners can be categorised by the principle of the distance measurement system. They continue by explaining that this is because the distance measurement system correlates to both the range and the resulting accuracy of the system. Fröhlich and Mettenleiter (2004) suggest associated ranges, accuracies and manufacturers of the different classifications of terrestrial laser scanners as based upon the technological principle of measurement (see Table 6.3). The three different technologies for range measurements that are used with laser scanners are listed as:
• The most popular measurement system for laser scanners is based upon the time of flight principle. This technique allows unambiguous measurements of distances up to several hundred metres. Long range scanning is usually associated with reasonable accuracy.
• The phase measurement principle represents the other common technique for medium ranges. The range is restricted to one hundred metres. Accuracy of the measured distances is possible to within a few millimetres.
• Several close range laser scanners with ranges up to few metres are available that use pro-optical triangulation to measure. However, they are more for use in industrial applications and reverse engineering (e.g. online monitoring as used in construction). Accuracies down to a few micrometers can be achieved with this technology.
As the term 'short-range' suggests, this approach to measurement is often restricted in terms of measurable distance and thus area of coverage, and many short-range, precise LiDAR scanning

Table 6.3 Classification of laser scanners based upon the technological principle of measurement (from Fröhlich & Mettenleiter, 2004).

Measurement technology	Range (m)	Accuracy (mm)	Manufacturers
Time of flight	<100	<10	Callidus, Leica, Mensi, Optech, Riegl
	<1000	<20	Optech, Riegl
Phase measurement	<100	<10	IQsun, Leica, VisImage, Zoller + Frohlich
Optical triangulation	<5	<1	Mensi, Minolta

instruments are confined to laboratories. At the smallest scale of measurement, *in-situ* laser scanners can measure very small items such as bones and small hand-held artefacts. Because such scanners are restricted in terms of practical deployment and range of measurement, their utility in the 'natural' environment is often restricted to lab-based studies on small-scale surface features.

Early forms of small-range 'scanners' were based upon the optical sensor being mobile within a fixed grid structure. Such equipment was successfully used to measure small scale soil surface topography both in laboratory (Bertuzzi *et al.*, 1990), and in field conditions (Huang & Bradford, 1990). In such techniques the motor-driven mobile sensor moves systematically within the grid taking multiple measurements perpendicularly to the grid level. This means that laser penetration into the 3D surface structure is limited to an upper surface as viewed vertically. More modern short-range scanners can be broadly categorised into two types: fixed optical sensor origin and mobile optical sensor origin. Fixed origin scanners operate and measure in the same way as standard terrestrial laser scanners from a fixed position per scan. Mobile optical sensor scanners utilise a dynamic arm, which allows the sensor to be moved around an object in three-dimensions. The LiDAR-based distance measurements are then corrected using positioning information as provided by situation electronics within the mobile arm. This type of measurement allows a more complete surface representation to be acquired due to the incidental measurement angle of the laser pulse changing in relation to the static positioning of the object that it being measured. This avoids shadowing affects and is theoretically similar to the multiple-scan positioning approach that is commonly employed when conducting medium-long range laser scan surveys using mobile fixed-measurement origin LiDAR scanners.

Terrestrial laser scanning in practice

In contrast to traditional geodetic instruments (e.g. total stations, GPS), most available laser scanners are not well specified in terms of accuracy, resolution and performance, and only a few systems are checked by independent institutes regarding their performance in relation to manufacturer specifications (Fröhlich & Mettenleiter, 2004). Boehler *et al.* (2003) explain how the accuracy specifications given by laser scanner producers in their publications and pamphlets are not comparable. They continue by stating that experience shows that sometimes the manufacturer-quoted specifications should not be trusted, and that the accuracy of these instruments which are built in small series can vary from instrument to instrument. Schultz and Ingensand (2004) state that specific procedures have been developed for the calibration and testing of total-station type equipment in recent years allowing instrumental errors, angular resolutions and distance accuracies to be defined. They continue by explaining that unfortunately, the mechanical design of laser scanners is different to total stations, meaning that many of these investigations cannot be applied to laser scanners. A further complicating factor is noted: that it is impossible to measure the exact same point in succession using a laser scanner. This situation has led to some innovative protocols being designed in tests on laser scanning equipment, by parties independent to the manufacturer. This has involved the production and measurement of geodetic instruments and targets, which have varied per instrument and are thus unstandardised (Schultz & Ingensand, 2004). Testing, and the improved understanding that this brings, is important for terrestrial laser scanning, because every point cloud produced by a laser scanner contains a considerable number of points that show gross errors (Boehler *et al.*, 2003). Boehler *et al.* (2003) continue to explain that this means that if a point cloud is delivered as a result of surveying, a quality guarantee, as is possible for other surveying instruments, methods and results, cannot be given.

Fröhlich and Mettenleiter, (2004) conducted tests using a number of scanners including a Riegl LMSZ360 under controlled conditions, which allowed comparisons to be made between the systems. They concluded that Riegl scanners are amongst the most advanced systems available on the market in that they allow a professional approach when measuring complex scenes such as in industrial environments. The results and technical specifications of the scanning tests of Fröhlich and Mettenleiter (2004) are shown in Table 6.4. The testing results shown in Table 6.4 compare favourably with the manufacturer-quoted specifications in all cases.

A comprehensive and important terrestrial laser-scanning testing regime was conducted and reported upon by Boehler *et al.* (2003). Boehler *et al.* (2003) designed their testing methodology based upon the methods and results that were delivered in previous research that was concerned with laser scanner accuracy (e.g. Balzani *et al.*, 2001; Johansson, 2002; Lichti *et al.*, 2000, 2002; Kern, 2003). A total of nine different terrestrial laser scanning systems were tested in order to elucidate information on five different factors that require consideration in terms of data accuracy and surface representation.

A very fast and easy check for the noise (accidental error) of range measurements was achieved by Boehler *et al.* (2003) when a plane target perpendicular to the observation direction was scanned and the standard deviation of the range differences of the points from an intermediate plane through the point cloud was computed. The results from this experiment for the various scanners revealed that this effect is negligible for most scanners apart from the Riegl LMSZ210. In the case of the Riegl LMSZ210, range accuracy improves from between 5 m up to about 20 m, but then gradually worsens until it is over 30 mm at a range of 90 m.

In order to examine the effects of surface reflectivity tests were designed by Boehler *et al.* (2003) based upon the theory that laser scanners have to rely on a signal reflected back from the object surface to the receiving unit in case of ranging scanners and to the camera in case of triangulation scanners. In either case, the strength of the returning signal is influenced (among other facts such as distance, atmospheric conditions, incidence angle) by the reflective abilities of the surface (albedo). White surfaces will produce strong reflections whereas reflection is weak from black or dark surfaces. The effect of coloured surfaces depends on the spectral characteristics of the laser (green, red, near infrared). Shiny surfaces are usually not easy to measure due to their non-diffusing characteristics. Boehler *et al.* (2003) observed that surfaces of different reflectivities result in systematic errors in range. For some materials, these errors were several times larger than the standard deviation of a single range measurement. Some scanners that allowed aperture adjustment showed errors in the first points after the laser spot had reached an area of a reflectivity that differed considerably from the previous area. It was suggested that more serious errors should be expected when a single object is formed of different materials, or differently painted or coated surfaces, and that this could be avoided if the object was to be temporarily coated with a unique and consistent material (Table 6.5).

Boehler *et al.* (2003) used the results of all of their tests (some of which are not mentioned within this chapter) to make a generic comparison and evaluation of all the scanners that were involved in the project. The results of these evaluations are shown in Table 6.6 and Table 6.7 in terms of the

Table 6.4 Technical information relating to different scanning systems and the testing results of Fröhlich & Mettenleiter (2004).

Manufacturer	System	Frequency (Hz)	Range (m)	Performance
Leica	HDS 3000	1000	>100	Accuracy of 6 mm @ 50 m
Mensi	GS200	5000	700	3 mm @ 100 m
Optech	ILRIS-3D	200	800	3 mm < 100 m 30 mm > 100 m
Riegl	LMSZ360	8000	800	5 mm constant
Zoller & Frolich	IMAGER 5003	500	52	5 mm constant

Table 6.5 Distance correction in mm due to different surface materials (as reported on by Boehler *et al.*, 2003). Positive sign = Distance is measured too short as compared to a white surface. The colours refer to spray-painted surfaces.

Type	White 90%	White 80%	Grey 40%	Black 8%	Metal paint	Aluminium foil	Blue foil	Orange cone
Callidus (1)	0	0	0	0	0	0 to −100	+6	−10
Callidus (2)	0	0	+4	+3	0 to −10	0 to −15	+5	−20
Cyrax (1)	0	0	0	0	0	0 to −10	+22	−40
Cyrax (2)	0	0	0	0	0	0	+17	−70
S25	0	0	0	0	0	0	0	0
GS100	0	0	0	+8.	0	0	n.a[a]	0
Riegl Z210	0	0	+13	+3	0 to −100	0 to −250	0	−100
Riegl Z420	0	0	0	0	0	0	0	−20
Zoller & Frolich	0	0	0	0	0	0 to +30	−18 m	−20

[a] Scanner did not record any points on this surface.

Table 6.6 Major advantages of the laser scanners as tested by Boehler *et al.* (2003).

Type	Advantage
Callidus	Very large field of view.
Cyrax2500	Good accuracy.
S25	Very high accuracy for short ranges.
GS100	Large field of view.
Riegl Z210	High ranges possible. Large field of view.
Riegl Z420i	Very high ranges possible. Large field of view.
Zoller and Frolich	Very high scanning speed. Large field of view.

Table 6.7 Major disadvantages of the laser scanners as tested by Boehler *et al.* (2003).

Type	Disadvantage
Callidus	Very coarse vertical resolution (0.25).
Cyrax2500	Small scanning window (40 x 40).
S25	Does not work in sunlight. Not suited for long ranges.
GS 100	Large noise.
Riegl Z210	Low accuracy.
Riegl Z420i	Large noise.
Zoller & Frolich	Low edge quality. Limited angular resolution.

respective advantages and disadvantages of each laser scanning system. It must be stressed that just because a certain scanner performed better than others, that does not necessarily mean that this scanner is the most appropriate instrument for a certain purpose.

Talaya *et al.* (2004) report on how they were able to utilise a dynamic Riegl LMSZ210 terrestrial laser scanner, which was able to measure urban scenes whilst attached to the roof of a mobile vehicle. The laser measurements were corrected using an integrated GPS and IMU (Inertial Measurement Unit) system. They report RMS accuracies of 0.18 m, 0.35 m and 0.13 m in eastings, northings and the vertical components respectively. These statistics have since been improved upon by the Streetmapper™ system (Redstall & Hunter, 2006). The Streetmapper system can accurately map coordinate points to within 38 mm in x and y dimensions and 50 mm in terms of elevation. This system can also be attached to an off-road quad bike and could thus be utilised in certain environments where the obtainable accuracies are acceptable to the project. It is increasingly becoming the norm for terrestrial laser scanners to be integrated with high-megapixel digital cameras. This combination of tools is very productive in the built environment as the colour information provided by many laser scanners does not provide the quality required to in order to detail surface textures and frescos (e.g. Guarnieri *et al.*, 2004; Jansa *et al.*, 2004; Kadobayashi *et al.*, 2004; Alshawabkeh & Haala, 2004; Beraldin, 2004).

Hetherington *et al.* (2007), show how terrestrial laser scanning can be used to map and monitor geomorphologically complex and dynamic

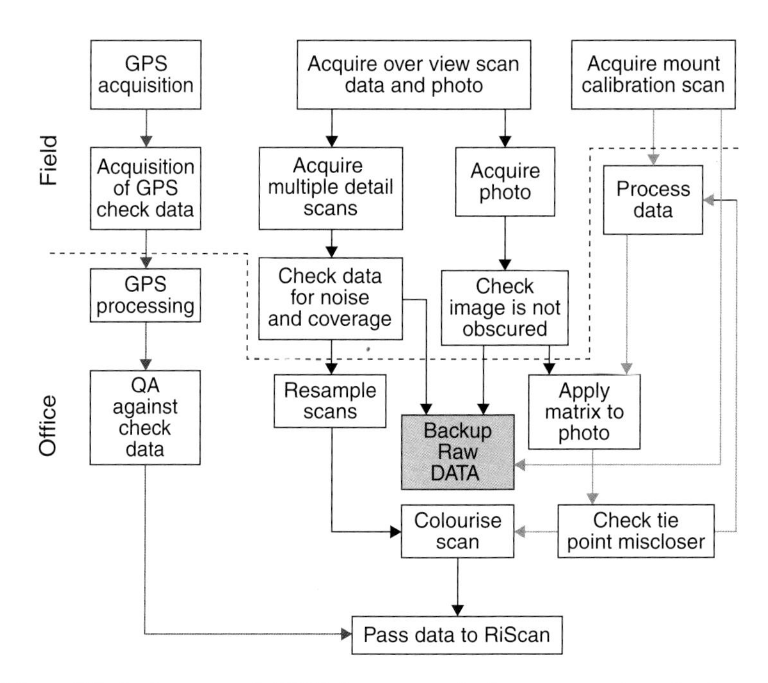

Fig. 6.2 The workflow of Hetherington *et al.* (2007) that was used when representing a fluvio-estuarine zone.

fluvio-estuarine zones. In this work, a Riegl LMSZ420i laser scanner was used along with a GPS-referenced control network, which allowed the easy global alignment of scans in the IMAlign component of the Polyworks™ suite of software. Initially, a local approach was taken to alignment, where IMAlign utilises an iterative process which starts by holding a single scan in a fixed position and then aligning it to a second scan. Once aligned, both scans are held fixed and a third scan aligned, and so on until all scans are aligned. Finally a 'global' approach is taken to scan alignment in that the alignment process is repeated for all scans with just one scan held fixed. It was found in early attempts at alignment that automated iterative scan alignment in IMAlign was not always successful. This is a common problem in natural environments due to the multi-scale 3D complexity of most surfaces. This complexity can result in pattern–point cloud matching software reaching a best-fit match that hazes small-scale variable surfaces in an attempt to better-match larger-scale variability. In order to improve the alignment process, certain scans were specifically acquired such that they contained hard, flat surfaces, specifically near buildings. The scan with the most surface regularity was chosen as the key locked-scan in the global alignment procedure in IMAlign. A schematic representation of the workflow of Hetherington *et al.* (2007) is shown in Figure 6.2. Scan merging affects are not relevant when surface analysis is conducted using a single scan. This approach was used by Heritage *et al.* (2006) to successfully elucidate information on grain size and roughness over a gravel bar feature.

The approach taken by Hetherington *et al.* (2007) allowed the scan data to be accurately merged, in relative terms, to within the inherent error bank of the terrestrial laser scanning equipment (0.01 m), whilst being globally accurate to within the expected error of the GPS control network (0.03 m). Table 6.8 provides details of the data and file size characteristics that can result after an intensive terrestrial laser scan survey regime and after their approach to post-processing.

Table 6.8 Example data characteristics resulting from a morphology-focused terrestrial laser scan survey over a 1.5 km chainage of a complex fluvio-estuarine environment (from Hetherington *et al.*, 2007).

Scan data stage	No. of scans	Coverage	No. of points	File size
Raw data	26	171,300 m^2	36 million	1.62 Gb (ASCII)
Processed data	26 (all used)	171,300 m^2	36 million	813 Mb (Binary)
Surfaced data	N/A	171,300 m^2	34.5 million (1.5 million classified as low, air or mirrored points)	80 Mb (20 cm Arc Grid)

Hetherington *et al.* (2007) make the point that terrestrial laser scanning station placement should be morphology-controlled if data quality is to be maximised. This approach relies on the surveyor choosing scan positions so that the deepest and most-obscured topography of the smallest scales of variation that are of importance in terms of project requirements are measured. A schematic representation of how TLS can be used as a tool for managing the natural and quasi-natural environment, based upon the levels of detection that are required is shown in Table 6.9.

Relatively more work has been conducted that focuses specifically on the measurement and representation of vegetation, and vegetative landscape parameters within the natural environment. Schmid *et al.* (2004) used TLS to measure soil roughness at a logged site to within a point accuracy of 4 mm, over a potential area of 240 m^2 per day. A summary of some key studies that have employed terrestrial laser scanning within the natural environment is shown in Table 6.10.

DATA INTEGRATION

Data from different laser scanning platforms can be complimentary and assist in completing the holistic representation of a scene by 'filling in the gaps', where objects are shadowed or out of view. Because of this, data integration is an important process if the benefits relating to different platforms having different perspectives of measurement are to be appreciated. Bohm and Huaala (2005) report on the issue of integrating aerial and terrestrial laser data, which is most useful for the representation of urban environments. This is due to each technique having differing perspectives and thus creating a final model with reduced shadow. They report that data can be linked to within approximately 0.24 m of each other, in terms of mean 3D point alignment in known locations. This issue is not as relevant in the many non-vegetated natural environments, as the scene is more typically associated with low-relief, single layer physical variation in topography.

Guarnieri *et al.* (2004) use a Riegl LMSZ360 laser scanner along side an independent photogrammetric survey with a texture resolution of 2 mm. This allowed them to texturise the terrestrial laser scan coordinate data with the more detailed information from the photogrammetric survey using the IMAlign module of the Polyworks™ suite of software. They report the final relative data accuracies of both surfaces as tested by an independent total station survey in Table 6.11. It is noted that the RMS errors as suggested by Guarnieri *et al.* (2004) do not correspond well to the 3.2 mm (@50m range) accuracy as quoted by Riegl for the LMSZ360 laser scanner.

Jansa *et al.* (2004) offer a good comparison of terrestrial laser scanners and data that can result from photo cameras; this is shown in Table 6.12. They report that using photogrammetric data along side TLS data (as acquired by a Riegl LMSZ360) can lead to an improvement of accuracy of laser scan data of a factor of three. Both Kabobayashi *et al.* (2004) and Alshawabkeh and Haala (2004) appreciate the benefits to accuracy and representation that combining terrestrial laser scanning with photogrammetric information can bring, but

Table 6.9 Detection of, and response to, various scales of spatio-temporal change (based on different levels of data quality and surface representation) in quasi-natural environments using terrestrial LiDAR (after Hetherington *et al.*, 2007).

Scale	Example feature	Size	Temporal change	Example spatial-change	Management
Micro	A single pebble, leaf or small volume of sediment.	0.01–0.05 m	Daily, as a response to 'every day' processes.	Single stone movement over a short distance, discreet sand removal/accretion, vegetation change.	Low priority.
Meso	Cobbles, individual rip rap, moderate volume of sediment.	0.05–0.20 m	Medium-term response to constant processes (weekly) or high a magnitude event (daily).	Deformation of protection measures, moderate movement or individual rip rap, moderate erosion/accretion.	Low to moderate priority. Increased levels of monitoring recommended.
Unit	Gabion cage, large individual rip-rap, substantial volume of sediment.	0.2–1 m	Daily as a response to a high-magnitude event or over 10s of years as a response to 'every day' processes and moderate magnitude events.	Displacement of individual large rip-rap, slumping of gabions, bank collapse and undercutting, small dune removal or accretion.	Moderate to high priority. Loss of structural strength. Worsened state possible. Monitoring vital.
Large	Rows of gabion cages, large sea walls, piles of rip-rap, complete slopes.	1–100 m	Daily response to an extreme event or Over 10s of years as a response to moderate and large events.	Complete slope failure, Large scale movement or removal of rip-rap or gabion cages. Damage and removal of large sea walls.	Immediate action and large scale remedial works.

Table 6.10 Summary applications and data characteristics for terrestrial laser scanning in the natural environment (from Heritage & Hetherington, 2005, 2007).

Environment and study type	Scan number/Survey duration	Survey area (m^2)	Overall accuracy/Point spacing	Authors
Barchan dune, measurement	11 scans/0.5 days	400	±6 mm/100 m^{-2}	Nagihara *et al.* (2004)
Glacial outwash plain, daily morphological change measurement	4 scans/Daily survey, 2 hours duration	4000	± 2 mm/2500 m^{-2}	Hetherington *et al.* (2005)
Felled forest, roughness estimation and volume balance measurement	80 m^2/day	240	±4 mm/50000 m^{-2}	Schmid *et al.* (2004)
Valley side, landslide measurement	3 scans	4000	±10–25 mm/1000 m^{-2}	Bitelli *et al.* (2004)
Glacial Lake/Valley, ice cliff evolution	14 scans/1 day	36000	±3–5 mm/1000–1400 m^{-2}	Conforti *et al.* (2005)
Cliff Face, stratigraphic modelling	Not specified	37500	±5 mm/>10000 m^{-2}	Bellian *et al.* (2005)
Pine Forest, Tree canopy structure measurement	3 scans/0.5 days	60	±5–15 mm/up to 100000 m^{-2}	Danson *et al.* (2006)

Table 6.11 Comparative accuracy of laser scan and photogrammetrically-acquired data against an independent total station survey (from Guarnieri *et al.*, 2004).

3-D Model	No. of check points	RMS x (m)	RMS y (m)	RMS z (m)
Image-based	22	0.017	0.025	0.020
Laser scanner	22	0.056	0.063	0.044

Table 6.12 A comparison of photographic and laser scanning sensors (from Jansa *et al.*, 2004).

Attribute	Laser scanner	Photo cameras
Spatial resolution	High	Very high
Spatial coverage	Very good	Good
Intensity/Colour	Limited	Very good
Illumination	Active	Passive (& active)
3D point density	High	Depends on texture
Depth accuracy	High	High
Acquisition procedure	Dynamical	Moment shot
3D reconstruction effort	Medium	High
Texture reconstruction	None or very limited	Very good
Instrument costs	High	Low

do not quantify this specifically. Beraldin (2004) reiterates the statements of Hall and Llinas (1997), stating that multi-sensor data-fusion techniques combine data from multiple sensors and related information from associated databases, to achieve improved accuracies and more specific inferences than could be achieved by the use of a single sensor alone. The combined approach could thus be an important area of research and application in environmental scanning, even when high levels of visual representation are not required.

CONCLUSIONS AND RECOMMENDATIONS

The approach to data acquisition for any given purpose should be driven by the size and complexity of the subject scene or object, and what level of measurement detail and representivity is required over the area in question. The project data requirements, in relation to the nature of the scene or object, will thus define what level of measurement can be classed as good quality. In general, those who require data have an increasingly varied suite of tools at their disposal in order to provide good quality data for a large and varied number of purposes. It is the responsibility of those who collect and manipulate data to ensure that the correct decisions are made from the outset in terms of selecting appropriate measurement techniques, and data are collected in the correct fashion (e.g. morphology-driven surveys). Data from both terrestrial and airborne platforms have their own sets of benefits, which can be maximised overall if data from both approaches are integrated. Integration with other techniques, such as photogrammetry, can also add benefits to data resulting from LiDAR surveys. Recent developments in full-waveform and blue–green LiDAR technologies mean that data are now potentially available to environmental scientists and engineers for previously-hidden landscape components.

REFERENCES

Alshawabkeh Y, Haala N. 2004. Integration of digital photogrammetry and laser scanning for heritage documentation. *Proceedings of the XXth ISPRS Congress, 12–23 July 2004 Istanbul, Turkey*, Commission 5. ISSN 1682–1750. 424–430.

Balzani M, Pellegrinelli A, Perfetti N, Uccelli F. 2001. A terrestrial 3D laser scanner: Accuracy tests. *Proceedings of the 18th International CIPA Symposium*. 445–453.

Bellien JA, Kerans C, and Jennette DC. 2005. Digital outcrop models: applications of terrestrial scanning LiDAR technology in stratigraphic modelling. *Journal of Sedimentary Research* **75**: 166–176.

Beraldin AA. 2004. Integration of laser scanning and close-range photogrammetry – the last decade and beyond. *Proceedings of the XXth ISPRS Congress, 12–23 July 2004 Istanbul, Turkey*, Commission 5. ISSN 1682–1750.

Bertuzzi P, Caussignac JM, Stengel P, Morel G, Lorendeau JY, Pelloux G. 1990. An Automated, non-contact laser profile meter for measuring soil roughness *in situ*. *Soil Science* **149**: 169–178.

Blair JB, Rabine DL, Hofton MA. 1999. The Laser Vegetation Imaging Sensor: a medium-altitude, digitisation-only, airborne laser altimeter for mapping vegetation and topography. *ISPRS Journal of Photogrammetry and Remote Sensing* **54**: 115–122.

Boehler W, Bordas Vicent M, Marbs A. 2004. Investigating laser scanner accuracy. *Proceedings of the XIXth CIPA Symposium*, 30th of September – 4th of October, 2003, Antalya. Turkey.

Bohm J, Haala N. 2005. Efficient integration of aerial and terrestrial laser data for virtual city modelling using laser maps. *Proceedings of the Working Group V/3 Workshop Laser Scanning 2005. 12–14 September 2005, Enschede, the Netherlands*. 192–197.

Bornaz L, Rinaudo F. 2004. Terrestrial laser scanner data processing. *Proceedings of the XXth ISPRS Congress, 12–23 July 2004 Istanbul, Turkey*, Commission 5, ISSN 1682–1750. 514–520.

Bryant RG, Gilvear DJ. 1999. Quantifying geomorphic and riparian land cover changes either side of a large flood event using airborne remote sensing: River Tay, Scotland. *Geomorphology* **29**: 307–321.

Charlton ME, Large ARG, Fuller IC. 2003. Application of airborne LiDAR in river environments: The River Coquet, Northumberland, UK. *Earth Surface Processes and Landforms* **28**: 299–306.

Cobby DM, Mason DC, Davenport IJ. 2001. Image processing of airborne scanning laser altimetry data for improved river flood modelling. *ISPRS Journal of Remote Sensing* **56**: 121–138.

Conforti C, Deline P, Mortara G, Tamburini A. 2005. Terrestrial Scanning LiDAR Technology applied to study the evolution of the ice-contact image lake (Mont Blanc, Italy). *Proceedings of the 9th Alpine Glaciological Meeting*, February 2005, Milan, Italy.

Cooper MAR. 1998. Datums, coordinates and differences, In: *Landform Monitoring, Modelling and Analysis*. Lane SN, Richards KS, Chandler JH (eds), John Wiley & Sons Ltd., Chichester. 21–36.

Danson FM, Hetherington D, Morsdorf F, Koetz B, Allgower B. 2006. Forest Canopy Gap Fraction from Terrestrial Laser scanning. *Geoscience and Remote Sensing Letters* **4**: 157–160.

Davenport IJ, Bradbury RB, Anderson GQA, Haymen GRF, Krebs JR, Mason DC, Wilson JD, Veck NJ. 2000. Improving bird population models using airborne remote sensing. *International Journal of Remote Sensing* **21**: 2705–2717.

Filin S. 2004. Surface classification from airborne laser scanning data. *Computers and Geosciences* **30**: 1033–104.

French JR. 2003. Airborne LiDAR in support of geomorphological and hydraulic modelling. *Earth Surface Processes and Landforms* **28**: 321–335.

Fröhlich C, Mettenleiter M. 2004. Terrestrial Laser Scanning – New Perspectives in 3D Surveying, *Proceedings of the ISPRS working group VIII, Freiburg, Germany, 03–06 October 2004, International Archives of photogrammetry*. Remote Sensing and Spatial Information Sciences Volume XXXVI, Part 8/W2, Thies M, Koch B, Spiecker H, Weinacker H (eds), ISSN 1682–1750.

Geist T, Lutz E, Stotter J. 2004. Airborne laser scanning technology and its potential for application in glaciology. *Proceedings of the ISPRS working group III, Dresden, Germany, 08–10 October 2004, International Archives of photogrammetry*, Remote Sensing and Spatial Information Sciences Volume XXXVI, Part 3/W13, Maas HG, Vosselman G, Streilein A (eds), ISSN 1682–1750.

Gerthsen C, Vogel H, 1993, Physik (17), Verbesserte und erweiterte, Springer-Werlag, Berlin-Heidelberg.

Gilvear DJ, Waters TM, Milner AM. 1995. Image analysis of aerial photography to quantify changes in channel morphology and instream habitat following placer mining in interior Alaska. *Freshwater Biology* **34**: 389–398.

Gilvear DJ, Waters TM, Milner AM. 1998. Image analysis of aerial photography to quantify the effect of gold placer mining on channel morphology, interior Alaska. In: *Landform Monitoring, Modelling and Analysis*. Lane SN, Richards KS, Chandler JH (eds), John Wiley & Sons Ltd., Chichester. 311–339.

Gomes Pereira LM, Wicherson RJ. 1999. Suitability of laser data for deriving geographical information: A case study in the context of management of fluvial zones. *ISPRS Journal of Photogrammetry and Remote Sensing* **54**: 105–114.

Guarnieri A, Vettore A, El-Hakm S, Gonzo L. 2004. Digital Photogrammetry and laser scanning in cultural heritage survey. *Proceedings of the XXth ISPRS Congress. 12–23 July 2004 Istanbul, Turkey*, Commission 5. ISSN 1682–1750. 154–159.

Hall DL, Llinas J. 1997. An introduction to multisensor data fusion. *Proceedings of the Institute of the Electrical and Electronics Engineers* **85**: 6–23.

Heritage GL, Hetherington D. 2005. The use of high-resolution field laser scanning in mapping surface topography in fluvial systems. *Proceedings of Symposium S1 (Sediment Budgets) held during the Seventh IAHS Scientific Assembly at Foz do Iguaçu, Brazil, April 2005. IAHS Publ.* 291.

Heritage GL, Hetherington D. 2007. Towards a protocol for laser scanning in fluvial geomorphology. *Earth Surface Processes and Landforms* **32**: 66–74.

Heritage GL, Hetherington D, Large ARG, Fuller IC, Milan DJ. 2006. Terrestrial laser scanner-based grain roughness quantification using a random field approach. *RSPSoc Annual Conference Proceedings*. 5–8 September 2006, Cambridge, UK.

Hetherington D, Heritage GL, Milan DJ. 2005. Daily fine sediment dynamics on an active Alpine glacier outwash plain. *Proceedings of Symposium S1 (Sediment Budgets) Held During the Seventh IAHS Scientific Assembly at Foz do Iguaçu*, Brazil, April 2005. IAHS Publ. 291.

Hetherington D, German SE, Utteridge M, Cannon D, Chisholm N, Tegzes T. 2007. Accurately representing a complex estuarine environment using terrestrial LiDAR. *RSPSoc Annual Conference Proceedings*. 11–14 September 2007, Newcastle Upon Tyne, UK.

Hodgson ME, Jensen JR, Schmidt L, Schill S, Davis B. 2003. An evaluation of LiDAR and IFSAR- derived digital elevation models in leaf-on conditions with ESGS Level 1 and Level 2 DEMs. *Remote Sensing of Environment* **84**: 295–308.

Hopkinson C, Chasmer LE, Zsigovics G, Creed IF, Sitar M, Treitz P, Maher RV. 2004. Errors in LiDAR ground elevation and wetland vegetation height estimates. *Proceedings of the ISPRS working group VIII/2, Freiburg, Germany, 03–06 October 2004, International Archives of photogrammetry*, Remote Sensing and Spatial Information Sciences Volume XXXVI, Part 8/W2, Thies M, Koch B, Spiecker H and Weinacker H (eds), ISSN 1682–1750.

Huang C, Bradford JM. 1990. Portable laser scanner for measuring soil surface roughness. *American Journal of Soil Science* **54**: 1402–1406.

Hug C, Ullrich A, Grimm A. 2004. Litemapper 5600 – A Waveform-digitizing LiDAR terrain and vegetation mapping system. *Proceedings of the ISPRS working group VIII/2*, Freiberg, Germany, October 2004.

Hyde P, Dubayah R, Peterson B, Blair JB, Hofton M, Hunsaker C, Knox R, Walker W. 2005. Mapping forest structure for wildlife habitat analysis using waveform LiDAR: Validation of montane ecosystems. *Remote Sensing of Environment* **96**: 427–437.

Jansa J, Strudnicka N, Forkert G, Haring A, Kager H. 2004. Terrestrial laser scanning and photogrammetry – acquisition techniques complimenting one another. *Proceedings of the XXth ISPRS Congress, 12–23 July 2004 Istanbul, Turkey*, Commission 5. ISSN 1682-1750.

Johanson M, 2002, Explorations into the behaviour of three different high-resolution ground-based laser scanners in the built environment. *Proceedings of the 6th International Working Group of CIPA*. Workshop on Cultural Heritage Recording.

Kadobayashi R, Kochi N, Otani H, Furukawa R. 2004. Comparison and evaluation of laser scanning and photogrammetry and their combined use for digital recording of cultural heritage. *Proceedings of the XXth ISPRS Congress, 12–23 July 2004 Istanbul. Turkey*. Commission 5. ISSN 1682-1750. 401–407.

Keckler D. 1995. *Surfer for Windows – User's Guide*, Golden Software Inc., Colorado.

Kern F. 2003. Automatisierte Modellierung von Bauwerksgeometrien aue 3D-Laserscannerdaten. *Geodatische Schriftenreihe der Technischen Universitat Braunschweig*, Nr 19. ISBN 3-926146-114-1.

Lane SN. 1998. The use of digital terrain modelling in the understanding of dynamic river channel systems. In: *Landform Monitoring, Modelling and Analysis*, Lane SN, Richards KS, Chandler JH (eds), Wiley. Chichester. 311–339.

Lane SN. 2000. The measurement of river channel morphology using digital photogrammetry. *Photogrammetric Record* **16**: 937–957.

Lane SN. 2001. The measurement of gravel-bed river morphology. In *Gravel-Bed Rivers V*, Mosley MP (eds), New Zealand Hydrological Society, Wellington. 291–320.

Lane SN, Westaway RM, Hicks DM. 2003. Estimation of erosion and deposition volumes in a large gravel-bed, braided river using synoptic remote sensing. *Earth Surface Processes and Landforms* **28**: 249–271.

Latypov D. 2002. Estimating relative LiDAR accuracy information from overlapping flight lines. *ISPRS Journal of Photogrammetry and Remote Sensing* **56**: 236–245.

Legleiter CJ, Roberts DA, Marcus WA, Fonstad MA. 2004. Passive optical remote sensing of river channel morphology and in stream habitat: Physical basis and feasibility. *Remote Sensing of Environment* **93**: 493–510.

Lemmens M. 2007. Terrestrial Laser Scanners – Product Survey. *GIM International Magazine*, August 2007. Vol 21 Issue 8.

Lichti DD. 2004. A resolution measure for terrestrial laser scanners. *Proceedings of the XXth ISPRS Congress, 12–23 July 2004 Istanbul, Turkey*, Commission 5. ISSN 1682–1750.

Lichti DD, Gordon SJ, Stewart MP, Franke J, Tsakiri M, 2002. Comparison of digital photogrammetry and laser scanning, *Proceedings of 6th CIPA International Working Group*.

Lichti DD, Stewart MP, Stewart MP, Franke J, Tsakiri M. 2000., Calibration and testing of a terrestrial laser scanner. *International Archives of Photogrammetry and Remote Sensing* **18** (B5): 485–492.

Mamon G, Youmans DG, Sztankay ZG, Morgan CE. 1978. Pulsed GaAs laser terrain profiler. *Applied Optics* **17**: 868–877.

Mark K, Bates P. 2000. Integration of high-resolution topographic data with floodplain flow models. *Hydrological Processes* **14**: 2109–2122.

McKean J, Roering J. 2004. Objective landslide detection and surface morphology mapping using high-resolution airborne laser altimetry. *Geomorphology* **57**: 331–351.

Nasset E. 2002. Predicting forest stand characteristics with airborne laser scanning data. *ISPRS Journal of Photogrammetry and Remote Sensing* **52**: 49–56.

Nasset E, Okland T. 2002. Estimating Tree height and tree crown properties using airborne scanning laser in a boreal nature reserve. *Remote Sensing of Environment* **79**: 105–115.

Nagihara S, Mulligan KR, Xiong W. 2004. Use of a three-dimensional laser scanner to digitally capture the topography of sand dunes in high spatial resolution. *Earth Surface Processes and Landforms* **29**: 391–398.

Popescu SC, Wynne RH, Nelson RF. 2002. Estimating plot-level tree heights with LiDAR: local filtering with a canopy-height based variable window size. *Computers and Electronics in Agriculture* **37**: 71–95.

Rango A, Chopping M, Ritchie J, Havstad K, Kustas W, Schmugge T. 2000. Morphological characteristics of shrub coppice dunes in desert grasslands of southern New Mexico derived from scanning LiDAR. *Remote Sensing of Environment* **74**: 26–44.

Redstall M, Hunter G. 2006. Accurate terrestrial laser scanning from a moving platform. *Geomatics World Magazine* **July/August**: 28–30.

Schultz T, Ingensand H. 2004. Terrestrial laser scanning – investigations and applications for high-precision scanning. *Proceedings of the FIG working week*, May 22–27, 2004. Athens, Greece. 1–14.

Talaya J, Alamus R, Bosch E, Serra A, Komus W, Baron A. 2004. *Proceedings of the XXth ISPRS Congress, 12–23 July 2004 Istanbul, Turkey*, Commission 5. ISSN 1682–1750.

Thoma DP, Gupta SC, Bauer ME, Kirchoff CE. 2005. Airborne laser scanning for riverbank erosion measurement. *Remote Sensing of Environment* **95**: 493–501.

Weinrebe W, Greinert J. 2002. Multibeam Bathymetry: Sounding Ocean Floor Morphology at Continental Margins to Understand Geodynamic History. *Proceedings of the 13th Biennial International Hydrographic Symposium*, Kiel Germany, 8–10th Oct 2002. 12–22.

Winterbottom SJ, Gilvear DJ. 1997. Quantification of channel bed morphology in gravel bed rivers using airborne multispectral imagery and aerial photography. *Regulated Rivers Research and Management* **13**: 489–499.

White SA, Wang Y. 2003. Utilizing DEMs derived from LIDAR data to analyze morphologic change in the North Carolina coastline. *Remote Sensing of Environment* **85**: 39–47.

7 Terrestrial Laser Scanning to Derive the Surface Grain Size Facies Character of Gravel Bars

NEIL S. ENTWISTLE[1] AND IAN C. FULLER[2]

[1]School of Environment and Life Sciences, University of Salford, Manchester, UK
[2]School of People, Environment and Planning, Massey University, Palmerston North, New Zealand

INTRODUCTION

Particle size measurement is dependent upon sampling of river-bed gravels where sample size, operator bias and surface heterogeneity can greatly affect the accuracy of the result. An alternative approach to such sampling is offered by terrestrial laser scanning. Scans provide the opportunity for measurement of a far greater number of clasts and may also detect subtle pebble clusters. This potentially yields useful information in quantification of channel roughness. Derivation of surface grain size is necessary for a variety of applications in river engineering, geomorphology and river ecology, including sediment transport, habitats and flow hydraulics. In geomorphic research, assessment of morphological response to flood flows has been informed using reconstructions based on reach D_{84} information (e.g. Bartholdy & Billi, 2002) and threshold flows for sediment entrainment based on reach averaged grain size parameters (e.g. Fuller *et al.*, 2002). Water flow level in river channels is moderated by the interaction with the roughness of the surface over which it flows. Characterisation of surface grain-size is notoriously problematic due to patchiness (Crowder & Diplas, 1997; Buffington & Montgomery, 1999), incompatibility between sampling approaches (Diplas & Sutherland, 1988; Fraccarollo & Marion, 1995), operational bias (Wohl *et al.*, 1996; Marcus *et al.*, 1999), and the sheer range of grain sizes present in a gravel-bed river from silt (<0.063 mm) to boulders in excess of 500 mm (Verdu *et al.*, 2005). Grain size can vary at different spatial scales, for example between bedforms such as pools and riffles (Milan *et al.*, 2001), within a bar (Church *et al.*, 1987), and show gradual downstream changes (Buffington & Montgomery, 1999), as well as cross-channel changes (Bartholdy & Billi, 2002). Bartholdy and Billi (2002) found a reduction in grain size in both downstream and cross-stream directions, the latter related to an increasing deposition of overbank fines on the surface of a gravel bar, away from the coarser channel. Sedimentologically, this produces a complex facies assemblage. Channel beds therefore comprise several physiographic units

Laser Scanning for the Environmental Sciences,
1st edition. Edited by G.L. Heritage and A.R.G. Large.

with specific morphological and sedimentological features, which are variable spatially and temporally (Billi & Paris, 1992). Skin roughness can also be influenced by the presence of micro-scale bedforms such as pebble clusters (Wittenburg & Newson, 2005). Fundamental to the accuracy of the particle size approaches is the sampling of river-bed gravels where sample size (Wolman, 1954; Church *et al.*, 1987; Wohl *et al.*, *1996*), operator bias (Hey & Thorne, 1983), particle shape (Milan *et al.*, 1999) and surface heterogeneity (Nikora *et al.*, 1998) can greatly affect the result. Rice and Haschenburger (2004) suggest that even in relatively homogeneous units, very large numbers of particles must be sampled to provide an accurate dataset. This concurs with Mosley and Tindale's (1985, p. 465) observation nearly twenty years previously that, 'accurate characterisation of bed sediment in gravel-bed rivers is very demanding'. Despite these problems, a standard surface sample of 100 clasts (Wolman, 1954) remains the accepted method for grain-size characterisation amongst scientists and engineers concerned with channel hydraulics. One hundred clasts are usually sampled over a patch of gravel comprising a single patch or population, 'a zone or area considered homogenous' (Dunne & Leopold, 1978), with sampling being conducted over a grid or transect (Wolman, 1954).

The intermediate 'b' axis of gravel clasts is usually measured. The median (D_{50}) size of the cumulative frequency distribution of the 100 clast sample is usually employed to estimate roughness. However, an index grain diameter (D_i) used to identify critical discharge for entrainment of sediment size D_i, suggests D_{16} as an appropriate value for D_i (Bathurst, 1987). However, Billi and Paris (1992) indicate a higher degree of variability in D_{16} values compared with D_{84} values. Here the D_{84} values on a riffle showed a better relation with the hydraulic characteristics expressed in terms of critical discharge (Billi & Paris, 1992). Thus, a question arises as to whether the D_{50} measure is appropriate, especially due to the effects of imbrication, and microscale bedforms (clusters). Imbrication effectively decreases the roughness height from the bed, whilst pebble clusters tend to provide locally increased roughness height. The former provides an argument for using the c-axis as a roughness measure (Gomez, 1993) whilst the latter suggests some form of multiplier/scale factor be applied to the grain size value (Hey, 1979; Wiberg & Smith, 1991; Clifford *et al.*, 1992).

Surface grain size has also been measured using photographic methods (Kellerhals & Bray, 1971; Ibbeken & Schleyer, 1986). Grain-sizes can either be measured on the photograph at grid-selected positions, or a count of visible grains can be made and converted to a mean grain size via a calibration relation (Church *et al.*, 1987; Rice, 1995; Petts *et al.*, 2000). Two problems arise from photographic assessment of grain size; first, the partial hiding of the clast by fines, shadow or another clast and second, the imbrication angle. Imbrication angle can vary with clast size, shape and sphericity (Church *et al.*, 1987). As a result of these effects, Church *et al.* (1987) found a mean negative bias of 22% (phi units) when comparing D_{50} values obtained from photographs with those obtained from grid-by-number sampling. Direct measurement of micro-roughness using a mini-Tausendfuessler device (De Jong, 1992) gives an exact measure, but over a very small area of the channel. De Jong (1992) has used this in conjunction with photo-sieving to improve grain roughness estimation, but this remains limited in spatial extent and restricted to a cross-section or long profile approach. Latulippe *et al.* (2001) suggest that the usual way of characterising bed sediment is visually to delimit substrate facies boundaries. However, they acknowledge that this is highly subjective and have proposed an alternative process of visual characterisation linked to bulk sampling. Limited sampling is still required and extensive operator training is required to reduce subjectivity between operators.

Remote sensing and image processing techniques have recently been used to assess surface grain size (Butler *et al.*, 2001; Lane, 2002). The method uses a filtering program to identify individual clasts and measure their b-axes. Alternatively, high-resolution photogrammetry has been used to generate digital elevation models (DEMs) from which grain size characteristics have been derived relatively accurately (McEwan *et al.*, 2000). Whilst accurate representations of grain scale roughness can be obtained, the approach is limited to relatively small areas. Camera calibration with respect to focal length from the gravel bed is also an important consideration (Lane, 2002), and images may require significant post-processing. Verdu *et al.* (2005) indicate the feasibility of using geostatistics in conjunction with image analysis to measure grain size characteristics accurately across larger reaches. Carbonneau *et al.* (2004, 2005) have obtained D_{50} measures from imagery with pixel resolutions of between 3 and 10 cm using a similar geostatistical approach. Errors in deriving D_{50} and D_{84} associated with these combined imagery and geostatistical approaches remain relatively high (>30%) (Verdu *et al.*, 2005). Furthermore, the approach is computationally intensive.

Surface roughness has also been measured using a random field of spatial elevation data (Gomez 1993; Nikora *et al.*, 1998). The success of this approach has been tempered by the lack of high-resolution topographic data covering all roughness scales, however, improved data point resolution is now achievable using terrestrial laser scanning technology (Heritage & Hetherington, 2005). Developments in the measurement of surface character now allow detailed digital elevation models (DEMs) to be constructed of complex topography; and these may be used to further investigate roughness character and heterogeneity in an attempt to improve quantification of surface roughness.

LASER SCANNING

Surfaces measured using laser scanning technology have been proved to be extremely detailed (capturing between 1000 and 10,000 points/m^2), highly accurate (3D location errors of less than 0.02 m) and may be efficiently collected utilising a reflector based, or a total station, EDM tiepoint system, to mesh individual scan-cloud data. The resultant DEMs thus map all roughness elements from the grain-scale upwards across the entire scanned area, revealing their 3D structure and spatial distribution (Heritage *et al.*, 2005; Milan *et al.*, 2007). Subsequent data analysis yields fundamental information for assessment of sediment transport, habitat and hydraulics.

Facies characterisation is important because patchy bed surfaces affect both physical and biological processes within a reach (Buffington and Montgomery, 1999). This chapter describes a method to reliably quantify the population grain-size distribution of natural gravel surfaces using random field terrestrial laser scanner *x-y-z* data. In addition, we generate maps of surface grain size, from which large-scale facies assemblages are identifiable. Lambley provides the focus of the first aim, whilst the two contrasting sites of Sharperton and Kingsdale Beck are used to identify and compare broad facies assemblages.

FIELD SITES AND METHODOLOGY

A 180 m^2 gravel point bar located on the River South Tyne at Lambley, Northumberland, UK was chosen as the study site, due to the excellent visibility afforded across the surface from bridge and valley-side laser scan vantage points (Figure 7.1). More challenging locations in the River Coquet at Sharperton in Northumberland and the Kingsdale Beck in North Yorkshire were also scanned (for facies identification only), where higher elevation observation points were not available. Data collection was therefore at a more oblique angle and across a much larger bar area of 10,600 m^2 in the Coquet and a 50 m reach of the dry riverbed of the Kingsdale Beck. At each site a Riegl LMS Z210 field laser scanner collected random field data across the entire bar/exposed bed surfaces. This instrument uses a pulsed eye-safe infrared laser source (0.9 m wavelength) emitted in precisely defined angular directions controlled by a spinning mirror arrangement. A sensor recorded the time taken for light to be reflected from the incident surface, with the LMS Z210 collecting a series of

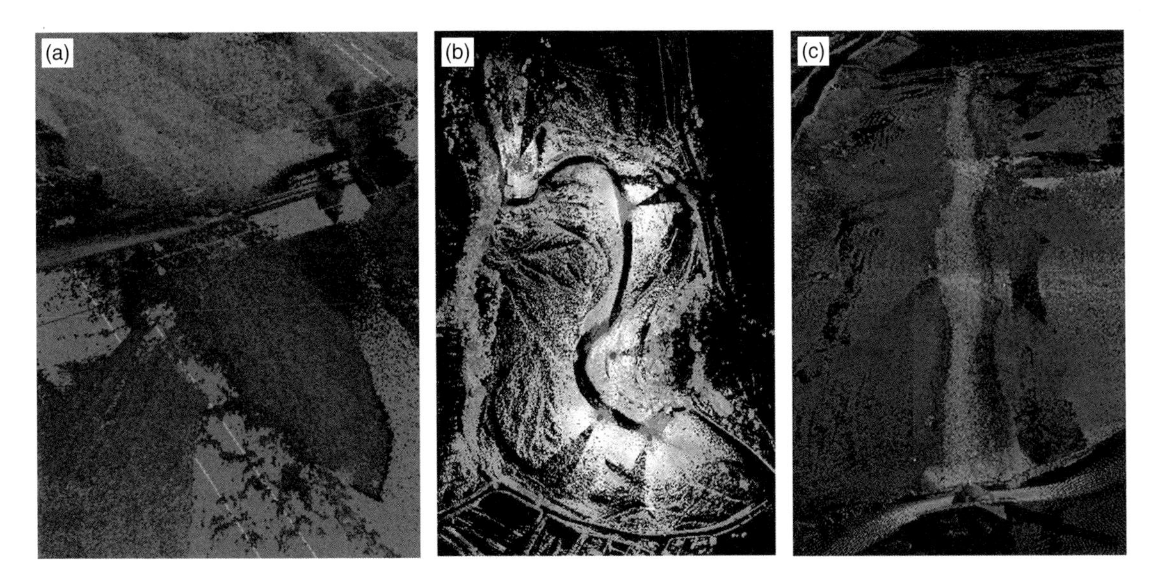

Fig. 7.1 Combined laser scan cloud image of the point bar study site at (a) the River South Tyne at Lambley, (b) the River Coquet at Sharperton and (c) Kingsdale Beck, North Yorkshire. See Plate 7.1 for a colour version of these images.

independent datasets recording range distance, relative height, surface colour and reflectivity. The scanner unit was mounted on a tripod and was capable of scanning through 80° vertically and 330° horizontally, stepping 0.072° to 0.36° depending on the resolution required and the time available for scanning. Vertical scanning rates varied between 5 and 32 scans s^{-1}. Angular measurements were recorded to an accuracy of 0.036° in the vertical and 0.018° in the horizontal. Range error was 0.025 m to a radial distance of 350 m (technical data from Heritage & Hetherington, 2005). Scans were generally restricted to 240° in front of the scanner, and were collected with substantial overlap, ensuring that the surface of the study reach was recorded from several directions. The effect of this approach was to increase the point resolution across the surface and to reduce the possibility of unscanned areas due to the shadowing effect of roughness elements along the line of each scan.

Four to six repeat scans from up to six independent locations were meshed using retro-reflectors linked to a common EDM theodolite based coordinate system. Problems are encountered when attempting to survey submerged bed topography with a laser scanner, due to reflection of the shorter wavelengths of light from the water surface. Milan *et al.* (2007) have however demonstrated that using standard deviation of errors between the laser derived surface and true surface as a measure of uncertainty can yield laser returns from the bed (< 0.2 m submerged depth are possible, and usable providing an appropriate level of detection is applied to the DEM specific to the topic of investigation).

Technical data

Spacing statistics for the data points in the final datasets are given in Table 7.1. The Lambley dataset contained 3.8 million points with a mean spacing of 0.012 m. At Sharperton, 0.8 M data points were spaced at a mean distance of 0.04 m. In the Kingsdale Beck, the final dataset comprised 4.9 million points with a mean spacing of 0.009 m. Seventy-five percent of the data points were spaced at less than 0.016 m, 0.053 m and

0.009 m apart for the Lambley, Coquet and Kingsdale sites respectively. Surface points were on average accurate to ±0.009 m when compared with 113 independently surveyed EDM theodolite validation points measured at Lambley only (Table 7.2). An example 2 cm resolution DEM of a bar surface derived from Lambley data alongside an oblique photograph is given in Figure 7.2.

Table 7.1 Laser data point spacing statistics. All values in metres.

	Lambley	Sharperton	Kingsdale
Mean	0.013	0.043	0.009
25%-tile	0.005	0.024	0.002
Median	0.009	0.035	0.004
75%-tile	0.016	0.053	0.009
Maximum	1.378	2.326	1.963
Interquartile Range	0.011	0.029	0.007

Table 7.2 Laser scan elevation error statistics based on 113 independent EDM points measured at Lambley (independent measurement not taken at Sharperton or Kingsdale).

Parameter	Metres
Mean	6.19E-16
Median	−0.009
Standard Deviation	0.053
Sample Variance	0.003
Kurtosis	1.931
Skewness	0.593

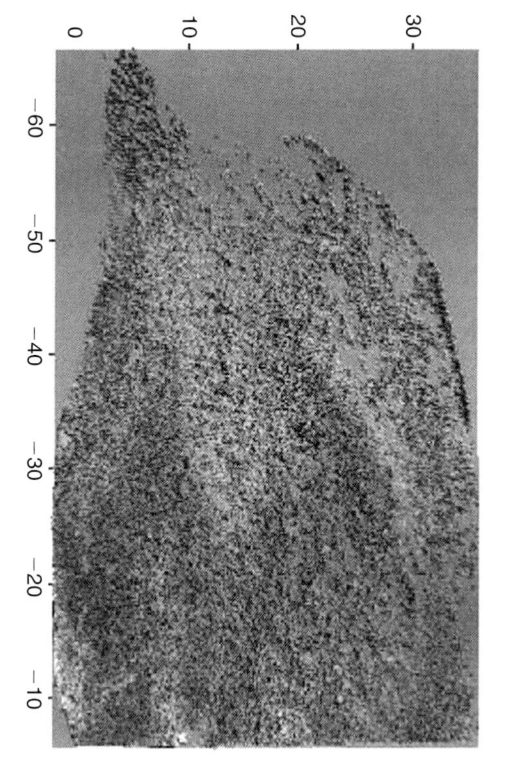

Fig. 7.2 Surface and photo of the point bar study site at the River South Tyne at Lambley. See Plate 7.2 for a colour version of this image.

RIVER SOUTH TYNE: GRAIN SIZE ANALYSIS

Laser scanning of the Lambley study site (Figure 7.2) was combined with detailed sampling of the bar surface to verify grain size data generated by the laser scan. Random field data were extracted for eight bar surface $2\,m^2$ sub-areas displaying distinct facies assemblages at Lambley. The local standard deviation of the elevation data were computed using a 0.3 m moving window, equivalent to the largest clast visible across the bar surface. Each standard deviation measure was then multiplied by a factor of two to generate the effective clast height (Nikora *et al.*, 1998, Gomez, 1993) and a frequency distribution constructed for each sub-area (Figure 7.3). A conventional grid-by-number sampling of the long (a), intermediate (b) and short (c) axis of 100 clasts (Wolman, 1954) was also used to define particle size distribution data at each of eight $2\,m^2$ sub-areas (Figure 7.4). Direct comparison of the particle size, a, b, c axis and random field frequency distributions were made revealing excellent linear relationships at this site (Figure 7.5).

It is to be expected that the laser based measurements would show some relationship with the clast c-axis due to the flow orientating the primary axis in the streamwise direction exposing the shortest axis to the flow. The linear regression (Table 7.3) indicates that the effective clast height measure as suggested by Nikora *et al.* (1998) and Gomez (1993) underestimates the c-axis slightly. The excellent relation also obtained for the a- and b-axis is surprising as these measures have no direct link to surface elevation change. It is suggested that autocorrelation is occurring in this instance, as many of the South Tyne clasts have a similar shape (discoid, cf. Table 7.4). This relationship may not hold where there is a greater diversity of clast shape.

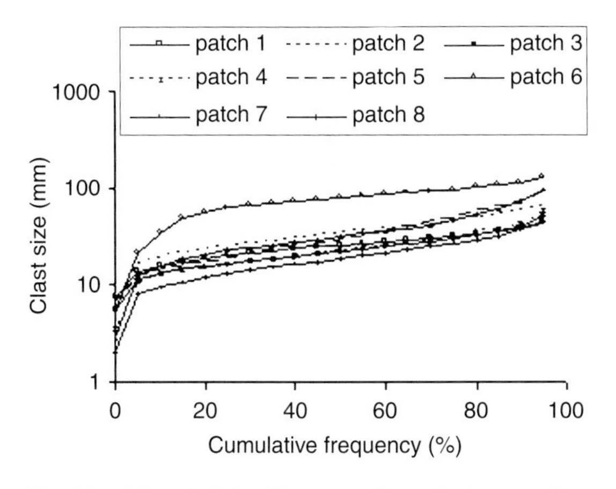

Fig. 7.3 Clast height (2 × moving window surface standard deviation) frequency across the point bar study site at the River South Tyne at Lambley.

Application of the random field approach across the entire bar surface generated an effective sample of 120,000 clasts. These may be visualised as a clast size surface based on the 0.3 m moving window (Figure 7.6). Visual comparison reveals an excellent link between textural facies and the surface roughness DEM with the larger sediments on the inside of the point bar and the two ridge lines of coarser material being identified. Further details on these findings and justification of this approach are given in Heritage *et al.* (2007).

SURFACE FACIES ASSESSMENT: RIVER COQUET AND KINGSDALE BECK

The validation of laser scanning as a means of measuring grain size characteristics at Lambley suggests an accurate and objective characterisation of surface sediment parameters is feasible. Scanning at Sharperton and Kingsdale was not accompanied by conventional grain size measurement, because the scans can be used to assess surface grain size distribution, patterns or patches based on identification of D_{16}, D_{50} and D_{84} grain sizes. An 11×11 moving window was run over the standard deviation surface produced from the laser scan data of the two river reaches, and a local median ($surface_{50}$), upper ($surface_{84}$) and lower quartile ($surface_{16}$) relief height were calculated from the 121 data points. These statistics were then multiplied by a factor of two (after Gomez 1993) to obtain equivalent grain frequency values.

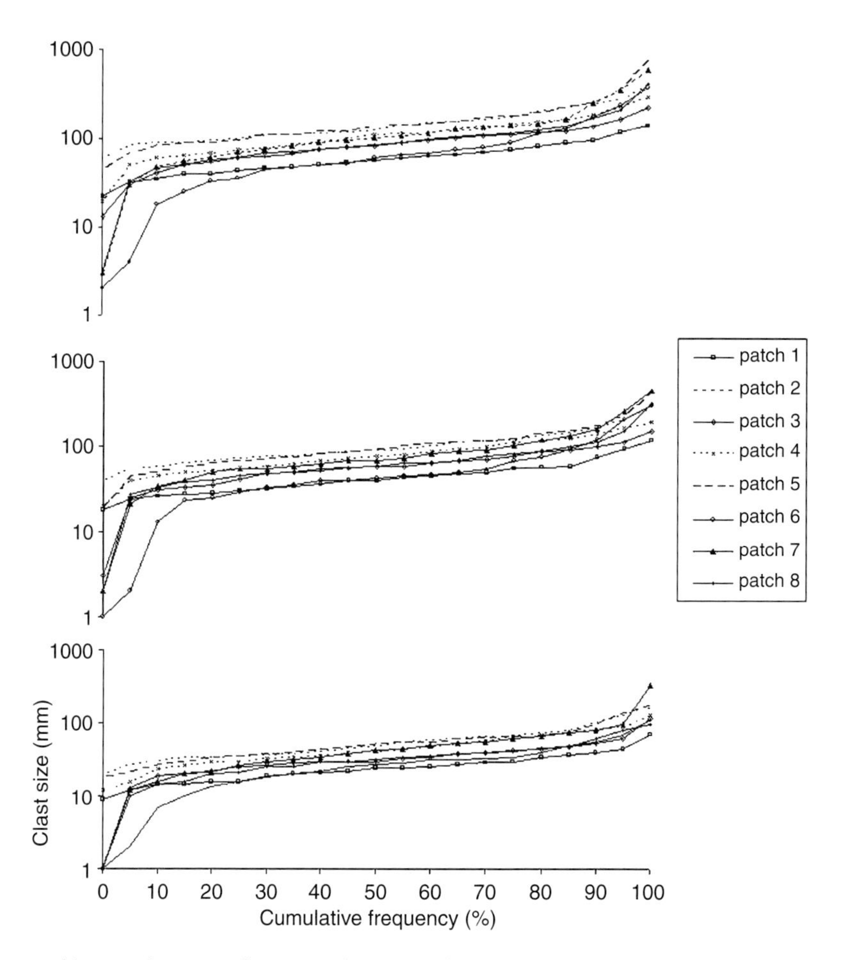

Fig. 7.4 Surface grid-by-number stone frequency (x-axis) and particle a, b and c-axis diameter (y-axis) for Lambley. Top graph: a-axis, middle: b-axis, bottom: c-axis.

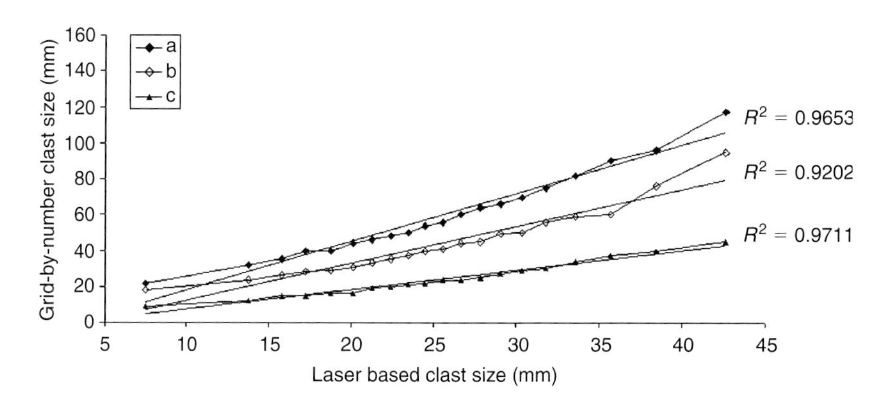

Fig. 7.5 Empirical relation between grid-by-number sampled clast dimensions and laser based local standard deviation for patch 1, which is the first of eight 2 m^2 areas constituting the bar surface surveyed.

Table 7.3 Empirical relation $y = mx + c$ between clast c-axis (y) and effective clast height (x) (Nikora *et al.*, 1998; Gomez, 1993) for Patch 1.

Clast axis	m	c	r^2
a	1.28	1.71	0.96
b	2.52	−0.06	0.97
c	3.54	0.77	0.98

Table 7.4 Percentage clast size distribution in 8 patches at Lambley on the River South Tyne.

	%Blade	%Rod	%Disc	%Sphere
Patch 1	13	14	54	19
Patch 2	20	19	48	13
Patch 3	22	14	50	14
Patch 4	17	13	58	16
Patch 5	18	20	55	7
Patch 6	20	21	41	18
Patch 7	17	17	48	18
Patch 8	22	13	48	17
Average	**18**	**16**	**50**	**15**

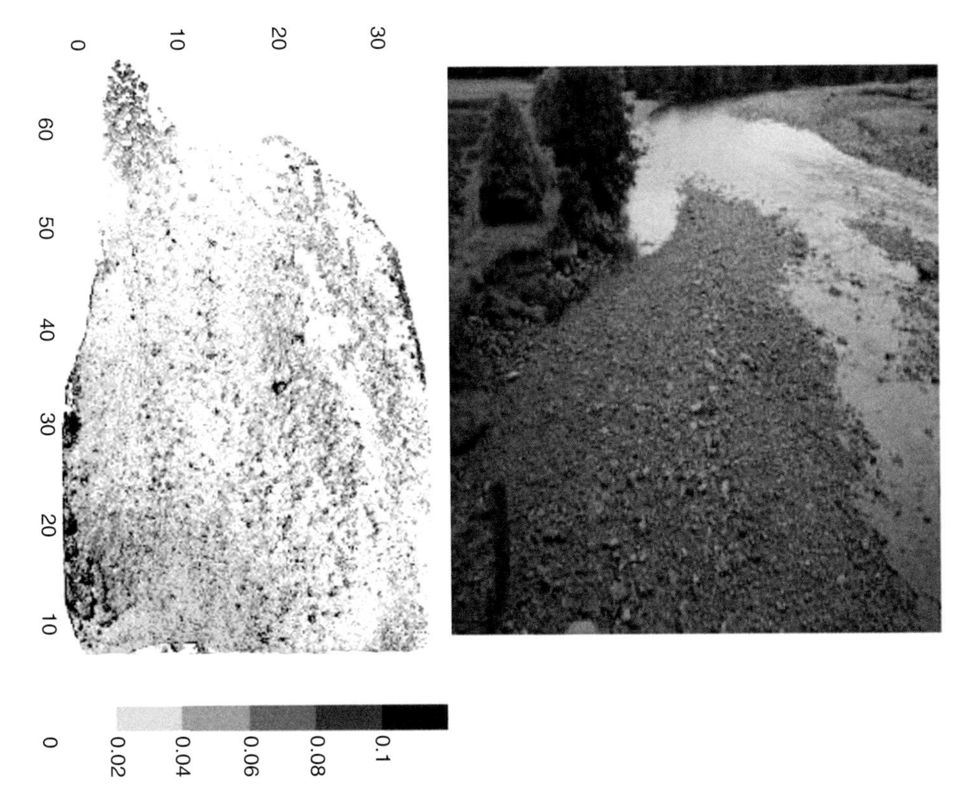

Fig. 7.6 Comparison of point bar surface roughness (2 × local standard deviation) and textural facies visible on an oblique photograph.

Coquet facies

Figure 7.7 shows the distribution of D_{16}, D_{50} and D_{84} across the bar surface at Sharperton, which permits detection of discrete facies associated with bar morphology at this site. This shows distinct patches of coarser grained sediment. Each of the index grain sizes demonstrates a consistency in identification of discrete substrate facies/patches across the bar surface. Morphologically, these patches reflect lobes and channels in the bar surface associated with accretion of the bar (see Fuller *et al.*, 2003). Coarser D_{84} values reflect sediment lobes, whilst the finer patches reflect both chute channels bisecting the bar, together with a fining towards the active channel (left of the images). This is in contrast with the findings of Bartholdy and Billi (2002), where coarsening towards the channel was observed. However, Bridge (1977) also identified the presence of a coarse lobe on a bar platform at the bar head, as is evident at Sharperton, such features are not uncommon in gravelly rivers (e.g. Bluck, 1979; Campbell & Hendry, 1987), and may be associated with downstream progression of bedload sheets. Each facies assemblage will reflect the precise combination of processes responsible for the formation of the discrete morphological unit (Bridge 1993). At Sharperton, lobate accumulation of coarse sediment was associated with a series of floods that were also responsible for repeated avulsion of the channel, which has subsequently left chute channels abandoned on the bar surface (Fuller *et al.*, 2003). The facies identified here are exclusively gravel, and as such reflect the importance of coarse grained sediment in bar construction at this site, contrasting with the greater abundance of sandy facies identified in a much bigger wandering gravel bed river by Desloges and Church (1987). The precise facies assemblage will naturally reflect the unique array of sediment sources in a catchment. Laser scanning at this site permits a clear identification of the coarse sediment characteristics associated with bar morphology and objectively identifies discrete sediment patches.

Kingsdale facies

Figure 7.8 shows the distribution of D_{16}, D_{50} and D_{84} across the bar surface at Kingsdale Beck. Grain size distribution is less patchy in Kingsdale Beck than Sharperton and suggests a greater uniformity. This is unsurprising, as this often dry channel experiences a flashy regime with a rapid falling limb and is more uniform topographically than the complex barform at Sharperton. Morphological facies analysis of Kingsdale Beck demonstrates characteristic high gravel imbrication, where a series of longitudinal ribs and pebble clusters are visible. Further investigation of visible pebble clusters indicates wake and stoss deposits of smaller substrate. Clear riffle-pools sequences are visible within the reach where riffle sediments are both coarser and better sorted

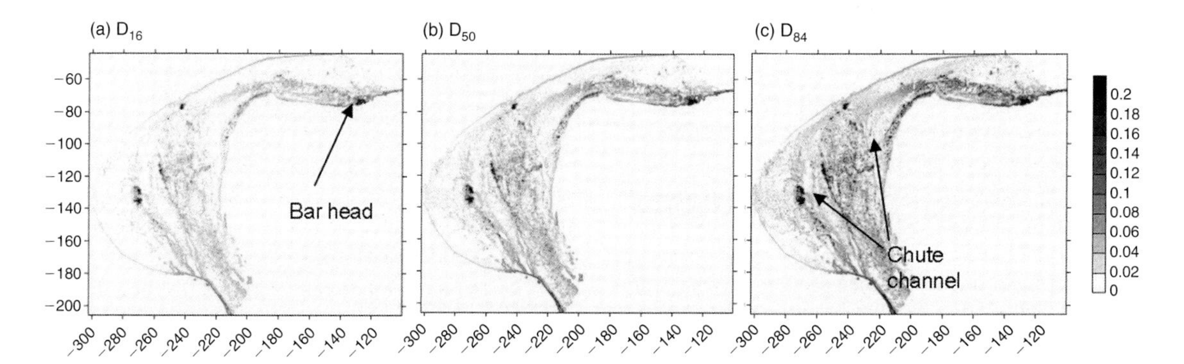

Fig. 7.7 Grain size characterisation of a lateral bar at Sharperton, River Coquet. All scales are in metres.

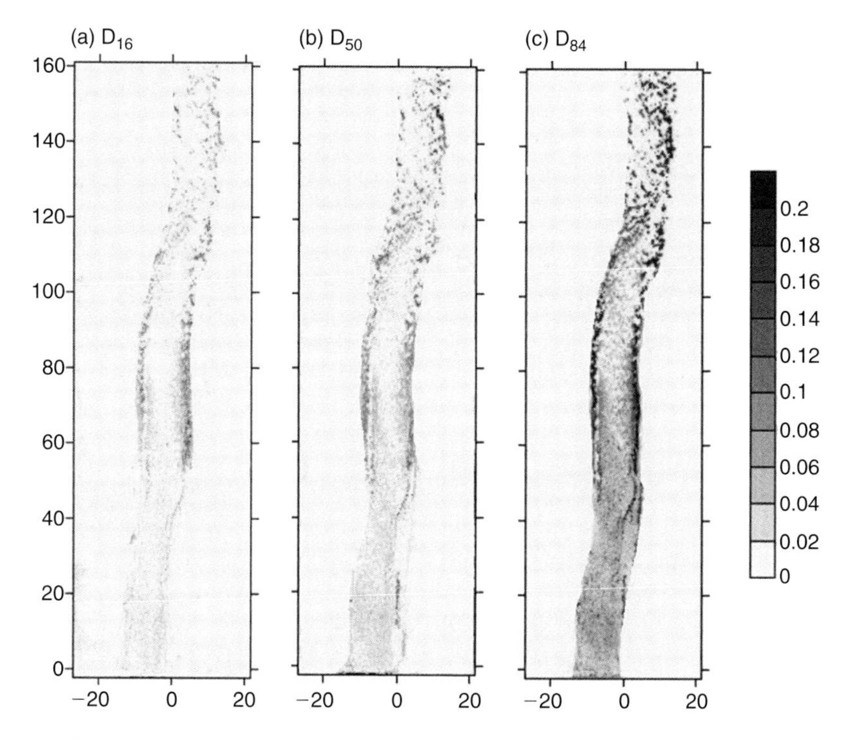

Fig. 7.8 Grain size characterisation of the dry river bed at Kingsdale. All scales are in metres.

than the pool sediments, with a larger percentage of D_{84} sediment clasts.

Facies synthesis

Buffington and Montgomery (1999) argue that grain size analysis of a given patch should encompass the entire surface area of the patch. Laser scanning readily achieves this. Furthermore, scanning whole patch assemblages overcomes the problem of gradation between discrete patches. Gravely beds tend to exhibit continuous spatial gradients of grain size and sorting, making boundary identification problematic (Buffington & Montgomery, 1999). In addition, laser scanning provides a means of accurately quantifying grain sizes in those patches, without the need for labour-intensive sampling, thus speeding up the whole process of surface characterisation. However, the drawback in this application of laser scanning revolves around its application to dry surfaces only. Sub-aqueous patch characterisation still requires either labour-intensive sampling and/or photogrammetry (e.g. Butler *et al.*, 2001).

CONCLUSIONS

The results of a random-field laser scan exercise to determine the grain-size characteristics of heterogeneous gravels indicate that the method is able to characterise the surface sediment with a high degree of confidence. Application of the random field approach across the entire bar surface at Lambley effectively generated a sample of up to 120,000 clasts.

The surface roughness recorded by the laser scanner very closely relates to the minor (c-) axis of the clasts on the point bar surface at Lambley. We argue therefore that the random-field approach should replace grid-by-number sampling wherever practicable in order to reduce the inherent errors associated with conventional grain-size sampling.

Facies maps based on D_{16}, D_{50} and D_{84} demonstrate the utility of laser scanning in the precise and objective identification of substrate facies/units/patches across dry surfaces of gravel-bed rivers. Wake and stoss deposits surrounding pebble clusters and longitudinal ribs were found at Kingsdale Beck, downstream fining patterns on point bars at Lambley and scroll bar morphology at Sharperton were observed. The results show close association between discrete facies and previously identified morphological units, as well as calculated grain size distributions. There is more work to be done in mining the datasets on patches for information equivalent to that presented, for example, by Buffington and Montgomery (1999). However, that is not within the scope of this chapter, which has sought to provide a first assessment of the utility of terrestrial laser scanning to characterise grain sizes on exposed (dry) gravel bar surfaces.

ACKNOWLEDGEMENTS

Thanks to George Heritage, Dave Milan and Andy Large for assistance in the field, data processing and analysis. Fieldwork was supported by funding from Massey University's overseas duties scheme in 2005.

REFERENCES

Bartholdy J, Billi P. 2002. Morphodynamics of a pseudomeandering gravel bar reach. *Geomorphology* **42**: 293–310.

Billi P, Paris E. 1992. Bed sediment characterisation in river engineering problems. In: Bogen, *Erosion and Sediment Transport Monitoring Programmes in River Basins*. J, Walling DE, Day T (eds), IAHS Publication **210**: 11–20.

Bluck BJ. 1979. Structure of coarse grained braided stream alluvium. *Transactions of the Royal Society of Edinburgh* **70**: 181–221.

Bridge JS. 1977. Flow, bed topography, grain size and sedimentary structure in open channel bends: a three-dimensional model. *Earth Surface Processes* **2**: 401–416.

Bridge JS. 1993. Description and interpretation of fluvial deposits: a critical perspective. *Sedimentology* **40**: 801–810.

Buffington J, Montgomery DR. 1999. A procedure for classifying textural facies in gravel-bed rivers. *Water Resources Research* **35**: 1903–1914.

Butler JB, Lane SN, Chandler JH. 2001. Automated extraction of grain-size data for gravel surfaces using digital image processing. *Journal of Hydraulic Research* **39**: 519–529.

Campbell JE, Hendry HE. 1987. Anatomy of a gravelly meander lobe in the Saskatchewan River, near Nipawin, Canada. In: *Recent Developments in Fluvial Sedimentology*, Ethridge FG, Flores RM, Harvey MD (eds), Society of Economic Paleontologists and Mineralogists, Special Publication **9**: 179–189.

Carbonneau PE, Lane SN, Bergeron NE. 2004. Catchment-scale mapping of surface grain size in gravel bed rivers using airborne digital imagery. *Water Resources Research* **40**: Article No. W07202.

Carbonneau PE, Bergeron NE, Lane SN. 2005. Automated grain size measurements from airborne remote sensing for long profile measurements of fluvial grain sizes. *Water Resources Research* **41**: Article No. W11426.

Church MA, McLean DG, Wolcott JF. 1987. River bed gravels: sampling and analysis. In: *Sediment Transport in Gravel Bed Rivers*. Thorne CR, Bathurst JC, Hey RD (eds), John Wiley: Chichester, 43–87.

Clifford NJ, Robert A, Richards KS. 1992. Estimation of flow resistance in gravel-bedded rivers: a physical explanation of the multiplier of roughness length. *Earth Surface Processes and Landforms* **17**: 111–126.

Crowder DW, Diplas P. 1997. Sampling heterogeneous deposits in gravel-bed streams. *Journal of Hydraulic Engineering* **123**: 1106–1117.

De Jong C. 1992. Measuring changes in micro and macro roughness on mobile gravel beds. In: *Erosion and Sediment Transport Monitoring Programmes in River Basins*, Bogen J, Walling DE, Day T (eds), IAHS Publication **210**: 31–40.

Desloges JR, Church M. 1987 Channel and floodplain facies in a wandering gravel-bed river. In: *Recent Developments in Fluvial Sedimentology*, Ethridge FG, Flores RM, Harvey MD (eds), Society of Economic Paleontologists and Mineralogists, Special Publication **9**: 191–196.

Diplas P, Sutherland AJ. 1988. Sampling techniques for gravel sized sediments. *Journal of Hydraulics Division, ASCE* **114**: 484–499.

Dunne T, Leopold LB. 1978. *Water in Environmental Planning*, W.H. Freeman and Company, New York, 666p.

Fuller IC, Large ARG, Milan DJ. 2003. Quantifying channel development and sediment transfer following chute cutoff in a wandering gravel-bed river. *Geomorphology* **54**: 307–323.

Fuller IC, Passmore DG, Heritage GL, Large ARG, Milan DJ, Brewer PA. 2002. Annual sediment budgets in an unstable gravel bed river: the River Coquet, northern England. In: *Sediment Flux to Basins: Causes, Controls and Consequences*, Jones S, Frostick LE (eds), Geological Society Special Publication **191**: 115–131.

Fraccarollo L, Marion A. 1995. Statistical approach to bed material sampling. *Journal of Hydraulic Engineering* **121**: 540–545.

Gomez B. 1993. Roughness of stable, armoured gravel beds. *Water Resources Research* **29**: 3631–3642.

Heritage GL, Hetherington D. 2005. Notes on the performance of side scanning LiDAR across varied terrain. *International Association of Hydrological Scientists Red Book Publication* IAHS Publication **291**: 269–277.

Heritage GL, Milan D, Johnson K, Entwistle N, Hetherington D. 2007. Detecting and mapping microtopographic roughness elements using terrestrial laser scanning. *Proceedings of the Remote Sensing and Photogrammetry Society Annual conference*.

Hey RD. 1979. Flow resistance in gravel bed rivers. *Journal of the Hydraulics Division ASCE* **105**: 365–379.

Hey RD, Thorne CR. 1983. Accuracy of surface samples from gravel bed material. *Journal of Hydraulic Engineering* **109**: 842–851.

Ibbeken H, Schleyer R. 1986. Photo-sieving: a method for grain-size analysis of coarse-grained, unconsolidated bedding surfaces. *Earth Surface Processes and Landforms* **11**: 59–77.

Kellerhals R, Bray DI. 1971. Sampling procedures for coarse fluvial sediments. *Journal of Hydraulics Division, ASCE* **97**: 1165–1179.

Lane SN. 2002. The measurement of gravel-bed river morphology. In: *Gravel Bed Rivers V*, Mosley MP. (ed.) Wellington: New Zealand Hydrological Society, pp.291–311.

Latulippe C, Lapointe MF, Talbot T. 2001. Visual characterization technique for gravel-cobble river bed surface sediments; validation and environmental applications contribution to the Programme of CIRSA (Centre Interuniversitaire de Recherche sure le Saumon Atlantique). *Earth Surface Processes and Landforms* **26**: 307–318.

Marcus WA, Ladd S, Stoughton J, Stock JW. 1995. Pebble counts and the role of user dependent bias in documenting sediment size distributions. *Water Resources Research* **31**: 2625–2631.

McEwan IK, Sheen TM, Cunningham GJ, Allen AR. 2000. Estimating the size composition of sediment surfaces through image analysis. *Proceeding of the Institution of Civil Engineers-Water Maritime and Energy* **142**: 189–195.

Milan DJ, Heritage GL, Large ARG, Brunsdon CF. 1999. Influence of particle shape and sorting upon sample size estimates for a coarse-grained upland stream. *Sedimentary Geology* **128**: 85–100.

Milan DJ, Heritage GL, Large ARG, Charlton ME. 2001. Stage-dependent variability in shear stress distribution through a riffle-pool sequences. *Catena* **44**: 85–109.

Milan DJ, Heritage GL, Hetherington D. 2007. Application of a 3D laser scanner in the assessment of erosion and deposition volumes and channel change in a proglacial river. *Earth Surface Processes and Landforms* **32**: 1657–1674.

Mosley MP, Tindale DS. 1985. Sediment variability and bed material sampling in gravel-bed rivers. *Earth Surface Processes and Landforms* **10**: 465–482.

Nikora VI, Goring DG, Biggs BJF. 1998. On gravel-bed roughness characterization. *Water Resources Research* **34**: 517–527.

Petts GE, Gurnell AM, Gerrard AJ, Hannah DM, Hansford B, Morrissey I, Edwards PJ, Kollman J, Ward JV, Tockner K, Smith BPG. 2000. Longitudinal variations in exposed riverine sediments: a context for the ecology of the Fiume Taglilamento, Italy. *Aquatic Conservation: Marine and Freshwater Ecosystems* **10**: 249–266.

Rice S. 1995. The spatial variation and routine sampling of spawning gravels in small coastal streams. *Res. Br., B.C. Min. For., Victoria, B.C. Work. Pap.* 06/1995.

Rice SP, Haschenburger JK. 2004. A hybrid method for size characterisation of subsurface fluvial sediments. *Earth Surface Processes and Landforms* **29**: 373–389.

Verdu JM, Batalla RJ, Martinez-Casanovas JA. 2005. High resolution grain size characterisation of gravel bars using imagery analysis and grain size statistics. *Geomorphology* **72**: 73–93.

Wiberg PL, Smith JD. 1991. Velocity distribution and bed roughness in high gradient streams. *Water Resources Research* **27**: 825–838.

Wittenberg L, Newson MD. 2005. Particle clusters in gravel-bed rivers: an experimental morphological approach to bed material transport and stability concepts. *Earth Surface Processes and Landforms* **30**: 1351–1368.

Wohl EE, Anthony DJ, Madson SW, Thompson DM. 1996. A comparison of surface sampling methods for coarse fluvial sediments. *Water Resources Research* **32**: 3219–3226.

Wolman MG. 1954. A method of sampling coarse riverbed material. *Transactions, American Geophysical Union* **35**: 951–956.

8 Airborne Laser Scanning: Methods for Processing and Automatic Feature Extraction for Natural and Artificial Objects

CHRISTOPH STRAUB, YUNGSHENG WANG AND OCTAVIAN IERCAN

Department of Remote Sensing and Landscape Information Systems, University of Freiburg, Freiburg, Germany

INTRODUCTION

Airborne laser scanning (ALS), often also referred to as LiDAR (light detection and ranging), is a relatively new active remote sensing technology for the capturing of topographic data. The result of the measurement with an ALS system is a point cloud which provides 3D information relating to landscapes. For segmentation and classification procedures with image processing techniques, the irregularly distributed *x,y,z* coordinates have to be converted into a regular grid structure. Several algorithms have been developed during the last years for filtering and interpolation of the points to derive digital terrain models (DTM), which represent the bare earth, and digital surface models (DSM), which include vegetation, buildings, bridges etc. As a secondary product, a normalized digital surface model can be derived by subtracting the DTM from the DSM. In forests it is often described as canopy height model (CHM). Due to the fact that the values in the DSM represent the object heights for each *x,y* position of a study area it can form the basis for a range of different analyses.

Here we supply an introduction to processing, segmentation and classification techniques for extraction of information from 3D data. This is of relevance to many different disciplines, for example, forestry, nature conservation, agriculture, landscaping and urban planning. We first give an overview of existing methods for filtering and interpolation of the raw data to derive digital terrain models and digital surface models. Following this we describe approaches for automatic feature extraction for natural as well as for artificial (man-made) objects. Here, we focus on segmentation and classification of forest areas, road extraction and building extraction.

DATA PROCESSING – AN OVERVIEW OF EXISTING METHODS FOR FILTERING AND INTERPOLATION

ALS provides 3D point clouds. In order to obtain a terrain model (a model of the bare earth) as well as a digital surface model (a model of the earth with objects like trees and buildings) this data has to be 'filtered'. Several automatic filtering

Laser Scanning for the Environmental Sciences,
1st edition. Edited by G.L. Heritage and A.R.G. Large.
 ISBN 978-1-4051-5717-9

methods exist, some of which are described here. Elsewhere, Sithole and Vosselman (2003) provide a useful comparison of several filtering techniques. Generally speaking, ALS data can be defined as irregularly distributed points in 3D, although some algorithms resample the data into a regular grid structure before filtering. Filters frequently work on a local region, and may be classified according to their concept, that is their hypothesis of the bare earth, and in relation to their classification criteria. Filtering algorithms compare and classify (e.g. 'earth', 'object') two or more points at a time.

The reader is directed to Sithole and Vosselman 2003) who provide a comprehensive comparison of filtering algorithms: (a) robust interpolation filters (e.g. Kraus & Pfeifer, 2001); (b) progressive TIN densification (e.g. Axelsson, 1999, 2000); (c) adaptive slope based filters (e.g. Sithole & Vosselman, 2001); (d) active contours (e.g. Elmqvist, 2001); (e) slope-based filters (e.g. Roggero 2001); (f) progressive TIN densification/ regularization method (e.g. Sohn, 2002); (g) hierarchical modified block minimum (e.g. Wack & Wimmer, 2002) and (h) spline interpolation (e.g. Brovelli, 2002). The algorithms are compared on the basis of their performance in detecting outliers (i.e. measurement errors), object complexities, detached objects, vegetation and discontinuity (e.g. steep slopes etc.) and are described in more detail below. The authors identified above are used in Table 8.1 to differentiate between algorithms for the point of comparison.

Robust interpolation filters

This algorithm was developed for the calculation of DTMs and for point classification into terrain and off-terrain points (Kraus & Pfeifer, 2001). The algorithm is based upon the linear prediction of Kraus and Mikhail (1972), and is designed to iteratively approximate a terrain surface. Initially the terrain surface is computed as a rough

Table 8.1 Qualitative comparison of filters. PTD = Progressive TIN densification, SI = Spline interpolation, AC = Active contours, RIF = Robust interpolation filters, SBF=Slope-based filters, ASBF = Adaptive slope-based filters, TDR = Progressive TIN densification/regularization, HMBM = Hierarchical modified block minimum. Comparison Scale: 3 = Good, 2 = Fair, 1 = Poor, 0 = Removed. (*Source*: Sithole, 2005).

	PTD	SI	AC	RIF	SBF	ASBF	TDR	HMBM
Outliers								
High points	3	3	3	3	3	3	3	3
High points influence	3	3	3	3	3	3	3	3
Low points	2	3	3	3	2	2	2	3
Low points influence	1	3	3	3	3	1	3	3
Object complexity								
Large objects	3	3	3	3	3	3	3	3
Small objects	3	2	2	3	3	2	2	2
Complex objects	2	2	2	2	2	2	2	2
Low objects	2	3	1	2	3	2	1	3
Disconnected bare earth	2	2	2	2	2	2	2	2
Detached objects								
Building on slopes	3	2	3	3	2	2	2	2
Bridges	2	3/0	3/0	3/0	3/0	3/0	3/0	3/0
Ramps Vegetation	1	1	1	1	1	1	1	1
Vegetation	3	3	3	3	3	3	3	3
Vegetation on slopes	2	2	3	3	2	2	3	2
Low vegetation	2	2	3	3	2	2	2	3
Discontinuity								
Preservation	2	1	1	2	1	1	1	2

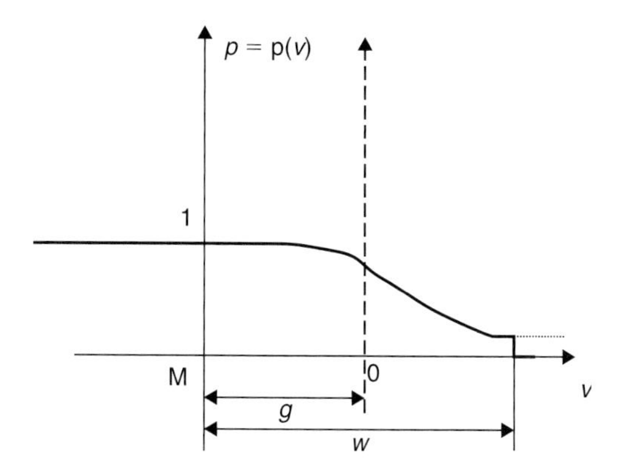

Fig. 8.1 The weight function *p* of the Robust Interpolation Filter. Here, P = the weight function, *g* = the shift value and *v* = the residual.

approximation – an averaging surface between terrain and off-terrain points. The vertical distances (residuals) of the laser points to this averaging surface are computed. Low points (terrain points) will have negative residuals and higher points (e.g. reflections on top or within vegetation) will have small negative values or positive residuals. The residuals are used in a weight function (Figure 8.1) to assign individual weights to each point. For terrain approximation the idea is to assign a small weight to points above the surface and a high weight to those points below the averaging surface. For ALS data the authors modified the original weight function for robust estimation by introducing a so called shift value (*g*).

As can be seen in Figure 8.1, the (weight) function *p* gives the value 1 to residuals equal to *g* and is maintained at 1 for all values smaller than *g*, which means that points with negative residuals (terrain points) will obtain the maximum weight. In the opposite direction, if residuals are equal or greater than *w*, the weight is set to 0. The determination of *g* is possible through three different methods, all of which use the analysis of the residual histogram. For new iterations new *g* values have to be determined. This is accomplished by selecting from the candidates supplied by the three methods of determination. The weights for each individual point are used to recalculate the surface. Due to the fact that points with large positive residuals (off-terrain points) will have a low weight, the surface will be attracted to the terrain points. This process is iterated several times (Kraus & Pfeifer, 1998). This method was further developed under the name hierarchical robust interpolation (Kraus & Pfeifer, 2001) and can be used for forested areas as well as for cities and urban agglomerations.

The progressive TIN densification filters

This algorithm iteratively densifies a TIN which in itself is a raw representation of the Earth's surface. The TIN is computed from neighborhood minima. The densification is achieved by adding laser points into the (raw) TIN. However, those points have to fulfill defined criteria with respect to the triangle that contains it: first, the point has to be within a defined range to the nearest triangle node; and second, the angle between the triangle's normal and the line joining the point and the nearest node has to fit within specific limits. In a new iteration two entities are computed: the TIN and the threshold constructed on the data and the algorithm conditions. The determination of the threshold is made by using a histogram for the two conditions of the algorithm. This algorithm is recommended for use in urban areas (Sithole, 2005) because of its surface modelling properties, even in regions with discontinuities. For more information, the reader is directed to Axelsson (1999, 2000, 2001).

Adaptive slope based filters

This filtering method is a modified version of the slope based filter developed by Vosselman (2000, 2001). It essentially uses the same methodology, but the principal filtering parameters are better elaborated for steep terrain. The filter approximates the bare earth with the help of a structuring element. The structuring element can have the shape of an inverted bowl which is centered (horizontally) on a point in the point cloud and is raised until it touches the point. The point is only accepted as belonging to the bare earth if there is no neighbouring point found beneath the structuring element. Otherwise it will be classified as off-terrain point. The slope of the structuring element is determined using a training dataset. In the modified version of the filter, the shape of the structuring element is altered with the slope of the terrain (Sithole, 2005).

Active contour model

This filter is a modification of the active shape models theory for ALS data filtering. The method has its origin in the computer graphics and image processing and it was meant to be used in detecting contours in images (Cohen 1993). The algorithm of this method has three levels. The first methodological level is the data sampling in which laser points are resampled into a regular grid. The lowest point is selected within each cell. The second level of the algorithm's methodology is the optimization of the active contour surface. The shape of the active contour surface is expressed by an energy function that takes into account internal and external forces (Elmqvist, 2000, 2002). The terrain approximation can be imagined as an elastic surface floating up from underneath the point cloud. The surface is 'attracted' to the points that are considered to be terrain points. The elastic surface is also stopped by the energy balance to extend to points that are not considered as terrain points. The third methodological level is the selection of terrain points within a tolerance zone around the final surface.

Slope based filter

This filtering method is a modified version of the slope based filter developed by Vosselman (2000, 2001). The filter is based on a two level algorithm. In the first level a raw DTM is approximated using a local minimum criterion to eliminate elevated points. A local operator is used to compute the minimum height and ground points are determined if their heights are compatible with a local slope (the variance of the local data is considered in a local regression criterion). The second level uses a threshold that will eventually classify the points in ground and non-ground points. The classification process is based on the vertical distance between a point and the DTM (bare earth) approximation obtained in the first level (Roggero, 2001). The entire process has to be performed twice in order to reach the desired accuracy of the terrain model.

Progressive TIN densification/ regularization method

This algorithm is based on the *divide et impera* (divide and conquer) principle. It starts by generating an initial terrain surface model (ITSM) and continues by following a downward *divide et impera* triangulation process and an upward *divide et impera* triangulation process. The ITSM is initialized using a rectangle which has four points assigned as terrain points. These corner points of the ITSM define the domain of the whole LiDAR dataset. The height values of the corner points are determined by averaging the height of the neighbouring laser points of each of the corners. The ITSM will be used as initial construction of a terrain surface in the following computation steps. The downward *divide et impera* is based on two suppositions (Sohn & Dowman, 2002). First, any point cannot be located underneath a reconstructed terrain surface model, and second, if supposition one is invalid within a local terrain surface model, a point with the maximum negative distance is selected as the most reliable terrain point and is used to recompute a TIN. This process is repeated for each triangle of the TIN until no more points can be found below the model. The resulting TIN is a first representation of the bare earth. The upward *divide et impera* is quite analogous to the 'Axelsson TIN densification'. First a buffer for each of the TIN's triangles is created on the basis of the downward *divide et impera*. The points included in the buffer of each triangle are passed through a series of steps and tested with the Minimum Description Length Model. If the points are determined as being part of the flattest tetrahedral they will be integrated into the TIN.

Hierarchical modified block minimum

This algorithm uses a hierarchical approach in combination with a weight function to differentiate ground surface from other objects. The algorithm has two phases; the first is the creation of raster elements and the second is the classification of those into ground or non-ground elements. In the first phase, a DTM with 9 m resolution is computed from the raw data. The lowest height from 99% of all height values within each raster element is assigned as raster value. Thus the influence of outliers is reduced. The usage of 9 m raster elements is explicitly useful in regions with large buildings and dense forested areas where the estimation of the ground is difficult. To get an accurate estimation of the heights of the ground points, gradient information of the DTM is calculated (to detect the lowest point perpendicular to the terrain). In addition a maximum allowed height deviation is defined in order to eliminate regions that strongly deviate from the DTM.

In the second phase a classification is performed, which will eventually lead to an estimation of ground and non-ground elements. For this a Laplacian of Gaussian (LoG) operation is used on the 9 m resolution DTM. This new 9 m DTM is used as basis for further calculations of a 3 m resolution DTM. The gradient information of the already determined 9 m DTM is used for the estimation of the height of each element of the 3 m DTM. A threshold is used in order to eliminate errors due to high buildings or trees. The false height values are replaced with values from the 9 m DTM. A weight function is used for the resolution of 3 m (and less) because edges in the terrain can be falsely classified as non-terrain. DTMs with a resolution of one metre or even less can be calculated with this methodology (Wack & Wimmer, 2002).

Edge based clustering

The principle of the edge based clustering filter is that objects existing on the surface of the Earth viewed planimetrically have distinct edges. Points inside closed edges can be classified as objects and not bare earth. The algorithm of the edge based filter is based on two sequences. The first sequence deals with splines. This step is influenced by the resolution of the raw data and is characterized by the usage of the bicubic spline interpolation using the Tychonov regularization parameter for ensuring the continuity of the surface. Data is here divided into equal tiles, each containing 200 × 200 splines. In essence, differentiation between object and bare earth is done by the positioning of the points above or below a particular spline. When a point is above a spline it is considered as being part of an object, and if it is below a spline it will be considered as being part of the Earth's surface.

The second step of the algorithm detects objects edges; in other words sudden changes in height on a small horizontal area are clearly determined as object edges. Considering that the surfaces that have to be determined are irregularly distributed in the raster image, an approximation of the DSM has to be performed. The DSM approximation can be done using one of the following procedures: the *bilinear spline function* and the *bicubic spline function*, both of which use the Tychonov regularization. Setting thresholds is also very important and a sensible subject of this method, more details can be found in Brovelli (2004).

QUALITATIVE COMPARISON OF THE ALGORITHMS

The above mentioned algorithms were compared qualitatively and quantitatively by Sithole (2005). The comparison was performed visually and especially complex regions were selected for the task. In the comparison procedure, several criteria were selected such as outliers, object complexity, attached objects, vegetation, discontinuity, etc. which can be seen in Table 8.1. The general impression of the filters is that they are able to distinguish between the bare earth and vegetation or buildings as well as between bare earth and the majority of the outliers. However some recognition problems still persist, such as in the case of bridges where the structures were, sometimes, entirely ignored. Filtering methods are not universally applicable. They will work in most situations but will fail in others (Sithole *et al.*, 2003).

EXTRACTING FOREST CHARACTERISTICS USING AUTOMATED STAND UNIT DELINEATION OF LIDAR DATA

Airborne Laser Scanning has been investigated as an alternative technique to derive forest inventory variables and to analyse vertical and horizontal structures of forests. Point clouds with high point density can be used to extract individual tree variables such as height and crown diameter. Further parameters can be estimated like stem diameter and stem volume. Many different approaches were developed and verified for single tree extraction. Some make use of the surface model and apply different versions of watershed segmentations or region growing techniques to delineate crown polygons. As an example, local maxima can be extracted from the DSM to locate potential tree tops. The maxima are used as seed points for an expansion using a pouring algorithm (similar to a watershed algorithm). The input image is segmented by 'pouring water' over it to derive tree crown polygons (Weinacker *et al.*, 2004; Koch *et al.*, 2006). Other approaches try to extract tree crowns directly from the raw data points. First the height difference between a point and the terrain (DTM) height below it is calculated. The result is a 'normalized' point cloud in which the point height represents the absolute height of an object. The study area is subdivided into voxels (volume elements) using a regular grid in three dimensional space. Using the voxels of a certain height layer, a horizontal 2D projection image can be generated, and the number of normalized laser points found within each voxel is assigned as the gray value of the correspondent pixel in the image. This is done for a series of horizontal 2D projections at different height levels. Tree crown regions are extracted from the projection images for the creation of a 3D crown model (Wang, 2007). In general, only the trees in the upper canopy level can be identified. Therefore tree numbers are often underestimated. Beside single trees, tree clusters can be extracted from laser scanner data as units of analysis (Bortolot, 2006). Tree clusters are defined as homogeneous (stand) units with respect to height variations and forest type. The following paragraphs describe a procedure which can be used to partition a forest area into different meaningful regions (forest stand units) with certain properties, using different metrics derived from laser scanner data.

SEGMENTATION AND CLASSIFICATION OF A FOREST AREA INTO STAND OR SUB-STAND UNITS BASED ON VEGETATION CHARACTERISTICS EXTRACTED FROM ALS DATA

Forest stand maps are widely used for optimized forest management. Based on stand units, silvicultural recommendations and prescriptions are made for establishment, thinning, release and harvest (Wynne, 2006). According to the system used in Germany, as well as in other countries, the forest area is subdivided into relatively homogeneous (stand) units. Similar or comparable stands often have the same management type. During the mapping process, groups of trees with similar physical characteristics are identified and grouped together in a logical and consistent manner. Additionally, topographical lines (e.g. ridgelines and the forest road network) are generally used for delineation (Hildebrandt, 1996). Conventionally, forest stand boundaries are digitized manually with the help of analogue maps and/or aerial photographs. The delineation results are verified by field surveys. To show the potential of ALS data for the extraction of different vegetation characteristics, a procedure for automated delineation of stand and sub-stand units is described. One approach which is often used to derive vegetation characteristics by means of laser scanner data is to subdivide the study area into grid cells (e.g. squares with a side length of 20×20 m) as shown in Figure 8.2.

Different metrics are computed within each grid cell and then used for classification. Therefore the procedure can be seen as an object-oriented classification procedure. Compared with pixel-based

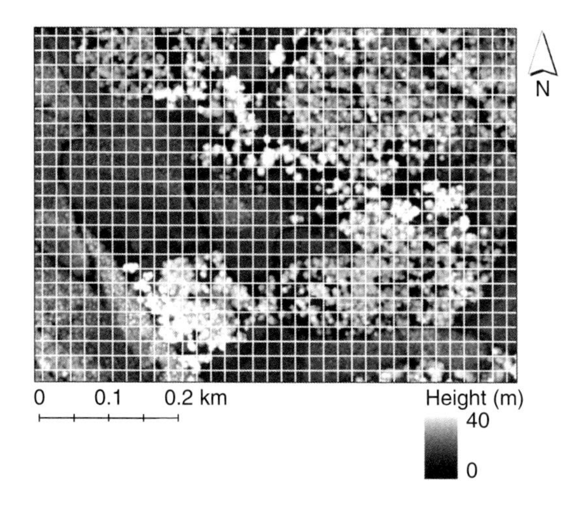

Fig. 8.2 Subdivision of a forest into squares (20 × 20 m) superimposed on a vegetation height model (nDSM) visualized as grey-scale image.

approaches, the main advantage of object-oriented methods is that groups of adjacent pixels are analysed to extract forest characteristics which cannot be extracted from individual pixels (Bortolot, 2006). The procedure can be subdivided into two main parts: (a) sorting of cells/objects into several classes and (b) grouping of classified objects into stand or sub-stand units. The following metrics are computed for each cell and are used to sort the cells into different 'stand classes':

- Percentage of conifers, deciduous trees and forest floor.
- Surface Roughness (with respect to variation of height values).
- Height Classes (assigned to developmental stages).

An overview of the processing steps is given in Figure 8.3.

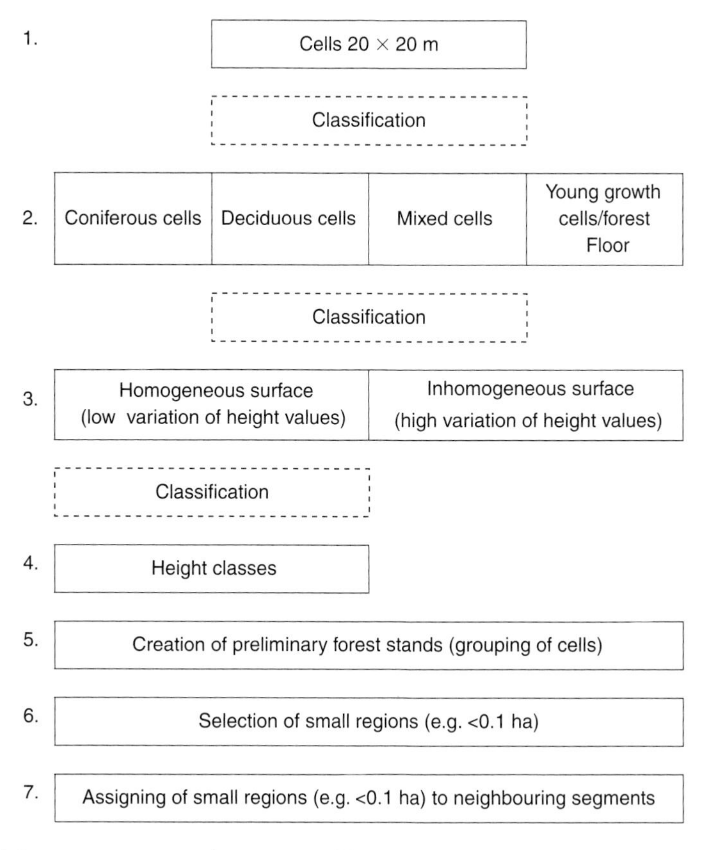

Fig. 8.3 Overview of the processing steps for automated forest stand delineation and classification.

CLASSIFICATION OF CONIFEROUS AND DECIDUOUS FOREST

ALS data obtained during leaf-off (winter) conditions can be used for classification into coniferous and deciduous forest. Figure 8.4a shows a profile of first echo and Figure 8.4b shows a profile of last echo data acquired during leaf-off conditions, for the same area with a deciduous and a coniferous forest stand. While in the first echo dataset the difference between the coniferous and deciduous stand is not directly visible, coniferous trees are clearly visible in the last echo dataset. Due to leaf-off conditions the laser penetrates to the ground in deciduous stands. In coniferous stands the penetration rate is much lower and still many reflections can be found within the tree crowns. To extract regions with conifers, a normalized digital surface model can be computed by only using the last echo points. The result is a height model of the conifers. Using a threshold operation (selection of grey values lying within a defined height interval) potential regions covered by conifers will be generated. Ground (forest floor) regions are thresholded from the nDSM computed from first echo data. The complement of coniferous and forest floor regions are assumed to be deciduous trees.

The percentage of coniferous, deciduous trees and ground are computed within each grid cell and the cells classified into the following categories (as shown in Figure 8.3):

1 'Coniferous Cells' (if percentage of coniferous trees >70%).

2 'Deciduous Cells' (if percentage of deciduous trees >70%).

3 'Young Growth Cells/Forest Floor' (if percentage of forest ground >70%).

4 'Mixed Cells' (cells which do not fulfill the criteria mentioned above).

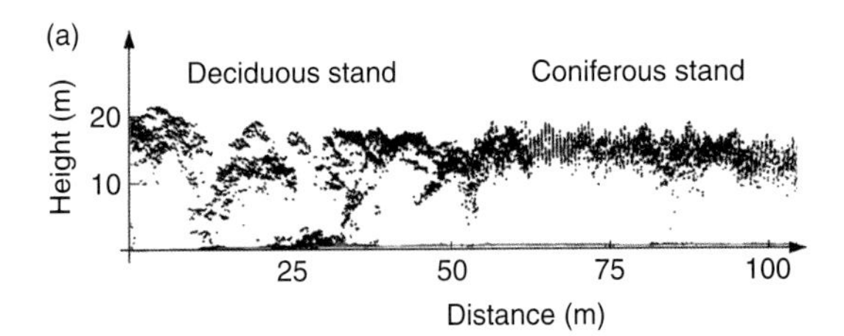

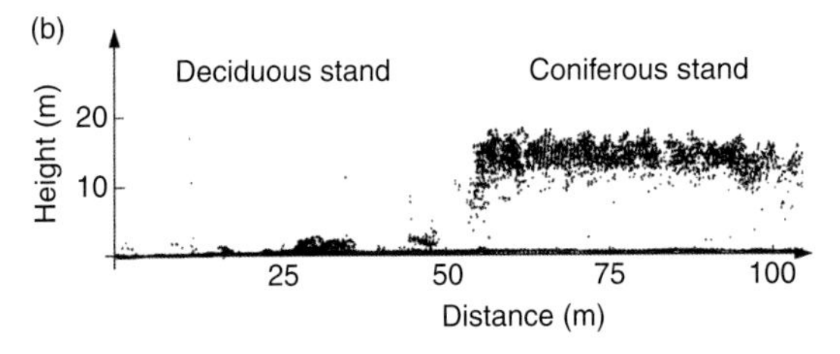

Fig. 8.4 (a) Profile of first echo data for a deciduous and a coniferous forest stand during leaf-off conditions; (b) profile of last echo data for a deciduous and a coniferous forest stand during leaf-off conditions.

CLASSIFICATION BASED ON 'SURFACE ROUGHNESS'

Cells can be further classified using the 'surface roughness' of a height model. As a measure for the 'roughness' the coefficient of variation of height values C_V within each cell is used – defined as the ratio of the standard deviation σ of height values to the mean μ:

$$C_V = \frac{\sigma}{\mu} \tag{8.1}$$

The result is a classification into 'even-aged forest' (homogeneous structure – low variation in height) and 'uneven-aged forest' (inhomogeneous structure – high variation in height).

METHODS FOR STAND HEIGHT COMPUTATION AND CLASSIFICATION USING HEIGHT CLASSES

Forest stand height is a crucial stand characteristic. An important stand height is the 'top height' which is often defined as the average height of the 100 (or 200) trees of largest diameter per hectare (Burschel & Huss, 1997). Due to the fact that the tree height is highly correlated to the stem diameter of a tree, the 100 highest trees per hectare have to be estimated. Several methods to estimate the height of forest stands based on LIDAR data have been developed and verified with field data during the last years. Early approaches computed a 'mean vegetation height' from the raw laser points for each forest stand. As an example, the arithmetic mean of largest laser heights within square grid cells with cell sizes of 15–30 m was used (Næsset, 1997). Several studies used multiple regression analysis with predictor variables derived from laser scanner data to develop models for height estimation. Various canopy height metrics and density measures were tested, like percentiles of the height distribution of laser pulses, variables related to canopy density (e.g. number of canopy hits divided by total number of transmitted pulses), maximum height, mean and median height etc. (Næsset & Bjerknes, 2001; Means *et al.*, 2000; Rieger, 1999). Once a suitable model for height estimation has been found a classification of the cells using height classes is possible. The classes can be assigned to developmental stages:

- 'juvenile' height < 3 m
- 'sapling' $3 \le h < 10$ m
- 'pole' $10 \le h < 15$ m
- 'mature trees' $15 \le h < 25$ m
- 'old trees' $h \ge 25$ m

GROUPING OF CELLS TO GENERATE MEANINGFUL STAND UNITS

After all processing steps described above, the cells are separated into different classes. To generate larger objects, neighbouring cells within the same class are merged into preliminary forest stand segments. To eliminate small regions, a minimum size for a forest stand unit can be defined (e.g. 0.1 ha). Small regions are 'dissolved' and are assigned to a neighbouring unit to which the minimum (top) height difference is calculated.

VEGETATION CHARACTERISTICS AND AIRBORNE LASER SCANNING DATA

Different vegetation characteristics can therefore be extracted from ALS data and can be used for delineation of stands and sub-stand units respectively. Statistical assessment in various studies has shown that stand parameters can be estimated accurately. Stand heights especially are estimated precisely with coefficient of determination (R^2) in the range of 0.74 and 0.94 (Rieger *et al.*, 1999; Means *et al.*, 2000; Næsset, 2000, 2002). The estimation of crown coverage proportion of coniferous trees from winter last echo data (which can be used for a classification into coniferous and deciduous forest) is in the range of 0.77 to 0.90 (Straub, 2007; Rieger *et al.*, 1999).

Natural surfaces (like tree height models) can be very complex and can have enormous variations in height values, which may influence the success of an automated delineation method. The more homogeneous the structure of the forest is (e.g. pure stands with relatively constant height values), the more the automated delineation result of the stand units will match the boundaries of existing forest stand maps. Moreover, the laser point density and the season have a high influence on the extraction of vegetation characteristics. Due to variable conditions accurate stand height estimations have to be calibrated with field data. The differentiation of deciduous and coniferous stand types is more reliable during winter (leaf-off) conditions due to different penetration rates in both stand types. Additionally the filtering and interpolation algorithm for DTM and DSM computation will influence the smoothness or roughness of the models and therefore the characterization of stand units. For that reason the computation method has to be adapted to the specific conditions of the area under investigation.

EXTRACTION AND MODELLING OF MAN-MADE OBJECTS

Extraction of forest roads based on LiDAR data

An accurate acquisition and documentation of the road and path network in the forest is of high importance for soil conservation and is a precondition for optimization of timber transport. At present, forest roads are digitized manually with the help of analogue maps and aerial photographs as well as information collected in field surveys. Digital terrain models (DTMs) with high resolution (e.g. 1m) from LiDAR data can be used to visualize and extract forest roads with high precision. Figure 8.5 shows a 3D view of a DTM in a mountainous terrain. The forest roads are clearly visible due to the low gradient compared to the steeper surrounding terrain.

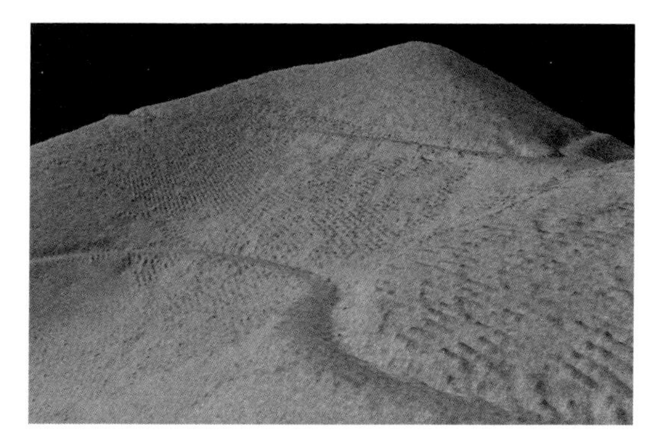

Fig. 8.5 Visualization of a digital terrain model in 3D of a forest area in a mountainous terrain – forest roads are clearly visible.

A slope model for break line extraction to detect the edges of roads was presented by (Rieger *et. al.*, 1999). Figure 8.6a shows a slope image derived from a LiDAR-DTM with 1 m resolution. The slope is calculated within a moving window and is defined as the maximum change in elevation (rise) over distance (run) for each point (x,y) of the DTM and its eight neighbours as degree value or a percentage slope (rise/run X 100) using the approach described in Horn (1981):

Height values within the moving window at position (x,y):

Z_1	Z_2	Z_3
Z_4	Z_5	Z_6
Z_7	Z_8	Z_9

Calculation of the gradient in east to west $\delta z/\delta x$ and south to north $\delta z/\delta y$:

$$\frac{\delta z}{\delta x} = \frac{(Z_3 + 2Z_6 + Z_9) - (Z_1 + 2Z_4 + Z_7)}{8 \cdot \delta x}$$

$$\frac{\delta z}{\delta y} = \frac{(Z_7 + 2Z_8 + Z_9) - (Z_1 + 2Z_2 + Z_3)}{8 \cdot \delta y}$$

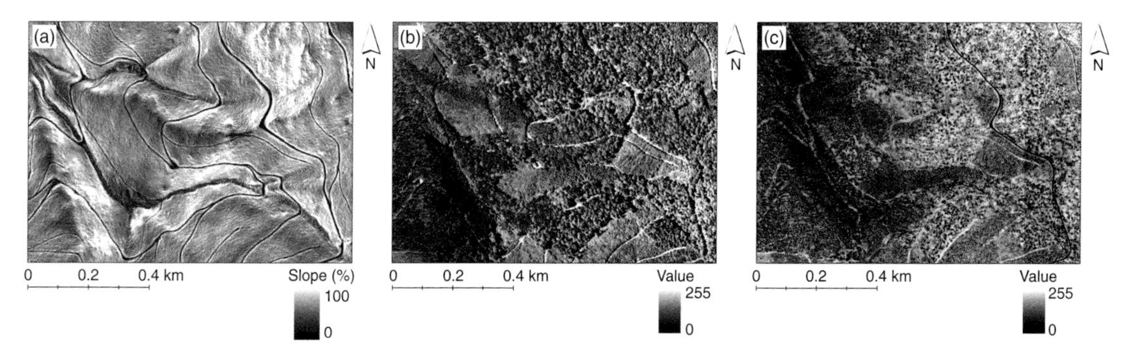

Fig. 8.6 (a) Slope image with 1 m resolution calculated from a LiDAR-DTM in a mountainous area. Forest roads are visible as dark lines, (b) Orthophoto with 25 cm resolution of the same area for comparison with the LiDAR data, (c) Intensity image with 1 m resolution calculated from the signal strength of the laser pulses that are emitted and reflected by a target.

Calculation of slope S:

$$\tan S = \sqrt{\left(\frac{\delta z}{\delta x}\right)^2 + \left(\frac{\delta z}{\delta y}\right)^2}$$

Dark regions in Figure 8.6a show low slope values and thus 'flat terrain' including roads, while bright values represent steep terrain. The comparison of the slope image to the orthophoto of the same area (Figure 8.6b) proves the high potential and advantage of LiDAR data for road extraction in mountainous terrain compared to other remote sensing data.

Approaches for automatic and semi-automatic road extraction from remote sensing data have been researched for many years and were mainly tested in urban and rural areas. The combination of height and intensity information from LiDAR data shows high potential for the detection of roads with bituminous surfaces. The assumption is that a road pixel will fulfil a series of criteria: the raw data points will lie near or on top of the DTM and will have a certain intensity value which represents the signal strength of the laser pulse that was reflected by the target. Additionally, a high local point density is assumed (Clode *et al.*, 2004). Figure 8.6c visualizes the intensity of the returned laser beam as grey scaled image (here: first pulse intensity). Even though the intensity image is very noisy, a road (with bituminous surface) appears as dark line on the right side of the image. Other approaches integrate information from laser scanning with topographic objects from cartographic databases like road centrelines to estimate detailed geometrical road parameters such as height, width, longitudinal and transversal slope as well as curvature (Hatger & Brenner, 2003), or parcel boundaries are used from a cadastral map for street surface reconstruction (Vosselman, 2003).

Due to the fact that most forest roads in Germany have an unpaved road surface with gravel, the assumption that a road will have certain intensity which is appropriate for urban areas will fail in many cases. Therefore road extraction techniques have to rely on height information only, for example by detecting abrupt height changes which are found at road edges in sloped terrain. The road surface itself will have a low gradient. Based on this assumption the following section describes an example for automated road extraction which was tested in mountainous forest area in Germany.

Method for automated forest road extraction

A terrain model with 1 m resolution s used as input into the processing chain. A median filter is used to reduce noise. The local slope of the smoothed DTM is calculated within a moving window as described above. Based on the slope model, in which roads become apparent as dark lines (see Figure 8.6), sub-pixel precise line extraction is utilized using an approach which was originally developed to extract lines from aerial images: 'For each pixel, the second order Taylor polynomial is computed by convolving the image with the derivatives of a Gaussian smoothing kernel. Line points are required to have a vanishing gradient and a high curvature in the direction perpendicular to the line' (Steger, 1996). The extracted lines (curvilinear structures) are returned with their centreline and width (potential edges of roads). In the next step all road segments shorter than 100 m will be deleted and the distances between the endpoints of the remaining road segments measured. If the distance is shorter than 10 m, adjacent segments are connected. In some cases, problems may occur with drainage lines which are interpreted as road features by the line extraction method. To delete those 'misclassified' road segments, hydrologic analysis algorithms are applied which make use of the DTM. For each cell of the DTM the flow direction of water is modelled. Water will flow in the direction with steepest descent starting from a selected cell to its eight neighbors. The flow direction grid is used to derive a flow accumulation grid. Each cell is given a value equivalent to the number of cells that flow to it (Jenson & Domingue, 1988). High values in the accumulation grid show areas of concentrated flow and are used to define a stream network. All contour points (xy coordinate pairs that describe a road segment) are checked if they are close to or on top of a drainage line. If more than 80% of the contour points are in the range of a drainage line, the respective road segment will be deleted. In a final step the road segments are smoothed using a method which projects the contour points onto a local regression line (a least-squares approximating line) fitted to a defined number of original contour points.

A result of the processing chain for road extraction can be seen in Figure 8.7a, while Figure 8.7b shows a comparison to GIS data used as reference. With the help of profiles perpendicular to the centreline, detailed geometrical information can be measured, for example the width and longitudinal and transversal slope. These parameters can be used to analyse if road segments are accessible for trucks (in general ≤12% slope). However the method is limited to sloped terrain. If there is no height or inclination discontinuity, no information can be extracted from the terrain model and other data has to be used (e.g. orthophotos).

3D building modelling

Reconstruction of 3D building models

3D Building models are playing a more and more important role in environmental management such as urban planning, surveillance and damage control of natural disasters, and virtual tourism, etc. As an efficient technology of 3D capture, LiDAR has gained a high popularity in the studies of 3D building reconstruction in recent years. Automatic or semi-automatic approaches that derive 3D building models from LiDAR data, with or without other accessorial information such as Aerial Image and 2D GIS ground plans, have been achieved during the past 10 years. In this section, several related concepts of 3D building models are illuminated.

Concepts of 3D building model

A 3D building model is a generalized, or rather abstract and scaled, virtual representation of the real building. The resolution of a model represents the approximation between the model and the real building. Figure 8.8 shows a definition of five different resolutions or namely LoDs (Levels of detail) for the 3D building model. Among them, LoD0 is a simple DSM (Digital surface model), LoD1 is block structure of buildings without information on roofs, LoD2 is block structures combined with roofs features, LoD3 gives more detailed information on façades and finally LoD4 provides the possibility to navigate inside the models. The reader is reminded that when we talk about 3D building models generated from

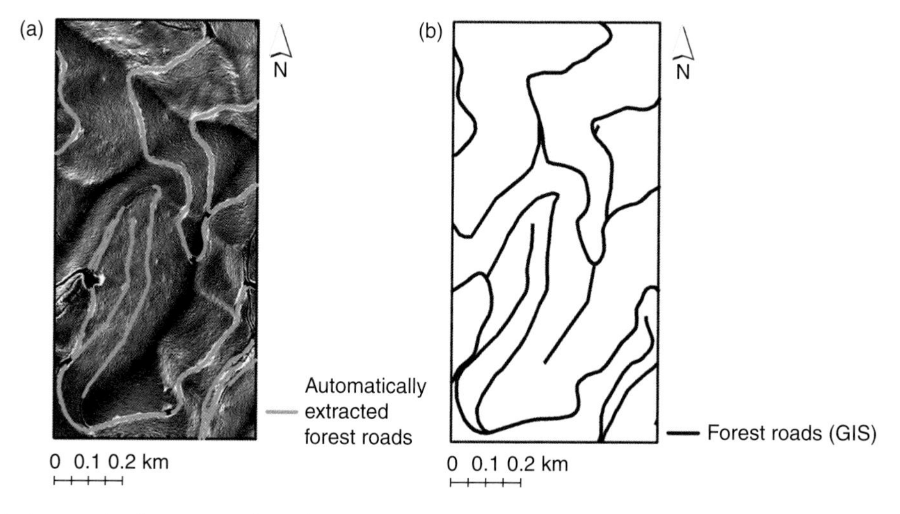

Fig. 8.7 (a) Automatically extracted road segments using image processing techniques (shown as red lines) superimposed on the slope image, (b) Reference data (road features from the GIS of the Department of Forestry). See Plate 8.7 for a colour version of these images.

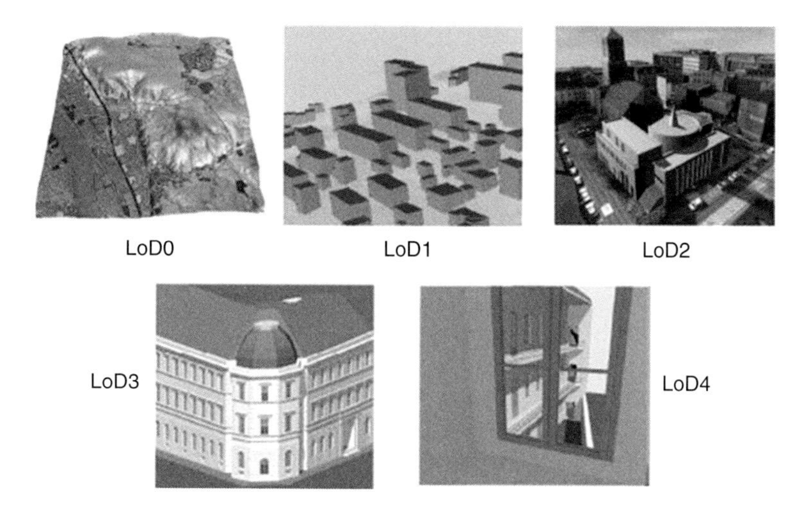

Fig. 8.8 Different resolutions of building models (Image source: Kolbe & Gröger, 2005).

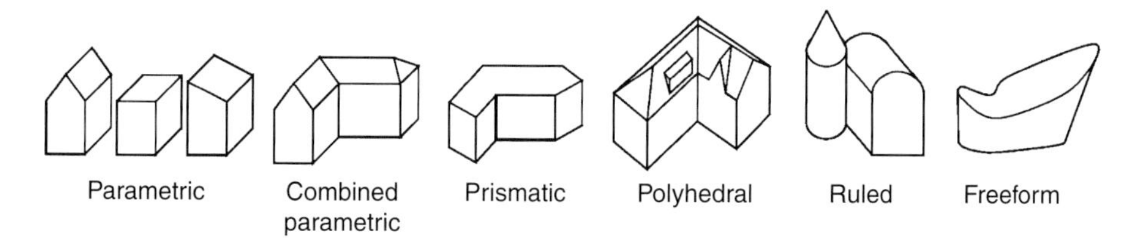

Fig. 8.9 Different classes of building models (Image source: Brenner, 2000).

LiDAR data, the LoD2 models are generally indicated as, for the most part, the majority of the points located on the roofs and information relating to façades is rarely available. In fact, derivation of the geometrical features of building roofs is the main focus of recent approaches to LiDAR – based 3D building modeling of buildings.

There is a more detailed definition of differing complexity of building models (Förstner, 1999), which is very important for analysing different LiDAR-based 3D building modelling processes. As shown in Figure 8.9, a prismatic model is actually a LoD1 model without roof descriptions. Both parametric and polyhedral models are of LoD2, a parametric model that reduces buildings to several predefined shapes, all the detailed structures such as dormers or balconies are eliminated, while a polyhedral model is more complex and has the highest approximation to the original building.

Building detection

The detection of buildings from the data is generally the first step for the reconstruction of 3D model. 2D ground plans are often used to solve the problem by offering the precise location of building outlines. Aerial Image is another useful supplement data resource for building detection, since plenty of algorithms are available for building detection on aerial images (Baillard, 1999; Heuel, 2000). nDSM (normalized Digital Surface Model) based methods are the most popular ones for extracting buildings from LiDAR data directly. Building regions can be easily detected by a height threshold on nDSM, which was calculated by subtracting DSM with DTM (Straub, 2001). For a more satisfactory result, local variance, morphological algorithms, texture classification and other edge detecting algorithms are been applied to improve the detected building borders (Elberink, 2000; Rottensteiner, 2002). To avoid the deformation of building structure on nDSM, buildings could also have been detected directly from DSM according to the second derivative characters of height changes in building regions on the DSM surface (Wang *et al.*, 2006). Besides the surface model based algorithms, some other approaches attempt to group 3D points of a single building directly from LiDAR point clouds by analysing local variety of point heights. The basic assumption of this is that there will be a high variety of point heights on trees, whereas the points of a single building tend to provide a regular change on heights (Verma *et al.*, 2006).

Building reconstruction

The approaches of 3D building model reconstruction can be roughly categorized into model-driven and data-driven. For the model-driven methods, several primitives of building models with typical roof structures such as flat, gable and hipped roofs, have been built up and fitted with the point clouds by changing the primitive parameters. The data-driven methods attempt to detect planes out of the point clouds and combine them to build models. Typical model-driven building reconstruction processes can be found in Brenner (2001), Maas (1999) and Verma (2006). As for the data-driven methods, Alharty (2004), Hofmann (2005), Rottensteiner (2002) and Vosselman (1999) have provided various possibilities.

Model-driven approaches

The basic presumption of model-driven approaches is that the building roofs are constructed by some prime structures, such as a flat plane, a gable or a hipped roof, thus a series of building primitives can be defined. There are several different expressions of the primitives. Figure 8.10a shows a parametric definition of building primitives given by Maas (1999), where (x_0, y_0) is the local origin of the building; θ is the angle between the main axis of the building to east which represents the direction of the building; X, Y, H are the length, width and height of the building; α is the slope of the roof plane.

A topographical expression of building roofs (Figure 8.10b) has been published in Verma (2006), where nodes *a*, *b*, *c*, *d* represent 4 different planes of the roof; lines between the nodes mean the two planes share an edge; O+ means that when the two planes are projected on the *XY* plane, they are orthogonal and point away from each other.

The most important process of a model-driven approach is to fit building primitives with the point clouds of a building. For the parametric primitives, the typical notion is to adjust the value of the parameters in order to get a best match between the primitive and the point clouds. This is accomplished by Haala (1999), with a least square procedure, and by Maas (1999) with analysis of invariant moments of point clouds. For the topographical primitives, this goal is normally achieved by an approximation evaluation between the primitive and a very crude model

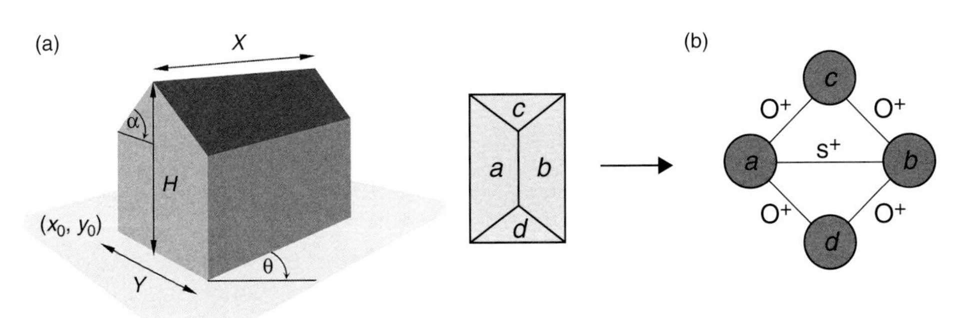

Fig. 8.10 Different expressions of building primitives (a) Parameter definition (Image source: Maas 1999), (b) Topographical definition (Image source: Verma *et al.* 2006).

calculated from the point clouds (Verma, 2006). The model-driven process is generally more efficient, but is relatively limited by the complexity of the primitives. To find several suitable primitives and combine them into a whole complex building remains a significant challenge.

Data-driven approaches

The extraction of roof planes from point clouds of a building is the dominant process of a data-driven approach. The most common method is to detect several segments of planes and group all the segments that have similar features (e.g. slope, orientation, etc.). An early research exercise for roof plane detection based on 3D Hough transformation was introduced in Vosselman (1999). Rottensteiner (2002) has presented an approach for getting roof planes by parameter (normal vector) based region growing; a similar process can be found in Alaharty (2004). Gorte (2002) and Hofmann (2005) have tried to use a TIN-mesh calculated from point clouds as a start point for plane detections; triangles are considered to belong to a same plane when the difference between their parameters (slope, orientation and distance from local origin) is within a tolerance. Unlike the others, Schwalbe (2005) has given another strategy to derive planes from point clouds. The points located on a single roof plane will be projected as a single line if observed from an orthogonal direction, thus the detection of planes can be reduced to a problem of line detection on a specific projection of point clouds. Figure 8.11 summarizes typical plane detection methods used in LiDAR

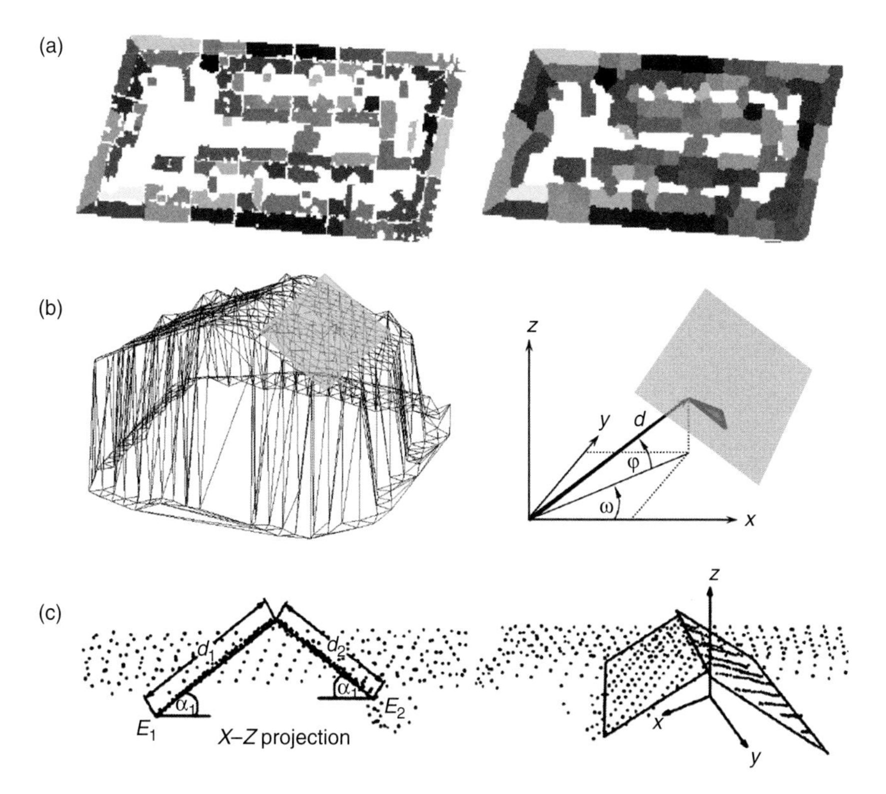

Fig. 8.11 Typical plane detection methods in data-driven approaches: (a) Left: Segments of planes detected from point clouds; Right: Voronoi diagram after merging coplanar segments (Image source: Rottensteiner 2002); (b) Left: 2.5D TIN structure generated from point clouds; Right: definition of the triangle parameters for triangle in TIN nets (Image source: Hofmann 2005); (c) Left: Lines representing roof faces; Right: Extracted roof planes (Image source: Schwalbe 2005).

data-driven approaches. The advantage of data-driven approaches is the high approximation between models and real buildings, but it always requires a high density of point clouds and is sometimes too sensitive to the errors in LiDAR data.

SUMMARY AND OUTLOOK

Different approaches for automatic feature extraction based on Airborne Laser Scanner data are described in this chapter. Both methods for natural and artificial (man-made) objects such as segmentation and classification of forest areas, forest road extraction and building extraction were explained. As shown different vegetation characteristics can be extracted from ALS data and can be used for delineation of stand and sub-stands units respectively. Due to the fact that the technique can be used for large-scale mapping of structural vegetation parameters, it can provide valuable information for environmental management. The transferability of the methods to ALS data, acquired with different sensor and flight parameters as well as during different seasons, is still a challenge. In general, the estimation of vegetation parameters has to be calibrated with field data.

An overview of different approaches for road extraction is given. So far, automatic road extraction techniques in forests based on terrain models try to detect height or inclination differences. Therefore those methods are limited to sloped terrain. To improve the extraction of roads in flat terrain other data sources have to be considered, such as othoimages, intensity data and vegetation height models. Based on ALS data the accuracy of existing forest road maps can be improved and additional attributes can be collected such as width, longitudinal and transversal slope.

Methods for 3D building reconstruction that are introduced in the chapter are all targeted to LoD2 building models. Due to the fact that ALS systems measure from above, all the mentioned studies concentrate on building roofs. Another big challenge to all the above-mentioned algorithms, are the non-plane surfaces on the roofs such as cambers and domes. High resolution images or photos for roofs have provided a good chance to improve the recent results. While terrestrial laser scanning data is now becoming a powerful instrument to gain information on building facades, there is a high potential to achieve LoD3 building models by combining the terrestrial data with the airborne data.

REFERENCES

Alharty A, Bethel J. 2004. Detailed building reconstruction from airborne laser scanner data using a moving surface method. *International Archives of Photogrammetry and Remote Sensing and Spatial Information Sciences* **34**B: 213–219.

Axelsson P. 1999. Processing of laser scanner data – algorithms and applications. *Journal of Photogrammetry and Remote Sensing* **54**: 138–147.

Axelsson P. 2000. DEM generation from laser scanner data using adaptive TIN models. *International Archives of Photogrammetry and Remote Sensing and Spatial Information Sciences* **33**: (B4/1), 110–117.

Axelsson P. 2001. Ground estimation of laser data using adaptive tin-models. *Proceedings of OEEPE Workshop on Airborne Laser Scanning and Interferometric SAR for Detailed Digital Elevation Models*, March 1–3, Stockholm, Sweden, No. 40, CD-ROM, 11p.

Baillard C, Schmid C, Zisserman A, Fitzgibbon A. 1999. Automatic line matching and 3D reconstruction of buildings from multiple views. *International Archives of Photogrammetry and Remote Sensing and Spatial Information Sciences* **32**: Part 3–2W5, 69–80.

Bortolot Z. 2006. Using tree clusters to derive forest properties from small footprint LiDAR data. *Photogrammetric Engineering and Remote Sensing* **72**: 1389–1397.

Brovelli M, Cannata M, Longoni U. 2002. Managing LiDAR data with GRASS. *Proceedings of the Open Source GIS-GRASS User Conference*, 11–13 September 2002, Trento, Italy.

Brenner C. 2000. Towards fully automatic generation of city models. *International Archives of Photogrammetry and Remote Sensing and Spatial Information Sciences* **34**: 169–174.

Burschel P, Huss J. 1997. *Grundriss des Waldbaus*, Parey, Berlin.

Clode S, Kootsookos P, Rottensteiner F. 2004. *The Automatic Extraction of Roads from LiDAR Data, XXth ISPRS Congress*, 12–23 July 2004, Istanbul, Turkey.

Cohen LD, Cohen I. 1993. Finite Element Methods for Active Contour Models and balloons for 2D and 3D Images. *IEEE Transactions on Pattern Analysis and Machine Intelligence*, PAMI-15.

Elberink SO, Maas HG. 2000. The Use of Anisotropic Height Texture Measures for the Segmentation of Airborne Laser Scanner Data. *International Archives of Photogrammetry and Remote Sensing and Spatial Information Sciences*, Vol. XXXIII, Amsterdam.

Elmqvist M. 2000. *Automatic Ground Modelling using Laser Radar Data*. Master Thesis, Linköping University.

Elmqvist M. 2002. Ground surface estimation from airborne laser scanner data using active shape models. *International Archives of Photogrammetry and Remote Sensing and Spatial Information Sciences*, Vol. 34, part 3A, September 9–13, Graz, Austria, 114–118.

Förstner W. 1999. *3D-City Models: Automatic and Semiautomatic Acquisition Methods*. Photogrammetrische Woche, Stuttgart.

Gorte B. 2002. *Segmentation of TIN-Structured Surface Models*. Joined Conference on Geospatial Theory, Processing and Applications, Ottawa, Canada.

Haala N, Brenner C. 1999. Extraction of buildings and trees in urban environments. *Journal of Photogrammetry and Remote Sensing* **54**:130–137.

Hatger C, Brenner C. 2003. Extraction of Road Geometry Parameters from Laser Scanning and Existing Databases. *International Archives of Photogrammetry, Remote Sensing and Spatial Information Sciences*, Vol. XXIV, Part 3/W13.

Heuel F, Lang F, Förstner W. 2000. Topological and Geometrical Reasoning in 3D Grouping for Reconstructing Polyhedral Surfaces. *International Archives of Photogrammetry and Remote Sensing and Spatial Information Sciences* **34**: 169–174.

Hildebrandt G. 1996. *Fernerkundung und Luftbildmessung für Forstwirtschaft, Vegetationskartierung und Landschaftsökologie*, Wichmann, Heidelberg.

Hofmann AD. 2005. *An Approach to 3D Building Model Reconstruction from Airborne Laser Scanner Data Using Parameter Space Analysis and Fusion of Primitives*. PhD Thesis, Institute for Photogrammetry and Remote Sensing, Technical University Dresden.

Horn BKP. 1981. *Hill Shading and the Reflectance Map*. Proceedings of the IEEE **69**: 14–47.

Jenson SK, Domingue JO. 1988. Extracting topographic structure from digital elevation data for Geographic Information System analysis. *Photogrammetric Engineering and Remote Sensing* **54**: 1593–1600.

Koch B, Heyder U, Weinacker H. 2006. Detection of individual tree crowns in airborne LiDAR data. *Photogrammetric Engineering and Remote Sensing* **72**: 357–363.

Kolbe TH, Gröger G. 2005. *A GML3 Application Profile for Virtual 3D City Models*. OGC TC Meeting New, York City.

Kraus K, Mikhail E. 1972. Linear Least Squares Interpolation. *Photogrammetric Engineering* **38**: 1016–1029.

Kraus K, Pfeifer N. 1998. Determination of terrain models in wooded areas with airborne laser scanner data. *Journal of Photogrammetry and Remote Sensing* **53**: 193–203.

Kraus K, Pfeifer N. 2001. Advanced DTM generation from LiDAR data. *International Archives of Photogrammetry and Remote Sensing and Spatial Information Sciences*, Vol. 34, 3W/4, WG IV/3. October 22–24, Annapolis (MD), USA, 23–30.

Maas HG, Vosselman G. 1999. Two algorithms for extracting building models from raw laser altimetry data. *Journal of Photogrammtry and Remote Sensing* **54**: 153–163.

Means JE, Acker SA, Fitt BJ, Renslow M, Emerson L, Hendrix CJ. 2000. Predicting forest stand characteristics with airborne laser scanning LiDAR. *Photogrammetric Engineering and Remote Sensing* **66**: 1367–1371.

Næsset E. 1997. Determination of mean tree height of forest stands using airborne laser scanner data. *Journal of Photogrammetry and Remote Sensing* **52**: 49–56.

Næsset E, Bjerknes K. 2000. Estimating tree heights and number of stems in young forest stands using airborne laser scanner data. *Remote Sensing of Environment* **78**: 328–340.

Næsset E. 2002. Predicting forest stand characteristics with airborne scanning laser using a practical two-stage procedure and field data. *Remote Sensing of Environment* **80**: 88–99.

Rieger W, Kerschner M, Reiter T, Rottensteiner F. 1999. Roads and buildings from laser scanner data within a forest enterprise. *International Archives of Photogrammetry and Remote Sensing and Spatial Information Sciences* **32**: 185–191.

Rieger W, Eckmüllner O, Müllner H, Reiter T. 1999. Laser-scanning for the derivation of forest stand parameters. Workshop of ISPRS WG III/2 & III/5: Mapping surface structure and topography by airborne and spaceborne lasers. *International Archives of Photogrammetry and Remote Sensing and Spatial Information Sciences* **32**: Part3-W14, 193–200.

Roggero M. 2001. Airborne laser scanning: Clustering in raw data. *International Archives of Photogrammetry and Remote Sensing and Spatial Information Sciences* Vol. 34, 3W/4, WG IV/3. October 22–24, Annapolis (MD), USA, 227–232.

Rottensteiner F. 2002. LiDAR activities at the Viennese Institute of Photogrammetry and Remote Sensing. *3rd International LiDAR Workshop, Mapping Geo-Surficial Processes Using Laser Altimetry*, October 7–9, Ohio State University, Columbus, USA.

Schwalbe E, Maas HG, Seidel F. 2005. 3D building model generation from airborne laser scanner data using 2D GIS data and orthogonal point cloud projections. *ISPRS WGIII/3,III/4,V/3 Workshop Laser scanning 2005*, Enschede, Netherlands.

Sithole G, Vosselman G. 2001. Filtering of laser altimetry data using a slope adaptive filter. *International Archives of Photogrammetry and Remote Sensing and Spatial Information Sciences*, Vol. XXXIV, 3/W4, Annapolis, MD, USA.

Sithole G, Vosselman G. 2003. Report ISPRS Comparison of Filters Online http://www.itc.nl/isprswgIII-3/filtertest/index.html (accessed August 2008).

Sithole G, Vosselman G. 2004. Experimental comparison of filter algorithms for bare-earth extraction from airborne laser scanning point clouds. *Journal of Photogrammetry and Remote Sensing* **59**: 85–101.

Sithole G. 2005. *Segmentation and Classification of Airborne Laser Scanner Data*. Publications on Geodesy of the Netherlands Commission of Geodesy **59**, Delft, 93–118.

Sohn G, Dowman I. 2002. Terrain surface reconstruction by the use of tetrahedron model with the mdl criterion. *Proceedings of the Photogrammetric Computer Vision, ISPRS Commission III, Symposium 2002*, September 9–13, Vol. 34 (3A), 336–344.

Steger C. 1996. Extraction of Curved Lines from Images. *13th International Conference on Pattern Recognition*, Volume II, 251–255.

Straub B, Gerke M, Koch A. 2001. Automatic Extraction of Trees and Buildings from Image and Height Data in an Urban Environment. *International Workshop on Geo-Spatial Knowledge Processing for Natural Resource Management*, Belward A, Binaghi E, Brivio PA, Lanzarone GA, Tosi G (eds), June 28–29 2001, University of Insubria, Varese (Italy), 59–64.

Straub C. 2007. *Erfassung des Rohholzpotentials und seiner Verfügbarkeit im Wald zum Zwecke der Energieholznutzung mit neuen Methoden der Fernerkundung*, Projektbericht, Freiburg.

Verma V, Kumar R, Hsu S. 2006. 3D building detection and modeling from aerial LiDAR data. *Proceedings of the 2006 IEEE Computer Society Conference on Computer Vision and Pattern Recognition* – Volume 2, 2213–2220.

Vosselman G. 1999. Building reconstruction using planar faces in very high density height data. *International Archives of Photogrammetry and Remote Sensing and Spatial Information Sciences* **32**: 87–92.

Vosselman G. 2000. Slope based filtering of laser altimetry data, *International Archives of Photogrammetry and Remote Sensing and Spatial Information Sciences*, Vol. XXXIII, Part B3, Amsterdam, 935–942.

Vosselman G, Maas HG. 2001. Adjustment and filtering of raw laser altimetry data. *Proceedings of OEEPE Workshop on Airborne Laser Scanning and Interferometric SAR for Detailed Digital Elevation Models*, March 1–3, Stockholm, Sweden. Official Publication No. 40. CD-ROM., 11 pages.

Vosselman G. 2003. 3-D reconstruction of roads and trees for city modelling. *International Archives of Photogrammetry, Remote Sensing and Spatial Information Sciences*, Vol. 34, Part 3/W13, Dresden, Germany, 231–236.

Wack R, Wimmer A. 2002. Digital terrain models from airborne laser scanner data – a grid based approach. *International Archives of Photogrammetry, Remote Sensing and Spatial Information Sciences*, Vol. 34 (3B, September 9–13, Graz, Austria), 293–296.

Wang Y, Weinacker H, Koch B. 2006. Automatic non-ground objects extraction based on Multi-Returned LiDAR data. *Photogrammetrie, Fernerkundung, Geoinformation*. Heft 2: 127–137.

Wang Y. 2007. Development of a procedure for vertical structure analysis and 3D-single tree extraction within forests based on LiDAR point clouds. *ISPRS Workshop on Laser Scanning 2007 and SilviLaser 2007*, September 12–14, 2007 Espoo, Finland.

Wehr A, Lohr W. 1999. Airborne laser scanning – an introduction and overview. *Journal of Photogrammetry and Remote Sensing* **54**: 68–82.

Weinacker H, Koch B, Heyder U, Weinacker R. 2004. Development of Filtering, Segmentation and Modelling Modules for LIDAR and Multispectral data as a Fundament of an Automatic Forest Inventory System. *Proceedings of the ISPRS working group VIII/2 Laser-Scanners for Forest and Landscape Assessment, Freiburg*, 3. bis 6. Okt. 2004, ISSN 1682–1750, 50–55.

Wynne RH. 2006. LiDAR Remote sensing of forest resources at the scale of management. *Photogrammetric Engeneering and Remote Sensing* **72**: 1310–1314.

9 Terrestrial Laser Scan-derived Topographic and Roughness Data for Hydraulic Modelling of Gravel-bed Rivers

DAVID J. MILAN

Department of Natural and Social Sciences, University of Gloucestershire, Cheltenham, UK

INTRODUCTION

Terrestrial Laser Scanning (TLS) now allows dry portions of gravel-bed rivers to be surveyed at the scale of a single particle, and thus has the potential to vastly improve estimates of bed roughness and produce high resolution models of the channel boundary. Flow modelling has traditionally utilised global estimates of roughness to numerically model velocity, however TLS now allows fully distributed data on roughness height to be incorporated. As yet, there has been little research that has attempted to incorporate high resolution roughness data into existing hydraulic models. This chapter investigates the effects of artificially degrading the resolution of roughness height and topographic data upon predicted velocity and boundary shear stress over three different discharges. As shown below, hydraulic simulations indicate a progressive improvement in predictions with a progressively finer grid resolution, with the 0.1 m grid yielding the best stream-wise (x) velocity matches (r^2=0.76). Furthermore, significant differences were found in the population of stream-wise velocities when comparing simulations that used a 0.15 and 0. 6 m resolution roughness height and topography grid. Increased spatial resolution of the roughness height and topography data revealed greater detail in the pattern of velocity and shear stress; with clast-scale shear stress patterns and flow vector deflection around clasts visible. Discharge-dependent variations in velocity and shear stress maxima were identified which were thought to relate to subdued pool-riffle morphology. The highest discharge exhibited the least spatial variability in velocity and shear stress, suggesting that the influence of both grain roughness and bedforms was lessened. Further development of CFD models to incorporate large topographic datasets obtained through terrestrial laser scan surveys is required.

MODELLING ROUGHNESS

Roughness is one of the fundamental parameters of the boundary over which fluids flow.

Laser Scanning for the Environmental Sciences,
1st edition. Edited by G.L. Heritage and A.R.G. Large.
 ISBN 978-1-4051-5717-9

Flow models require roughness values to optimise the agreement between measured and predicted velocities. Roughness coefficients such as Manning's *n*, the D'Arcy-Weisbach friction factor *f*, and the Chezy coefficient *C* are often used by hydraulic engineers to correct the relationship between flow and stage level, as part of a calibration process. Most Computational Fluid Dynamic (CFD) codes involving natural channels represent boundary roughness using the law of the wall, which requires specification of a roughness height (k_s) that is finer than that which can be represented through the topographic data. One of the main difficulties in applying wall functions within natural rivers involves the determination of an appropriate value for k_s. k_s in gravel-bed rivers is usually associated with a characteristic grain size percentile of the bed material (e.g. D_{50}, D_{84} or D_{90}).

k_s data is conventionally gathered using a grid-by-number approach, whereby the median grain size is derived from measurement of the intermediate b-axis of 100 clasts from distinct geomorphic units (e.g. Wolman, 1954), although some workers have preferred to use the c-axis as a roughness measure to account for the effects of partial burial and imbrication of surface clasts (e.g. Gomez, 1993). The use of a single percentile to represent k_s can be problematic due to the effects of packing/imbrication and the presence of microscale bedforms. Imbrication effectively decreases the roughness height from the bed, whereas pebble clusters tend to provide locally increased roughness height. To overcome this issue, some multiplier of grain size is often used. This can range between $0.4D_{50}$ and $3.5D_{84}$ depending on whether grain roughness alone or the combined effect of grain roughness and microtopographic roughness is required (e.g. Hey, 1979; Whiting & Dietrich, 1991; Ferguson & Ashworth, 1992; Wiberg & Smith, 1991; Clifford *et al.*, 1992; Gomez, 1993).

Furthermore, the spatial variability in surface k_s is rarely accounted for. Often an averaged k_s value is used across the whole of the model domain, although some workers (e.g. Booker *et al.*, 2001) have distinguished between riffles and pools. An alternative method of representing k_s is to treat the bed surface as a random field of surface elevations *Z(x,y,t)*, where *x* and *y* are longitudinal (stream-wise) and cross-stream coordinates, and *t* is time. Until recently the success of this approach has been tempered by the lack of high-resolution topographic data covering all roughness scales (e.g. Furbish, 1987; Robert, 1988, 1990, 1991; Clifford *et al.*, 1992). Improved data capture is now achievable using photographic and laser scanning technology (Nikora *et al.*, 1998; Goring *et al.*, 1999; Butler *et al.*, 2001; Aberle & Smart, 2003; Marion *et al.*, 2003; Nikora & Walsh, 2004; Smart *et al.*, 2004; Aberle & Nikora, 2006). Obtaining data on k_s at the grain scale in the field has been problematic, although some advances have been made using digital photogrammetry to produce millimetre and sub-millimetre precision digital elevation models (Butler *et al.*, 1998, 2002; Graham *et al.*, 2005).

Recent advances in Terrestrial Laser Scanning (TLS) permits measurements of bed topography to a very high resolution, as long as the bed surface is dry (Milan *et al.*, 2007). TLS has been shown to offer vertical accuracy of ±0.02 m (Milan *et al.*, 2007), and is used in the present study to obtain sub-grain resolution data, offering vastly improved k_s and topography data, and providing an opportunity to investigate the effects of artificially degrading the resolution of k_s and topographic data upon flow simulation. This study investigates the effect of the resolution of the k_s and topography grids upon velocity and turbulent flow simulations and attempts to identify whether there is any significant improvement in predictions through the use of a progressively finer grid.

APPLICATION OF THE TECHNOLOGY

In an effort to quantify its accuracy, and hence usefulness, in ephemeral streams, testing of TLS was carried out along a 45 m reach of Kingsdale Beck in North Yorkshire, UK (Figure 9.1). The

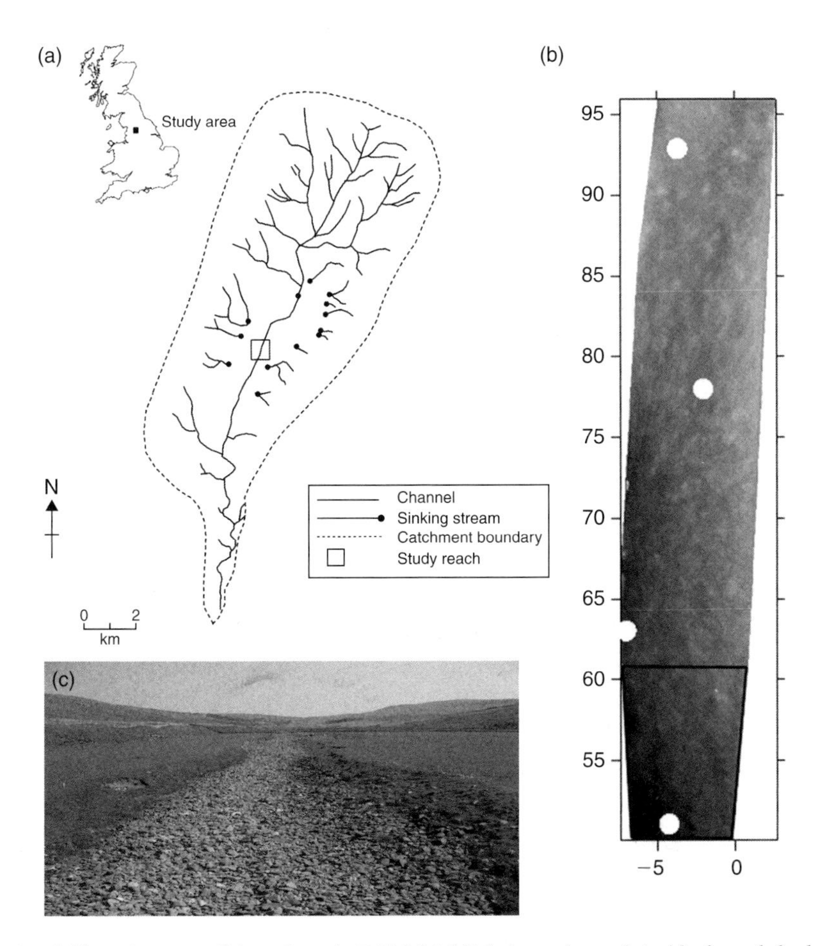

Fig. 9.1 Study site: (a) location map, (b) reach scale DEM, highlighting sub-reach in black, and the location of ADV measurements (white dots), (c) photograph of study reach when dry. View is upstream, taken from footbridge.

reach investigated was artificially straightened in the 1890s to improve agricultural access to the land. The reach has not been managed since and consequently has developed pool-riffle morphology. The catchment is underlain by Carboniferous limestone that results in a flashy flow regime. An upstream sinkhole captures base flow at the study reach, resulting in flow cessation at the surface during dry weather periods (Figure 9.1a). The channel consists of rounded and platy coarse gravel and cobble clasts. The median grain size (D_{50}) of the b and c-axis was 0.09 m and 0.04 m respectively whilst the upper 1st standard deviation statistics of the random field of bed elevations taken from the TLS (D_{84}) of the b and c-axis was 0.14 m and 0.06 m respectively (Heritage *et al.*, 2007). Microscale cluster bedforms are also present in the channel (Heritage *et al.*, 2007), which were also felt likely to influence near bed velocity vectors.

APPLYING THE TECHNOLOGY

Topography data

A Riegl LMS Z210 field laser scanner (see Heritage and Hetherington (2007) for technical specifications) collected random field data along a

45 m reach of the dry river bed. Data from four independent scan locations were meshed using retro-reflectors linked to a common EDM theodolite-based coordinate system. Median data point spacing was 0.0043 m with 75% of the data points spaced at less than 0.0092 m apart. One of the limitations with current CFD models is their inability to handle extremely large topographic datasets such as those generated by TLS. Instead of importing the 4,991,222 data points directly into the CFD model and then interpolating these raw data to produce a grid topography, the data had to be pre-processed. The raw data were imported as an ASCII *x,y,z* file into the SURFER® software package. A series of regular grid files were then produced using Delauney Triangulation as the interpolation algorithm, with resolutions of 0.1, 0.15, 0.3 and 0.6 m. These resolutions were used to produce four topographic DEMs. The 0.1 m resolution grid file contained 133,139 lines of data which was close to the processing limits of the CFD code employed in this investigation. The *x,y,z* data for each grid file was then re-interpolated beneath a computational mesh within the CFD software.

Surface roughness (k_s)

A 3×3 'moving window' approach was used to determine the local standard deviation (σ_z) of 0.09 m^2 sub-areas across the raw point cloud and the resultant data were multiplied by two, as suggested by Gomez (1993), to obtain the estimated clast c-axis measures at the equivalent 0.09 m^2 scale. Regular grid files were produced with 0.1, 0.15, 0.3 and 0.6 m node spacing, which were then used to produce four further DEMs of local $2\sigma_z$ for topographic surfaces of different grid resolutions. These $2\sigma_z$ DEMs were used to characterise surface roughness in the model.

Computational Fluid Dynamic (CFD) modelling

The CFD code used in this investigation was version 1 of SSIIM (Sediment Simulation in Intakes with Multiblock option; Olsen, 1996). The model has been applied previously to flow dynamics of gravel-bed rivers in the UK, in particular to address hydraulics and sediment transport processes through riffle-pool sequences (Booker *et al.*, 2001; Clifford *et al.*, 2005) and hydraulic modelling of fish habitat (Booker, 2003). The model solves the full 3D Navier–Stokes equations with a κ–ε turbulence closer model on a three-dimensional non-orthogonal grid. The SSIIM code uses Navier–Stokes equations for turbulent flow in a three-dimensional geometry to obtain the water velocity:

$$\frac{\partial U_i}{\partial t} + U_j \frac{\partial U_i}{\partial x_j} = \frac{1}{\rho} \frac{\partial}{\partial x_j} \left(-P\delta_{ij} - \rho \overline{u_i u_j} \right) \quad (9.1)$$

where the term on the left of the equation is the transient term and the next term is the convective term. Symbol notation is given in Table 9.1. The first term on the right of Equation (9.1) is the pressure term, and the second term on the right is the Reynolds stress term. Non-compressible, constant density flow is assumed. The k–ε model is used to calculate turbulent shear stress for the three-dimensional simulations within SSIIM. The eddy-viscosity concept υ_T with the k–ε turbulence closer model is used to model the Reynolds stress term:

$$\overline{u_i u_j} = \upsilon_T \frac{\partial U_i}{\partial x_j} + \frac{2}{3} k \delta_{ij} \quad (9.2)$$

In Equation (9.2), the first term on the right side of the equation forms the diffusive term in the Navier–Stokes equation, and the third term is incorporated into the pressure. The $k-\varepsilon$ model calculates the eddy-viscosity as:

$$\upsilon_T = c_\mu \frac{k^2}{\varepsilon} \quad (9.3)$$

where k is the turbulent kinetic energy, ε is its dissipation and c_μ is a constant. Further details of the modelling of k and ε are given in Olsen (2002). In this study the model was run with a free water surface, for steady-state solutions at three flows. The water surface was calculated using

Table 9.1 Notation used for SSIIM numerical symbols.

g = gravitational acceleration	u = fluctuating velocity
K = constant in wall function	u_* = shear velocity
k = turbulent kinetic energy (per unit mass)	ν_T = turbulent eddy viscosity
k_s = roughness height	z = coordinate in the vertical direction
Δh_{ij} =vertical movement for water surface calculation	ε = dissipation rate for k
l = difference in height of water surface at P_{ij} and P_{ref}	δ_{ij} = Kronecker delta
P = pressure (P_{ij} is extrapolated pressure at each cell, P_{ref} is the reference pressure)	ρ = fluid density
t = time	τ = boundary shear stress
U = average velocity	

the SIMPLE method (Patanker, 1980) for pressure correction. This was used to couple all cells except those closest to the surface and allowed calculation of a free water surface. The water surface is fixed at the downstream cross-section where the pressure, P_{ref}, is taken as a reference. A pressure deficit at each cell is then calculated by subtracting this reference pressure from the extrapolated pressure P_{ij} for each cell, which is then used to move the water surface (Olsen and Kjellesvig, 1998):

$$\Delta h_{ij} = \frac{1}{\rho g}\left(P_{ij} - P_{ref}\right) \qquad (9.4)$$

where l is the difference in the height between the water surface at the given and reference points.

Roughness can then be specified for each boundary cell within the computational mesh according to the law of the wall:

$$\frac{U}{u^*} = 2.303 \ln\left[\frac{30z}{k_s}\right] \qquad (9.5)$$

which is integrated over the bed cell with z equal to the distance from the wall to the centre of the cell.

SSIIM requires the specification of a Strickler value in order to generate an initial water surface, calculated using a 1D backwater model, knowing the downstream water level and the discharge. Water level was surveyed with a theodolite at the exit to the reach, in the same coordinate system as the bed, for each simulated flow. An n value of 0.03, and a bank roughness of 0.1 m, were found to give satisfactory correspondence between modelled and measured water elevations. Measured inflow velocity profiles were not available for all the flows presented here, but should ideally be used (Lane *et al.*, 1995). Following the technique of Booker (2003), an imposed uniform cross-stream velocity pattern with a vertical logarithmic profile, with zero cross-stream and vertical velocities was applied. Following the initial 1D backwater calculation, the 3D water surface was allowed to adjust.

Following Booker *et al.* (2001), boundary shear stress was estimated from:

$$\tau = 0.3\rho k \qquad (9.6)$$

Computational mesh

Two computational meshes were produced for this investigation, with mesh complexity limited to permit realistic computational time for convergence. The first covered the whole 45 m reach (323 m^2) and consisted of 157,500 cells with 450 in the stream-wise direction, 70 cross-stream and five in the vertical plane. Vertical spacing of grid nodes was 0, 10, 25, 75, and 100% from the bed. The second mesh concentrated on a 7 × 7 m sub-area of the first mesh towards the reach outlet (Figure 9.1c), and was constructed to permit a more detailed analysis of the influence of local roughness elements on turbulent flow. This second mesh had 56,000 cells with 50 in the stream-wise direction, 70 cross-stream, and had

16 vertical cells. Vertical spacing was at 0, 2, 4, 6, 8, 10, 15, 20, 30, 40, 50, 60, 70, 80, 90, and 100% from the bed. The *x,y* node spacing of 0.1 m for both meshes was approximately equivalent to the spacing of the finest scale topography DEM grid.

Coordinates for the four different resolution topographic DEMs were imported into SSIIM as a 'geodata' file and subsequently viewed beneath the computational mesh. These data points were then re-interpolated within SSIIM and *z* values allocated to each node at the boundary of the mesh. This data was stored in the 'koordina' file. Roughness height values were allocated to each node at the mesh boundary in a similar way. Instead of interpolating the topography data, the $2\sigma_z$ data was used instead. The roughness height data was stored in the 'bedrough' file. This data permitted a unique roughness height value to be associated with every node on the computational mesh. In comparison to previous similar investigations, the spatial resolution of the topographic and roughness height data entered into the model was greatly improved.

MODEL VALIDATION

Validation data was obtained through Acoustic Doppler Velocimeter (ADV) measurements, taken at four locations in the 45 m length study reach (Figure 9.1c). This instrument was used to obtain three-dimensional velocity information at 25 Hz for a measuring volume of approximately 0.01 m × 0.01 m × 0.08 m. Velocities were measured at four locations in the study reach and at each of these locations readings were taken at 25, 75 and 100% flow depth. Figure 9.2 shows plots of measured versus predicted stream-wise (*x*) velocity for the

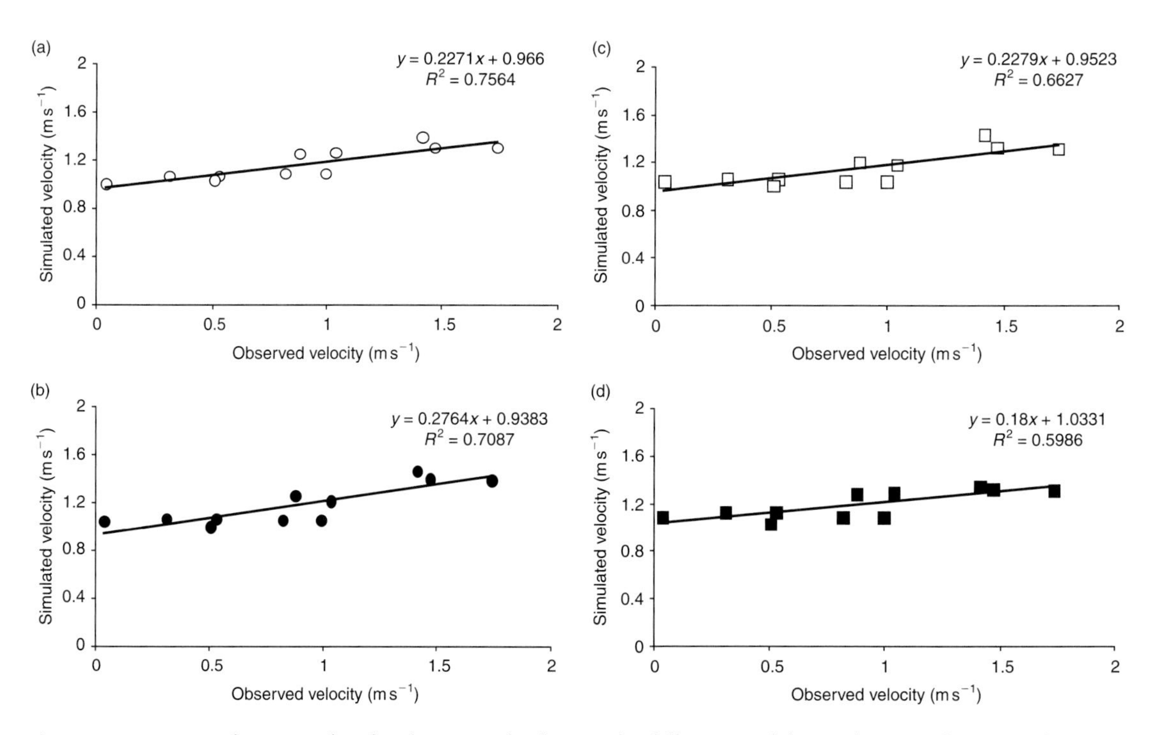

Fig. 9.2 Comparison been simulated and measured velocities for different model runs, (a) 0.1 m, (b) 0.15 m, (c) 0.3 m, and (d) 0.6 m.

four different model runs. Simulated velocities generally appear to be overestimated for values below 1 ms^{-1}, and underestimate values in excess of 1 ms^{-1}. The 0.1 m grid resolution appears to demonstrate the best velocity match according to the r^2 value of 0.76, while the lowest r^2 values (0.6) were found for the coarsest grid.

HOW USEFUL IS TLS IN THESE ENVIRONMENTS?

Figure 9.3a demonstrates shaded relief DEMs of topography for the four different grid resolutions. Gravel clasts are visible on the 0.1 m grid, whereas the DEM becomes much more pixelated for the 0.6 m grid. Figure 9.3b demonstrates the $2\sigma_z$ DEM as a shaded relief map for the four different grid resolutions. The final set of DEMs in Figure 9.3c demonstrates the results of the model runs for near bed velocity in the stream-wise (x) direction. The differences in velocity distribution shown are purely the result of differences in topography resolution and roughness height. Differences are apparent, but are subtle in nature. Much more detail is evident in the model run that employed the finest scale topography and roughness height grid (0.1 m). The zone of highest velocity appears along the right-hand bank, associated with the position of the thalweg. High velocities are also evident at the reach inlet. Lowest velocities are found towards the right bank in the middle of the reach, associated with shallower areas of flow depth.

To identify whether the differences in velocity distribution were significant, cumulative population distributions for velocities in x,y and z planes were generated for each of the four model runs, for (a) near bed (10% from bed) and for (b) 50% flow depth (Figure 9.4). For near-bed velocities it is only possible to distinguish x velocities from y and z, however at 50% flow depth it is possible to distinguish between x,y and z velocities. The effect of grid resolution appears to be minor, with most difference being shown for velocities in the x plane. A Kolmogorov-Smirnov test was used to identify differences in velocities in each plane between model runs. The only significant difference was between x near-bed velocities simulated using the 0.15 m grid resolution and the 0.6 m grid resolution ($p = 0.05$). For the velocity simulations at 50% flow depth, there is even more similarity in the velocity distributions for x,y and z velocities. Again, the greatest difference is seen for velocities in the x plane at 0.15 m and 0.6 m, however this difference is insignificant at the 0.05 level.

Hydraulic simulation for 3.34, 7.12 and 17.00 $m^3 s^{-1}$

Figure 9.5 compares (a) the spatial patterns in near-bed velocity for the three different flow events and (b) the spatial pattern of shear stress using the 0.1 m $2\sigma_z$ and topography grids only. The influence of grain roughness upon local variability in near bed velocity and shear stress is evident from inspection of the DEMs for 3.34 and 7.12 $m^3 s^{-1}$. For example, the patchy red areas on the shear stress DEMs indicate high shear stresses >40 Nm^2 on the stoss side of coarse clasts. Near-bed velocities tended to be greatest for the 7.12 $m^3 s^{-1}$ flow, where a peak velocity of 1.8 ms^{-1} of was simulated. Maximum velocity for the 3.34 $m^3 s^{-1}$ flow was 1.3 ms^{-1}, and for 17 $m^3 s^{-1}$ a velocity peak of 1.1 ms^{-1} was simulated. The reduction in near-bed velocity for the highest discharge may reflect ponding of water in the reach upstream of a footbridge situated at the tail of the reach. A spatial shift in the zone of maximum velocity and shear stress is evident with increasing discharge magnitude. Although morphology within the reach is relatively subdued in nature, there is a suggestion that pool-riffle morphology may be responsible for some of the spatial patterns simulated. The zones of relatively high shear stress and velocity are found to be associated with riffles at the lowest flow. At the intermediate discharge there is a zone of high velocity and boundary shear stress at the head of the reach and about 15 m downstream from the head of the reach, associated with the exit slope of a pool. In contrast, velocity and shear stress distribution is far more uniform at the highest discharge.

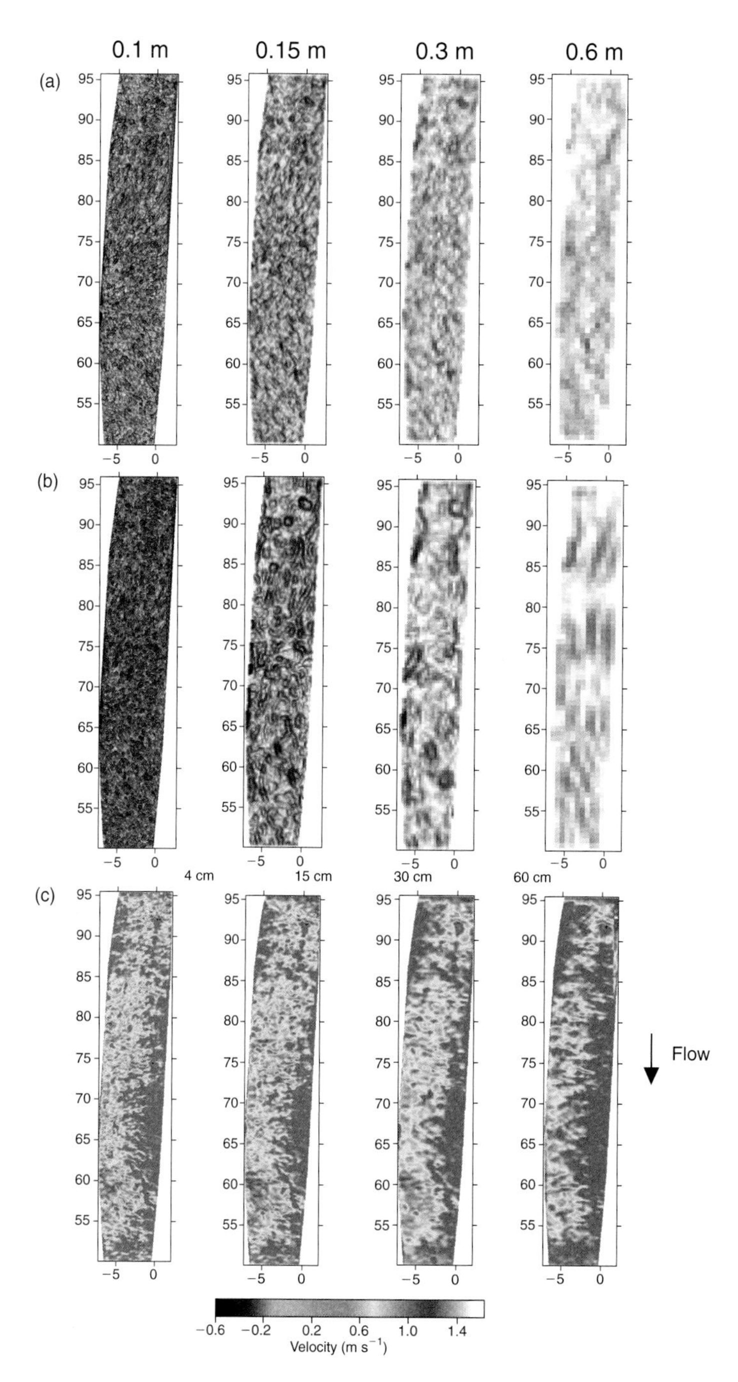

Fig. 9.3 (a) Shaded relief DEMs for grids of different resolution from 0.1 m (left) to 0.6 m (right), (b) shaded relief maps for standard deviation, (c) results for near bed (10% flow depth from the bed) stream-wise (x) velocity which have utilised various resolution topographic and roughness height information, for a discharge of $3.34\,m^3\,s^{-1}$. See Plate 9.3 for a colour version of these images.

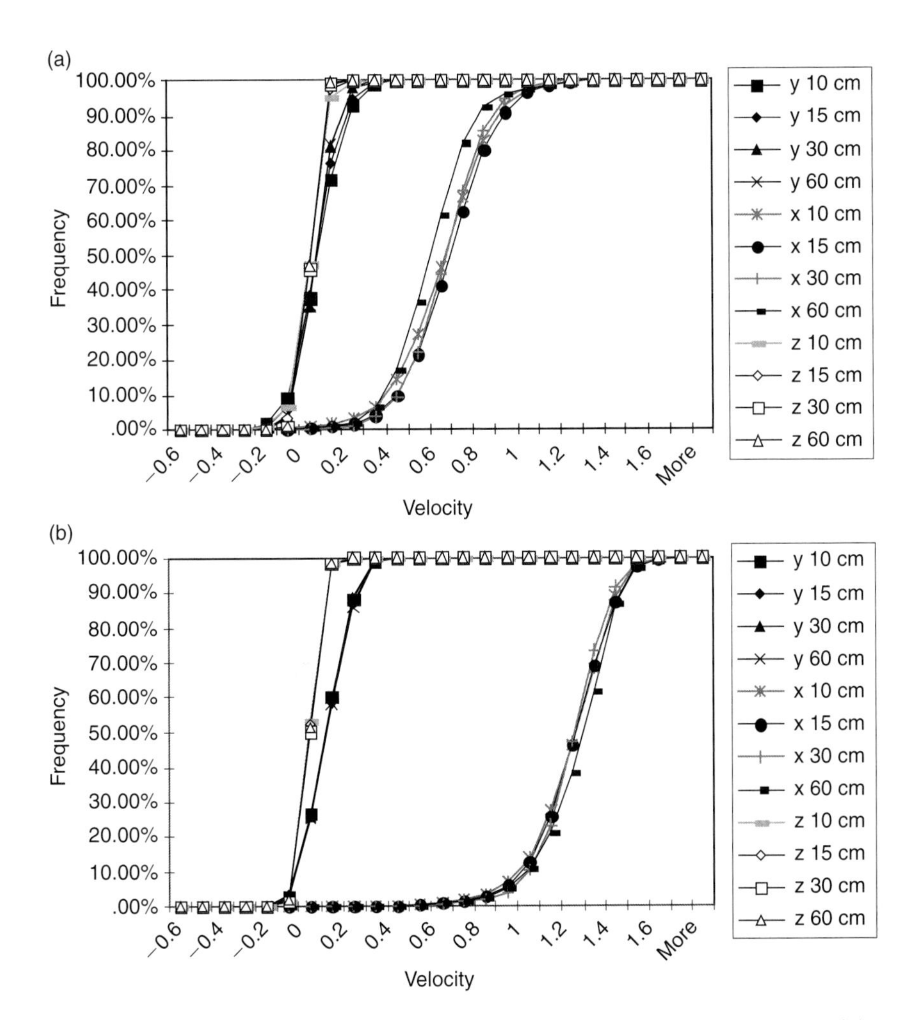

Fig. 9.4 Cumulative frequency population curves for simulated velocities in *x,y* and *z*, for model runs that have utilised different resolution topography and roughness height data; (a) near-bed, (b) 50% flow depth.

PATCH SCALE VELOCITY PATTERNS: FLOW DEFLECTION AROUND ROUGHNESS ELEMENTS

Figure 9.6 shows a planform view of flow vectors for the near bed cells (2% of the flow depth from the bed) for the 7 m × 7 m patch, with a roughness ($2\sigma_z$) contour map shown alongside. The 0.1 m resolution is compared with the 0.30 m resolution run. Differences in the resolution of the roughness height contour maps are clearly evident, as are the effects upon velocity vectors. Deflection of flow from the stream-wise direction is apparent, and appears to be dictated by the roughness of the bed/topography. There is evidence for flow divergence on the stoss side of clasts and convergence and flow strength reduction on the lee of clasts. The level of detail concerning flow vector deflection around roughness elements is lost in the 0.3 m grid. Also notable is that there is less variation in the velocities exiting the patch.

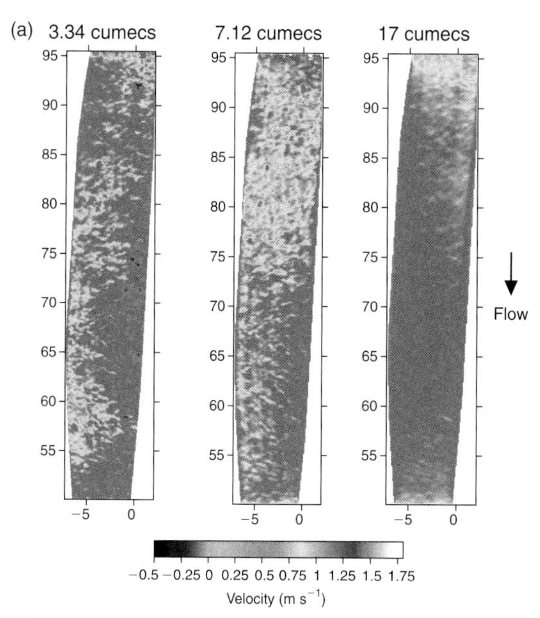

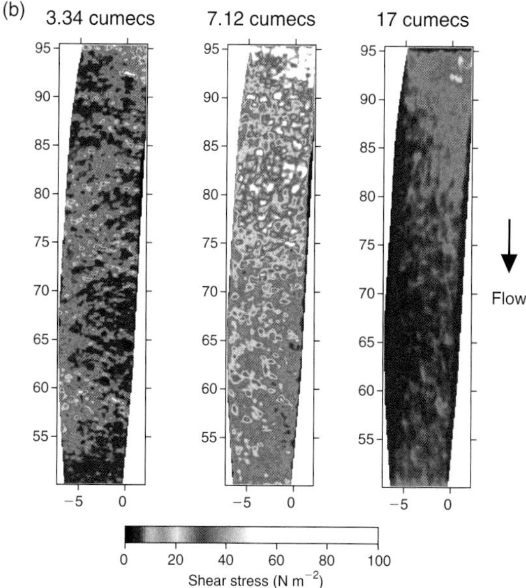

Fig. 9.5 Simulated (a) near-bed velocity and (b) shear stress for flows of 3.34 $m^3 s^{-1}$, 7.12 $m^3 s^{-1}$ and 17.00 $m^3 s^{-1}$. See Plate 9.5 for a colour version of these images.

DISCUSSION AND CONCLUSIONS

High resolution topographic data are essential to the future development of our understanding of flow and sediment transport interactions, and in the future management of fluvial systems for flood defence, sediment transfer and aquatic and riparian ecosystems. Furthermore, improved roughness and topographic data are key to the future development of flow and sediment transport modelling. Until recently the acquisition of high resolution field data for bed morphology and roughness was problematic. Heritage and Hetherington (2007) suggest that terrestrial laser scanning plugs the gap in the spatio-temporal sampling framework, allowing fluvial geomorphologists to gain data with high spatial resolution, and a broad temporal frequency required to survey and capture morphological change commonly experienced within fluvial systems (see Figure 16.1, this volume). From the exercise described in this chapter, it is clear that TLS does offer exciting possibilities for improving the quantification of bed roughness, with a concomitant improvement in flow prediction. Sub-clast resolution surveys permit the full roughness height distribution of exposed bar or channel surfaces to be quantified, and should in turn, with focused Research and Development, allow the development of the random field approach to surface clast size analysis, facies and surface roughness height determination (Nikora *et al.*, 1998; Aberle & Nikora, 2006; Heritage & Milan, in press). Spatial variability in k_s and topography may be attributed to each grid node, limited primarily by the computational mesh resolution, rather than the resolution of the raw random field data. Use of the distribution of k_s, may also be used to provide global measures of grain size for specific geomorphic units (Heritage & Milan, in press), avoiding the errors inherent in taking small samples (Hey, 1979). Furthermore, this may be used as input into hydraulic formulae used to predict flow velocity, such as the Colebrook–White formulae, which currently relies on roughness height determined from the Wolman grid approach.

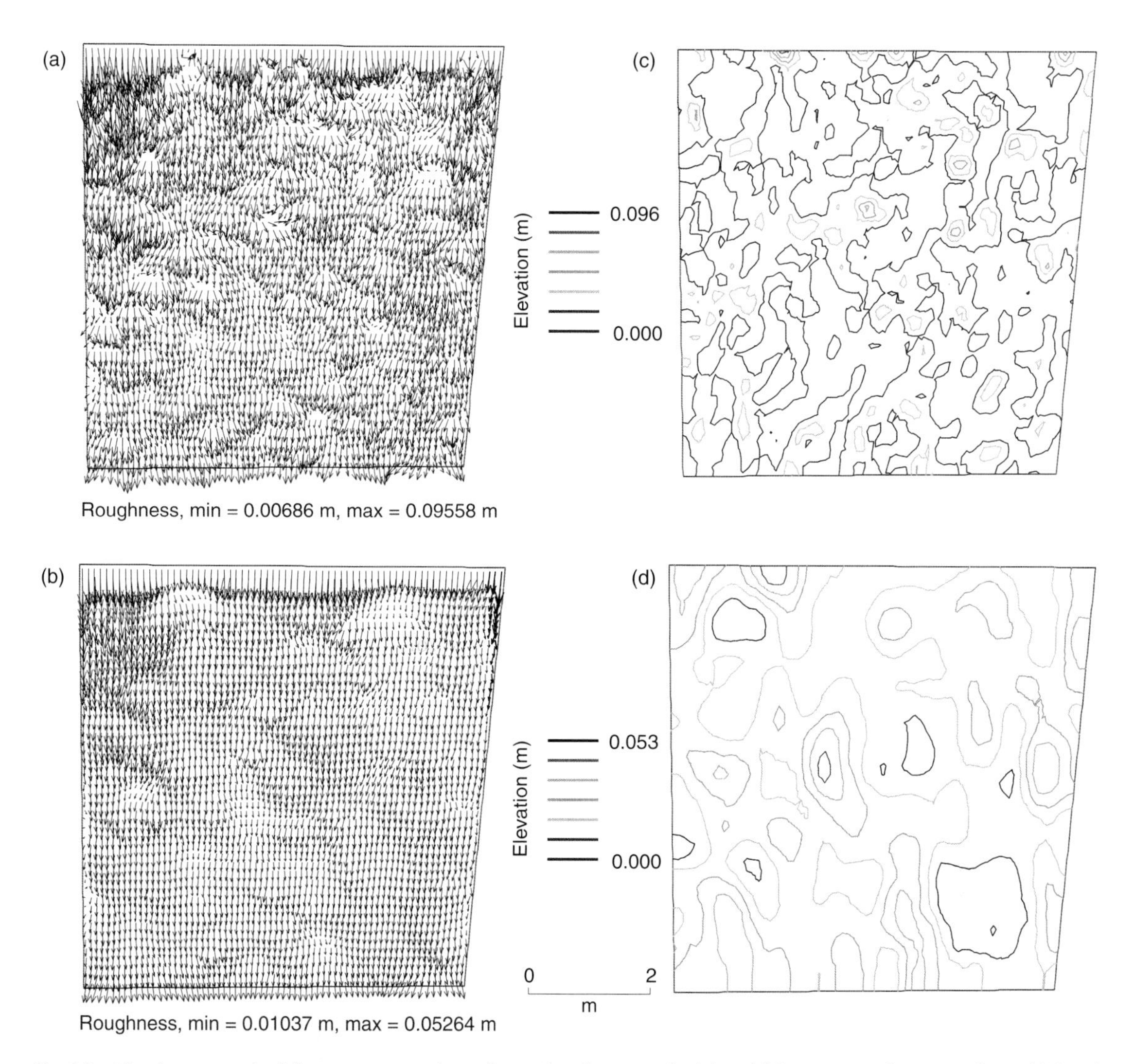

Fig. 9.6 Planform near-bed flow vectors and roughness height maps for (a) and (b) 10 cm resolution grids, and (c) and (d) 30 cm resolution grids. See Plate 9.6 for a colour version of these images.

It must be said that terrestrial laser scanning does have certain limitations. Due to reflection of the shorter wavelengths of light from the water surface and returns from particles within the water column, laser returns can only be guaranteed from dry surfaces, which limits its use to the survey of dry channels, bars, or rivers at low flow. Milan *et al.* (2007) demonstrated these problems for a proglacial river (Figure 9.7) and found that vertical errors increased with water depth, and ranged from +0·06 to −0·23 m, with the average error being −0·06 m (σ = 0·066 m). Bathymetric green wavelength LiDAR may provide a way to resolve this issue in the future (e.g. Brock *et al.*, 2007). A further limitation with TLS is that it is not possible to survey as large an area as airborne LiDAR; however TLS permits higher resolution and greater vertical accuracy compared to the

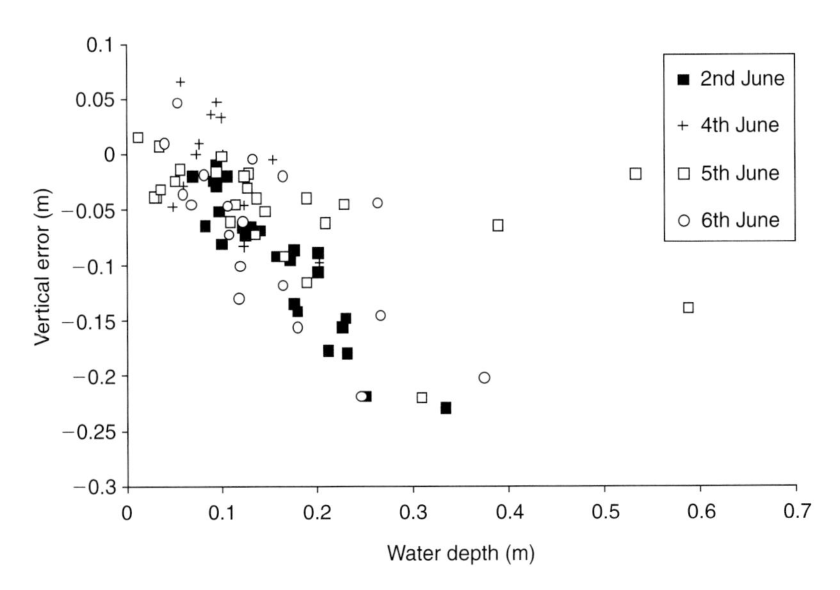

Fig. 9.7 Vertical error in submerged areas of reach. Vertical error was derived from a comparison between survey points surveyed using an EDM theodolite and TLS (after Milan *et al.*, 2007).

former. In addition the technique has the advantage over photogrammetry that post-processing effort and cost is vastly reduced.

Most CFD models are not yet capable of handling the large topographic datasets generated using TLS. Using the SSIIM model, there was a limit to the number of data points that could be interpolated, which necessitated pre-processing of the raw TLS point cloud. This restricted the length of the reach that could be modelled and meant that the raw topographic data had to be converted to a lower resolution regular grid file, prior to being re-interpolated within SSIIM. The ability of CFD codes to handle the large topographic datasets currently being obtained from TLS and LiDAR surveys is clearly an issue that needs to be overcome. Overall however, TLS has significant potential for research in gravel-bed rivers; through increasing the spatial resolution of the roughness height and topography data, much greater detail can be quantified in the pattern of velocity and shear stress. It is also possible using the technology to quantify clast-scale shear stress patterns and flow vector deflection around clasts when the finest resolution grids are applied. This has undoubted potential in aligning geomorphological research with its ecological implications and in addressing some of the outstanding scaling issues associated with major water legislation.

REFERENCES

Aberle J, Nikora V. 2006. Statistical properties of gravel-bed surfaces. *Water Resources Research* **42**: W11414.

Aberle J, Smart GM. 2003. The influence of roughness structure on flow resistance on steep slopes. *Journal of Hydraulic Research* **41**: 259–269.

Booker DJ. 2003. Hydraulic modelling of fish habitat in urban rivers during high flows. *Hydrological Processes* **17**: 577–599.

Booker DJ, Sear DA, Payne AI. 2001. Modelling three dimensional flow structures and patterns of boundary shear stress in a natural pool-riffle sequence. *Earth Surface Process and Landforms* **26**: 553–576.

Brock JC, Wright CW, Nayegandhi A, Woolard J, Patterson M, Wilson I, Travers L. 2007. USGS-NPS-NASA EAARL Submarine Topography – Florida Keys National Marine Sanctuary, USGS Open File Report 2007–1395 (DVD), URL: http://pubs.usgs.gov/of/2007/1395/

Butler JB, Lane SN, Chandler JH. 1998. Assessment of DEM quality for characterizing surface roughness using close range digital photogrammetry. *Photogrammetric Record* **16**: 271–291.

Butler JB, Lane SN, Chandler JH. 2001. Characterization of the structure of river-bed gravels using two-dimensional fractal analysis. *Mathematical Geology* **33**: 301–330.

Butler JB, Lane SN, Chandler JH, Porfiri K. 2002. Through-water close range digital photogrammetry in flume and field environments. *Photogrammetric Record* **17**: 419–439.

Clifford NJ, Robert A, Richards KS. 1992. Estimation of flow resistance in gravel bedded rivers: A physical explanation of the multiplier of roughness length. *Earth Surface Processes and Landforms* **17**: 111–126.

Clifford NJ, Soar PJ, Harmar OP, Gurnell AM, Petts GE, Emery JC. 2005. Assessment of hydrodynamic simulation results for eco-hydraulic applications: a spatial semivariance approach. *Hydrological Processes* **19**: 3631–3648.

Furbish DJ. 1987. Conditions for geometric similarity of coarse stream-bed roughness. *Mathematical Geology* **19**: 291–307.

Ferguson RI, Ashworth PJ. 1992. Spatial patterns of bedload transport and channel change in braided and near-braided rivers. In *Dynamics of Gravel-Bed Rivers*, Billi P, Hey RD, Thorne CR, Tacconi P. (eds), Chichester: John Wiley & Sons Ltd., 477–496.

Gomez B. 1993. Roughness of stable, armoured gravel beds. *Water Resources Research* **29**: 3631–3642.

Goring D, Nikora V, McEwan IK. 1999. Analysis of texture of gravel beds using 2-D structure functions. In *River, Coastal, and Estuarine Morphodynamics*, Proceedings of the IAHR Symposium Vol. 2, Seminara G, Blondeaux P, *et al.* (eds), New York: Springer, 111–120.

Graham DJ, Reid I, Rice SP. 2005. Automated sizing of coarse-grained sediments: image-processing procedures. *Mathematical Geology* **37**: 1–28.

Heritage GL, Milan DJ. in press. Terrestrial laser scanning of grain roughness in a gravel-bed river. *Geomorphology*.

Heritage GL, Hetherington D. 2007. Towards a protocol for laser scanning in fluvial geomorphology. *Earth Surface Processes and Landforms* **32**: 66–74.

Heritage GL, Hetherington D, Large ARG, Fuller I, Milan DJ. 2006. Terrestrial laser scanner based roughness quantification using a random field approach. In *Understanding a Changing World: Integrated approaches to monitoring, measuring and modeling the environment*. Proceedings of the Annual Conference of the Remote Sensing and Photogrammetry Society, University of Cambridge.

Heritage GL, Milan DJ, Johnson J, Entwhistle N, Hetherington D. 2007. Detecting and mapping microtopographic roughness elements using terrestrial laser scanning. Proceedings of the Annual Conference of the Remote Sensing and Photogrammetry Society, Newcastle University.

Hey RD. 1979. Flow resistance in gravel bed rivers. *Journal of the Hydraulics* Division ASCE, **105**: 365–379.

Lane SN, Richards KS, Chandler JH. 1995. Within-reach spatial patterns of processes and channel adjustment. In Hickin EJ (ed.), *River Geomorphology*, Wiley: Chichester, 105–130.

Milan DJ, Hetherington D, Heritage GL. 2007. Application of a 3-D laser scanner in the assessment of a proglacial fluvial sediment budget. *Earth Surface Processes and Landforms* **32**: 1657–1674.

Marion A, Tait SJ, McEwan IK. 2003. Analysis of small-scale gravel bed topography during armouring. *Water Resources Research* **39**: 1334.

Nikora VI, Goring DG, Biggs BJF. 1998. On gravel-bed roughness characterization. *Water Resources Research* **34**: 517–527.

Nikora V, Walsh J. 2004. Water-worked gravel surfaces: High-order structure functions at the particle scale. *Water Resources Research* **40**: W12601.

Olsen NRB. 1996. *A three dimensional numerical model for simulation of sediment movements in water intakes*. Dissertation for Dr. Ing. Degree, The Norwegian Institute of Technology, Division of Hydraulic Engineering, University of Trondheim, Norway.

Olsen NRB. 2002. A three-dimensional numerical model of sediment movements in water intakes with multiblock option – User's Manual version 1.1. Department of Hydraulic and Environmental Engineering, the Norwegian University of Science and technology.

Olsen NRB, Kjellesvig HM. 1998. Three dimensional numerical flow modelling for estimation of local scour depth. *Journal of Hydraulic Research* **33**: 571–581.

Patanker SV. 1980. *Numerical Heat Transfer and Fluid Flow*. New York: McGraw-Hill.

Robert A. 1988. Statistical properties of sediment bed profiles in alluvial channels. *Mathematical Geology* **20**: 891–225.

Robert A. 1990. Boundary roughness in coarse-grained channels, *Progress in Physical Geography* **14**: 42–70.

Robert A. 1991. Fractal properties of simulated bed profiles in coarse-grained channels, *Mathematical Geology* **20**: 205–223.

Schlicting H. 1979. *Boundary-layer Theory*. McGraw Hill: New York.

Smart GM, Aberle M, Duncan M, Walsh J. 2004. Measurement and analysis of alluvial bed roughness. *Journal of Hydraulic Research* **42**: 227–237.

Whiting PJ, Dietrich WE. 1991. Convective accelerations and boundary shear stress over a channel bar. *Water Resources Research* **27**: 783–796.

Wiberg PL, Smith JD. 1991. Initial motion of coarse sediment in streams of high gradient. Erosion and sedimentation in the Pacific Rim, *International Association of Hydrological Sciences Publication* **165**: 299–308.

Wolman MG. 1954. A method of sampling coarse river-bed material. *Transactions, American Geophysical Union* **35**: 951–956.

10 Airborne LiDAR Measurements to Quantify Change in Sandy Beaches

MICHAEL J. STAREK[1], K. CLINT SLATTON[1,2], RAMESH L. SHRESTHA[1] AND WILLIAM E. CARTER[1]

[1]Department of Civil and Coastal Engineering, University of Florida, Gainesville, USA
[2]Department of Electrical and Computer Engineering, University of Florida, Gainesville, USA

INTRODUCTION

In the United States the narrow coastal fringe that makes up only 17% of the nation's contiguous land area is home to more than half of its population (Crossett *et al.*, 2008). Similar population concentrations exist in other coastal areas around the world. Within the coastal zone, sandy beaches are a particularly valued natural resource providing vital economic and recreational benefits the world over. Beaches define a geomorphic barrier between the land and the sea, protecting vast investments in development and infrastructure to these regions. But sandy beaches are, by nature, highly dynamic landforms, constantly evolving with time and subject to the continual threat of erosion and coastal flooding. Loss of shoreline and damage to beaches can result in the loss of property, tourism and investment that can devastate local economies (Shrestha *et al.*, 2005). Monitoring beaches and studying the processes that govern their change are critical to the future sustainability of these environments and the economies that depend on them.

Changes in beach topography are a response to the forcing of waves, winds and currents and depend on the supply of sediment to the beach. This evolution constitutes a highly variable, three-dimensional process that can be loosely classified into three phenomena: rapid variations due to storm events, seasonal variations due to changes in wave and wind climate and long-term variations due to sea level rise, longshore sediment transport, anthropogenic influences, extreme storm events and climate change (Dean & Dalrymple, 2002). Because beaches constantly evolve with time, the challenge in coastal monitoring is to determine the persistence of the observed change. If a region of shoreline is undergoing growth (accretion) or retreat (erosion), it is important to determine whether this change is transitory, such as caused by seasonal influences, or the onset of a more permanent condition, such as an altering of the long-term sediment supply or sea encroachment. This becomes extremely important when assessing trends in shoreline change, recovery from storm impact, influences from anthropogenic structures, such as jetties and inlets, and the success of beach nourishment projects. By measuring the elevation of the beach repeatedly over time, changes in beach volume and shoreline position can be quantified providing insight into beach stability, storm impact and sediment transport.

Laser Scanning for the Environmental Sciences,
1st edition. Edited by G.L. Heritage and A.R.G. Large.
 ISBN 978-1-4051-5717-9

Standard methods for measuring beach change employ classical levelling and static GPS to survey two-dimensional profiles across the beach, which are oriented perpendicular to the waterline spaced at intervals of several hundred metres (Gutelius *et al.*, 1998; Shrestha *et al.*, 2005; Robertson *et al.*, 2007). By comparing a time series of profile measurements, changes in elevation and shoreline position with time can be quantified. Because of the labour intensive nature of these methods, it is not possible to sample at high enough spatial or temporal resolution to resolve the three-dimensional variability in beach change at large scales. The resulting profile data are limited in their utility for process modelling and only very approximate volume and shoreline change estimates can be expected. This undersampling problem is continually confronted by coastal planners who must make decisions about potential changes affecting a beach based on profiles separated by several hundred metres at temporal intervals of a year or greater. Furthermore, it is very difficult to perform rapid pre and post-storm mapping with these approaches, resulting in extensive delays before engineers have the information they need to rebuild or take other corrective actions in the aftermath of storm damage. Alternative survey methods include aerial photogrammetry and kinematic GPS (Gutelius *et al.*, 1998). But even with those methods it can be too costly, time consuming and simply not feasible to sample the beach surfaces densely enough to capture the high resolution variability at large scale. The limitations of these data collection methods limit our understanding of the three-dimensional evolution of beaches.

The advent of LiDAR technology with integrated GPS-inertial navigation systems in airborne platforms makes it possible to quantify three-dimensional change in beach topography at spatial scales needed to advance science and monitor erosion over long segments of coastline quickly, accurately and economically (Shrestha *et al.*, 2005). With the dense coverage of LiDAR data, it is possible to sample beach topography at much higher alongshore spatial frequencies than is practical with standard coastal surveying methods. Profiles can be extracted from the LiDAR data with metre-scale inter-profile spacing and shore-normal point spacing, limited only by the density of the data. These data can in turn be used for improved computational modelling. Volumes can be computed from profiles or surfaces extracted from LiDAR-derived digital elevation models (DEMs) to produce accurate erosion/deposition estimates for engineers, scientists and decision makers.

This revolution has been propelled both by topographic LiDAR systems that operate in the near-infrared portion of the optical spectrum and bathymetric systems that operate in the blue-green range of the spectrum. The focus in this chapter is on the application of small-footprint topographic LiDAR systems. Small-footprint systems enable beach and upland mapping with average spatial resolutions greater than 1 point per m^2, vertical accuracy (z) of 5–10 cm and horizontal accuracy (x,y) of 15–20 cm (Slatton *et al.*, 2007). However, the point density will vary locally depending on flight parameters, scan angle, beam divergence and pulse repetition rate (Axelsson, 1999; Baltsavias, 1999). In addition, modern discrete-return systems record multiple returns per transmitted pulse (including first and last). This multi-return capability is particularly useful for creating bare-earth DEMs because the last returns have a higher probability of reflecting from the true ground surface. Over minimally vegetated surfaces, such as beaches, accurate DEMs at high resolution can be generated to decimetre scales (Neuenschwander *et al.*, 2000).

Numerous studies have demonstrated the application of LiDAR data for quantifying beach change by comparing multiple datasets acquired over the same region (e.g. Tebbens *et al.*, 2002; Sallenger *et al.*, 2003; White & Wang, 2003; Woolard *et al.*, 2003; Zhang *et al.*, 2004; Shrestha *et al.*, 2005; Robertson *et al.*, 2007). In addition to changes in elevation, by mining the LiDAR data, many different features can be extracted where a feature in this context refers to a LiDAR measurable property of the sampled surface such as surface gradient or curvature (Luzum *et al.*, 2004). The objective of feature extraction is often to use the extracted features to discriminate between classes of objects in the data. For coastal LiDAR data, morphologic feature extraction can be used to 'characterize' change patterns observed in the LiDAR data (e.g. Brock *et al.*, 2004; Saye *et al.*, 2005; Starek *et al.*, 2007a).

In the following discussion, we present three case studies demonstrating the application of topographic LiDAR for characterizing beach change. Case Study 1 presents a framework to mine high-resolution LiDAR data over beaches through automated cross-shore profile sampling of LiDAR-derived DEMs. From the profiles, shoreline change can be quantified, erosion-prone zones detected and several different cross-shore features can be extracted for classifying different change patterns. Case Study 2 presents a simple example of calibrating an analytical diffusion model using LiDAR data to simulate the spatial spreading of newly deposited sand following a beach nourishment. Finally, Case Study 3 presents a brief example of quantifying changes in beach morphology due to hurricane impact using LiDAR data acquired shortly before and after the storm.

CASE STUDY 1: LIDAR DATA MINING FOR BEACH EROSION

Data mining and pattern classification techniques offer great potential to move the analysis of LiDAR data beyond visual interpretation and simple (first order) measurements made from the DEMs. This is particularly true for coastal monitoring with LiDAR data because of the importance of small-scale features in the DEMs. When acquired with sufficient temporal coverage, the high spatial-resolution information in the LiDAR data can reveal patterns in beach change and resolve non-stationary processes, such as localized zones of anomalous erosion. Data mining is the process of extracting relevant information from large data sets. We present a semi-automatic method to mine time-series LiDAR data and detect LiDAR-measurable morphologies (surface features) that are indicative of observed shoreline change patterns along a beach. These morphologic indicators can improve beach characterization providing insight into the change dynamics and for classification of high-impact zones. Several different features are extracted and then segmented into binary 'erosion' or 'accretion' classes. Class-conditional probability density functions (pdfs) are then estimated for each feature and the separation between class pdfs is ranked. The more separation provided by a feature, the greater its potential as an indicator for characterizing beach change. An example is presented demonstrating how the features can then be used to classify erosion-prone zones probabilistically.

Study area

As part of a costal monitoring program, the University of Florida (UF) Geosensing and Engineering Mapping (GEM) Centre acquired LiDAR data along the St Augustine Beach region of Florida, USA seven times between August 2003 and February 2007 (see Table 10.1). St Augustine Beach is located on the northeast coast of Florida in St. Johns County. A 10 km long stretch of the coastline comprises our study area. This region was selected because it contains a historical accretion zone in the southern portion and a highly erosive pier zone to the north, which is armoured with revetments and requires periodic beach nourishment to replace lost beach area. Effects from two nourishment projects conducted in 2003 and 2005 are observed in the data. Both nourishments focused on the around 2 km revetment region of the pier.

Table 10.1 LiDAR collection dates and data epochs covered by sequential acquisitions.

Time period	Dates	Length (years)
Epoch 1	16 August 2003 to 10 March 2005	~1.6
Epoch 2	10 March 2005 to 07 May 2005	~0.2
Epoch 3	07 May 2005 to 11 October 2005	~0.5
Epoch 4	11 October 2005 to 27 January 2006	~0.3
Epoch 5	27 January 2006 to 14 June 2006	~0.4
Epoch 6	14 June 2006 to 19 February 2007	~0.8

The surveys were conducted at approximately low tide using a topographic LiDAR system operating at a laser pulse rate of 33 kHz and a near-infrared wavelength of 1064 nm. First and last reflections were recorded for each return laser pulse. Most topographic LiDAR systems operate at near-infrared wavelengths, and thus can only record elevations from the surface above the water because near-infrared wavelengths do not penetrate the water column. As a result, LiDAR shoreline surveys are generally acquired during low-tide to capture more exposed land surface and diminish the potential effects from wave run-up on shoreline delineation. Nominal values for all flights included a flying height of 600 m above ground level, flight speed of 60 m/s, scan rate of 28 Hz and maximum scan angle from nadir of ±20°. These parameters resulted in a mean ground point density of 1.3 points/m^2 and swath width of approximately 440 m for each alongshore flight line.

Data processing

The raw LiDAR data consists of point clouds of irregularly spaced *x,y,z* values. Generally, only the last returns from each LiDAR shot are utilized when measuring surface change to minimize the probability of patches of vegetation or other land cover biasing the surface elevations. Those points must then undergo a series of pre-processing steps before the shoreline is delineated and morphological features are extracted. The three main preprocessing steps are (1) remove absolute position biases between flights that constitute the time series, (2) project elevations to an appropriate vertical datum and (3) filter out the LiDAR returns that are not from the ground surface.

Errors

Even with proper system calibration, a vertical bias in the LiDAR observations, due mostly to GPS induced trajectory errors, is sometimes present. These are handled by comparing the heights 'calibration lines' on hard surfaces in the study area, such as roads and runways, determined by ground-based GPS methods. Kinematic GPS profiles acquired on the ground can provide measurements of surface elevations with few-centimetre scale accuracy (Shrestha *et al.*, 2007). In this study, the offsets were removed by vertically shifting the laser data sets to match GPS profiles acquired along roads within the study area under the assumption that the road surfaces did not change during LiDAR and ground surveys. For the LiDAR system and flight parameters used in this work, along with proper calibration, experience shows that this correction generally yields vertical accuracies of 5–10 cm and horizontal accuracies of 15–20 cm relative to ground-survey measurements over slowly varying surfaces (Slatton *et al.*, 2007). For our seven LiDAR data sets, RMS height errors relative to the GPS control points fell within 5–8 cm after adjustment. Assuming a worst case elevation accuracy of ±8 cm deviation, we can thus expect our threshold in change detection to be roughly $\sqrt{8^2+8^2} = \pm 11.3$ cm deviation by propagating the error from the differencing of two elevations. Measured changes larger than this threshold can be considered statistically significant.

Vertical datum

For coastal applications, it is desirable to have LiDAR elevations referenced to a vertical datum when determining shoreline. The LiDAR points available for this study were output with respect to the NAD83 vertical datum projected in Universal Transverse Mercator (UTM). Therefore, the heights are initially referenced to the ellipsoid that defines the NAD83 datum. The ellipsoid heights must be converted using a geoid model into orthometric heights to relate the measured elevations to localized mean sea level (MSL). This allows a tidal elevation contour to be extracted as the shoreline. In this study, the horizontal (*x,y*) coordinates were projected to UTM zone 17 and the initial ellipsoid heights were transformed with respect to the NAVD88 vertical datum using the GEOID03 geoid model (Meyer *et al.*, 2004) .

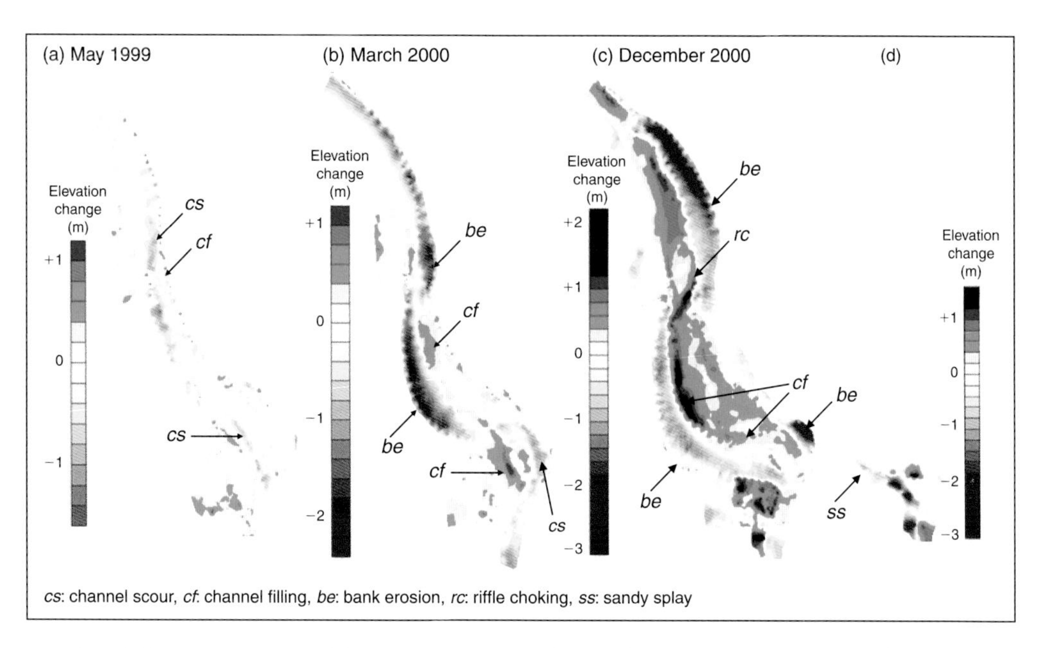

Plate 1.3 Theodolite repeat surveys of the River Coquet, Northumberland, UK at ~0.05 points m^{-2} resolution. Inset (d) shows a re-survey 7 days after a flood event caused avulsion at the lower end of the reach (after Fuller *et al.*, 2003).

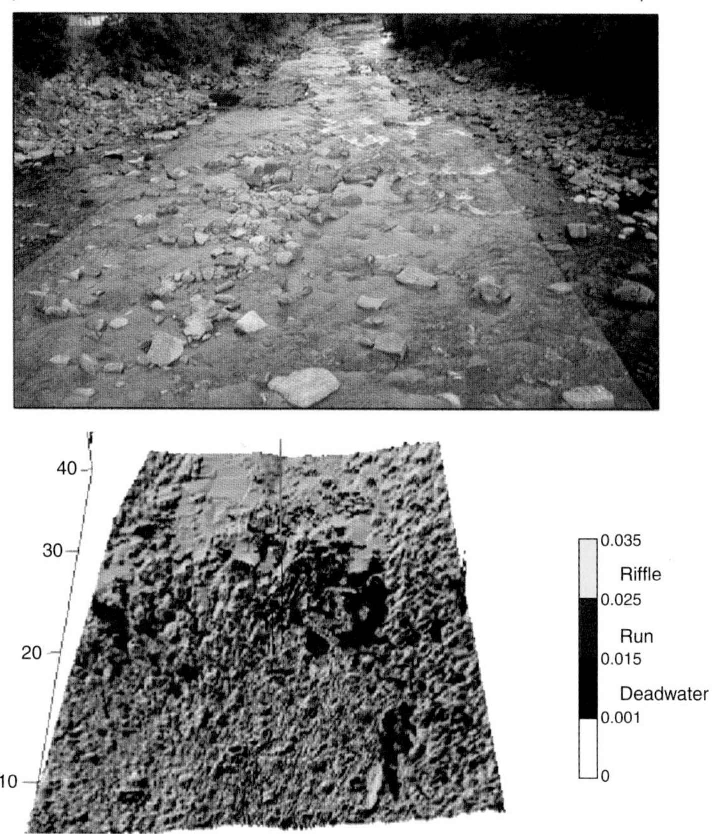

Plate 1.7 Spatial distribution of biotope types for the reach shown in Figure 1.6 defined by laser scan water surface roughness measurement (Large & Heritage, 2007).

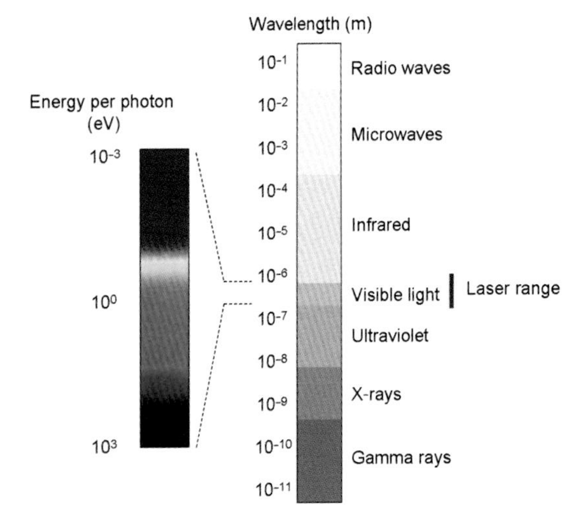

Plate 2.1 The electromagnetic spectrum and laser wavelengths.

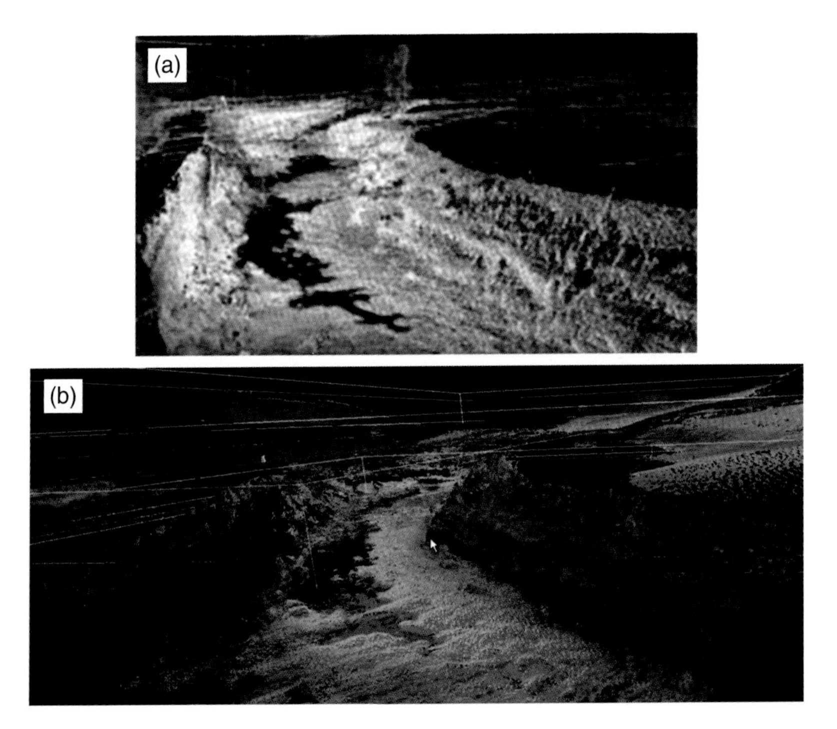

Plate 2.11 Example laser cloud model and colour-rendered surface for the River Wharfe at High Houses, North Yorkshire, UK.

Plate 4.2 The principle of airborne LiDAR operation. The laser illuminates a circular area (footprint) at the nadir point of the scan with a diameter dependent on the beam divergence. For moving platforms, the laser is integrated with a GPS and an inertial measurement unit (IMU) or inertial navigation system (INS) to provide the necessary orientation data.

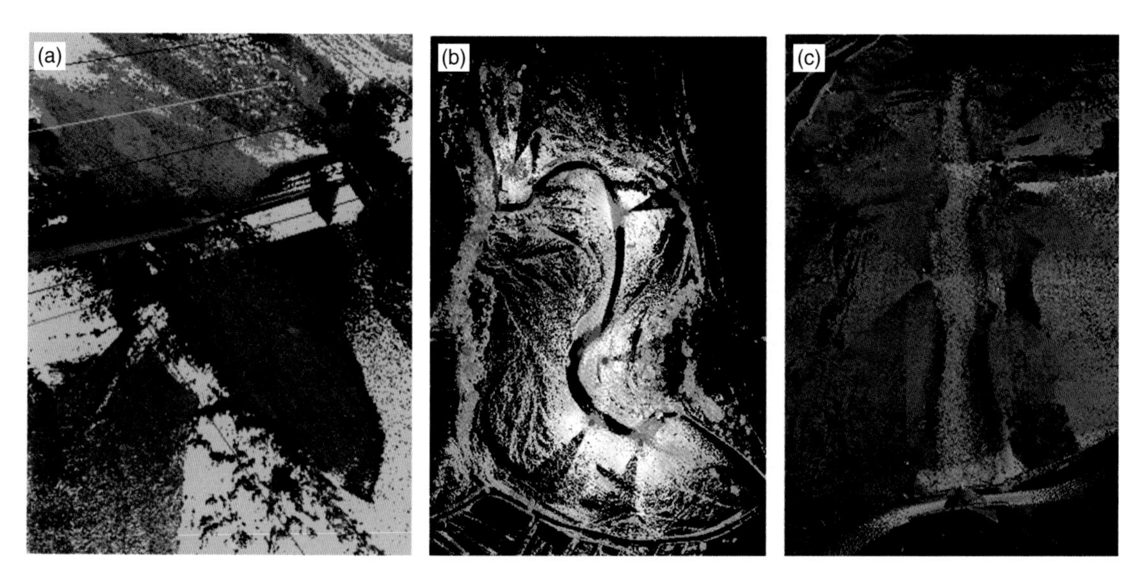

Plate 7.1 Combined laser scan cloud image of the point bar study site at (a) the River South Tyne at Lambley, (b) the River Coquet at Sharperton and (c) Kingsdale Beck, North Yorkshire.

Plate 7.2 Surface and photo of the point bar study site at the River South Tyne at Lambley.

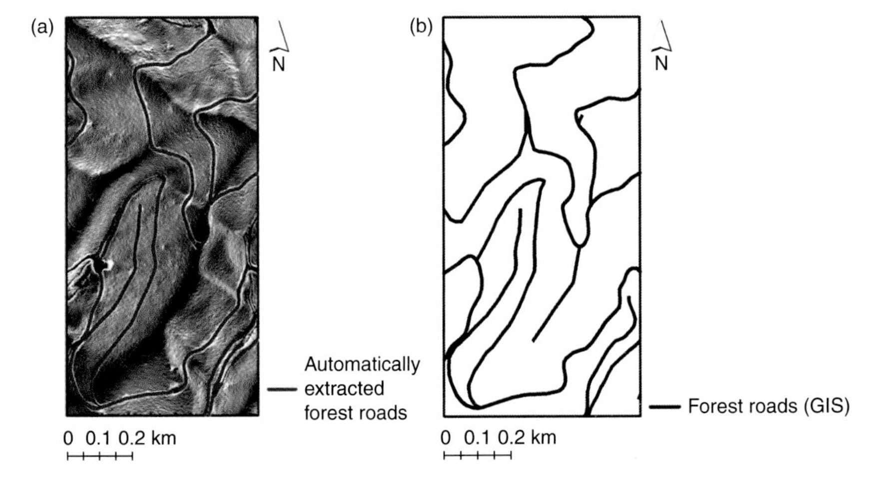

Plate 8.7 (a) Automatically extracted road segments using image processing techniques (shown as red lines) superimposed on the slope image, (b) Reference data (road features from the GIS of the Department of Forestry).

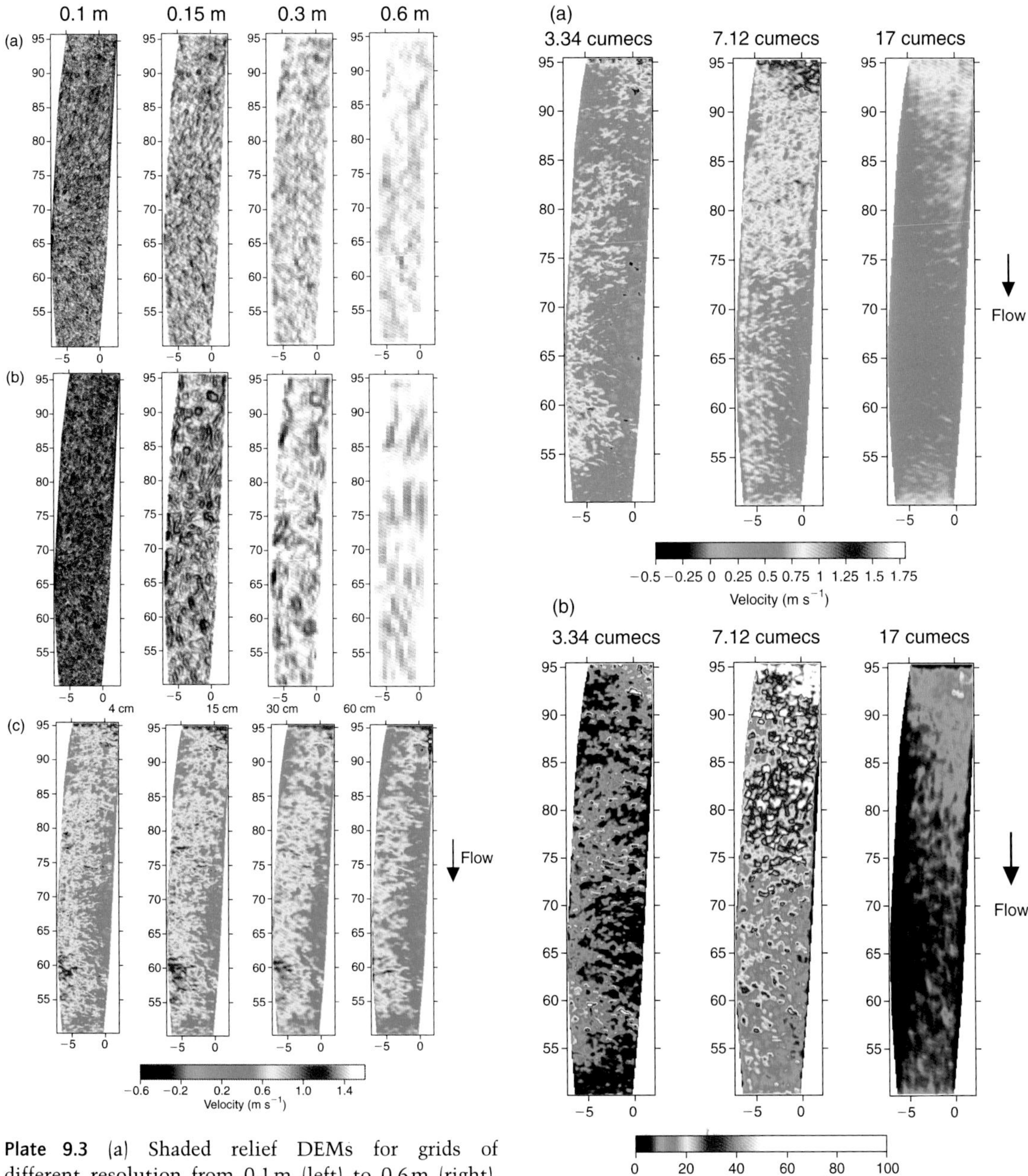

Plate 9.3 (a) Shaded relief DEMs for grids of different resolution from 0.1 m (left) to 0.6 m (right), (b) shaded relief maps for standard deviation, (c) results for near bed (10% flow depth from the bed) streamwise (x) velocity which have utilised various resolution topographic and roughness height information, for a discharge of $3.34\,m^3\,s^{-1}$.

Plate 9.5 Simulated (a) near-bed velocity and (b) shear stress for flows of $3.34\,m^3\,s^{-1}$, $7.12\,m^3\,s^{-1}$ and $17.00\,m^3\,s^{-1}$.

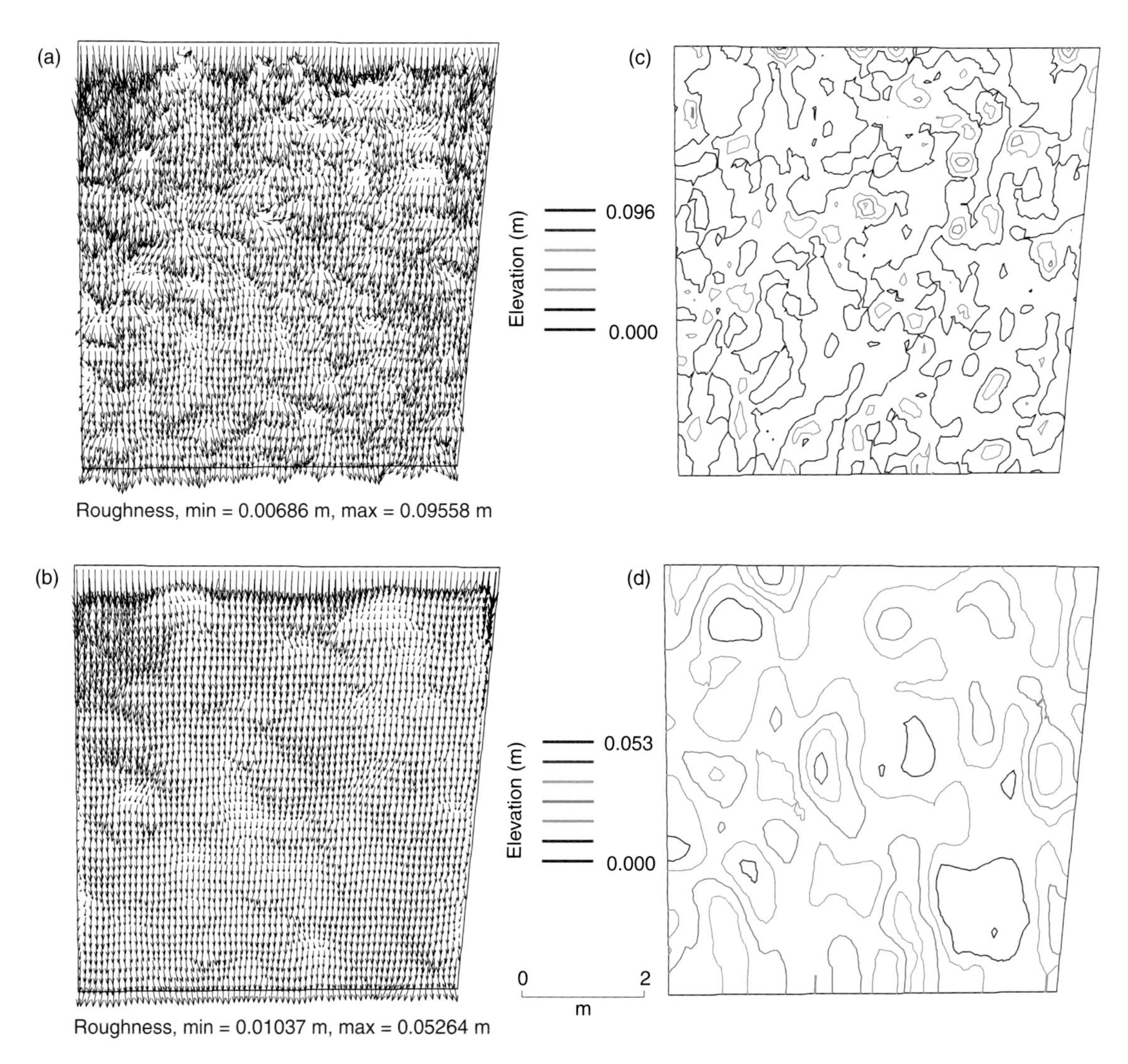

Plate 9.6 Planform near-bed flow vectors and roughness height maps for (a) and (b) 10 cm resolution grids, and (c) and (d) 30 cm resolution grids.

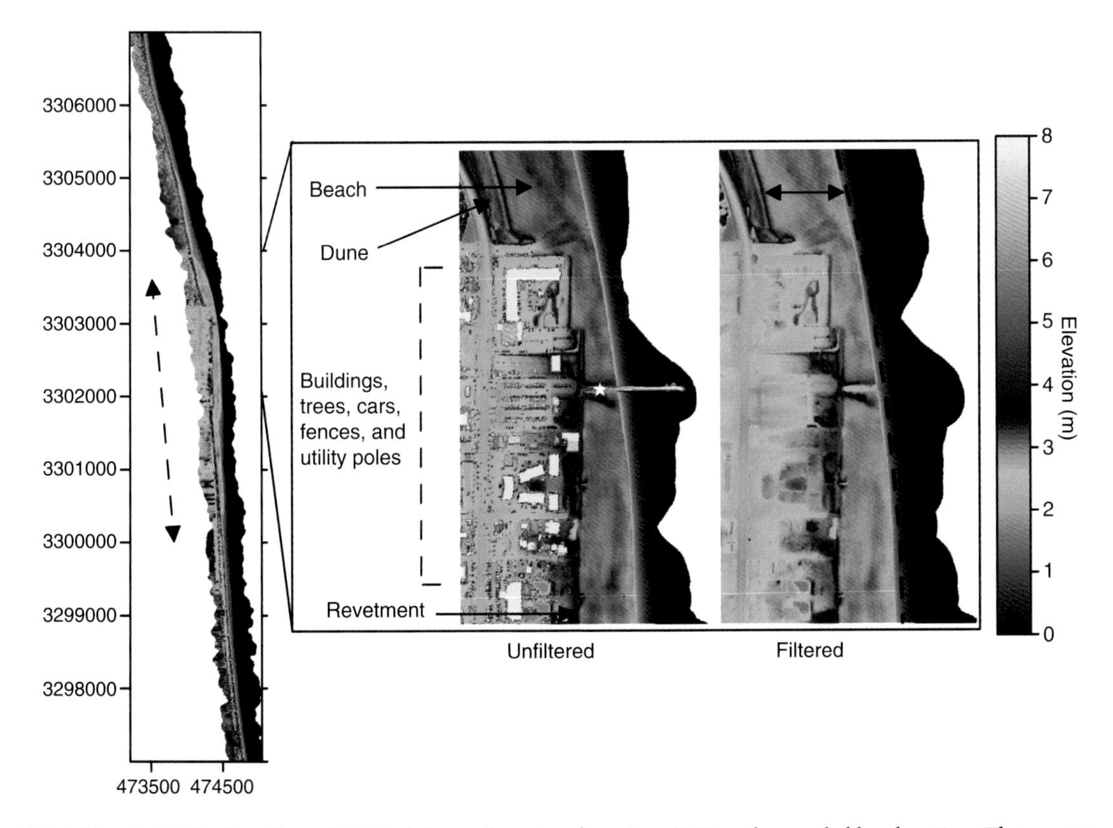

Plate 10.1 (Left) The 1 m filtered DEM of the study region from June 2006, colour coded by elevation. The *x*, *y* axes are in UTM Zone 17 metres, and the dashed line indicates the general direction of the flights. (Right) Zoomed in view of pier vicinity before and after filtering. The double-arrow on the right plot spans the beach from dune toe to shoreline (dashed line). The interpolated surface underneath the pier (indicated by the star) was excluded from the analysis since no LiDAR returns occur there. The filtering serves to remove LiDAR returns from non-surface objects on the beach, such as lifeguard towers, trash bins and people.

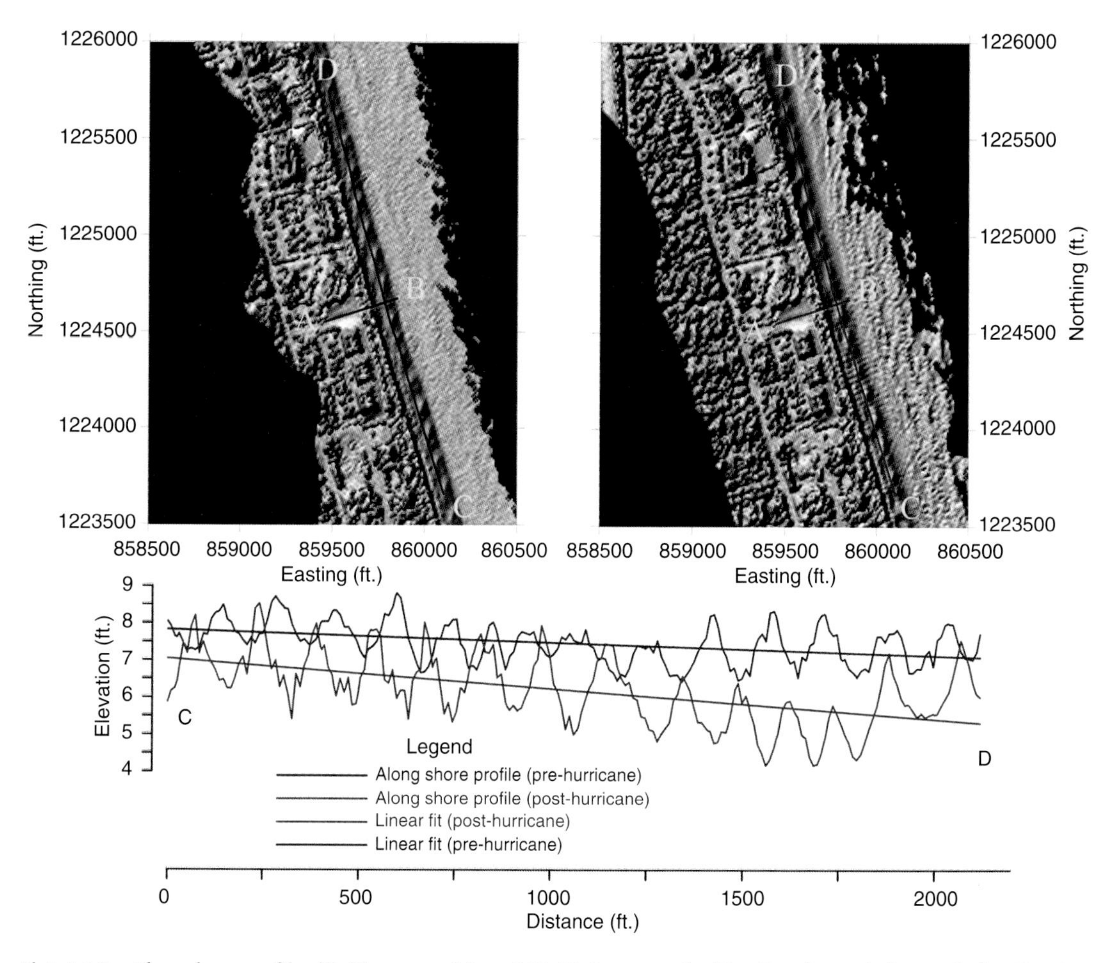

Plate 10.6 Alongshore profiles (D-C) extracted from LiDAR data over the Vero Beach area before and after Hurricane Irene. The left DEM is pre-hurricane and the right DEM is post-hurricane. In addition to a net reduction in elevation along the profile, modulation of the quasi-periodic topography is evident. Elevations in NAVD88 (ft). Reprinted from *ISPRS Journal of Photogrammetry and Remote Sensing*, 59, Shrestha RL, Carter WE, Sartori M, Luzum BJ and Slatton KC, Airborne laser Swath mapping: quantifying changes in sandy beaches over time scales of weeks to years, 222–232, 2005, with permission from Elsevier.

Plate 11.3 Projection plane view of the southwest face of the Nukhul syncline outcrop (see Figures 11.5–11.7 with a four times vertical exaggeration. Note the presence of a subtle anticline structure visible in the projection plane. This feature is not easily visible in the field, and is only clearly visible in vertically exaggerated data. The feature is also visible in the contour map in Figure 11.5.

Plate 11.4 Attribute analysis of LiDAR data. The scan data may be processed in order to extract more information. Parameters are (1) RGB colour, (2) Reflection Intensity, (3) Correlation between adjacent scan lines based on range from scanner (4) Surface roughness, (5) False colour intensity and (6) Dip of outcrop surface. Attribute analysis aids identification and analysis of subtle features hence aiding interpretation.

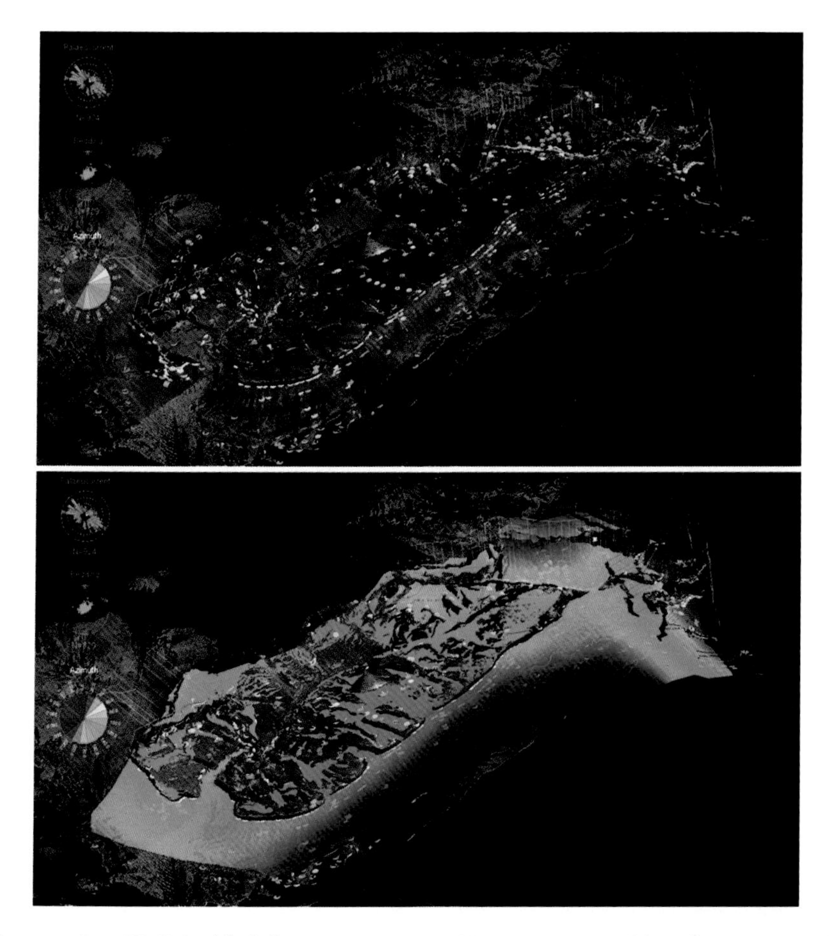

Plate 11.6 The complete Wadi Nukhul dataset comprising 85 scan stations, 924 palaeocurrents, 829 structural measurements and some 30 logs. The lowermost image shows the dataset with one of the hanging wall structural marker surfaces generated from the polyline and structural interpretations. Model is viewed from the north east.

Plate 12.8 Stixwould Priory and the possible causeway leading from the river (centre). LiDAR courtesy of Lincolnshire County Council: Source Environment Agency (March 2001).

Plate 12.10 Conventional aerial photograph of Welshbury Hillfort showing the dense tree canopy. Image courtesy of the Forestry Commission: Source Cambridge University Unit for Landscape Modelling (March 2004).

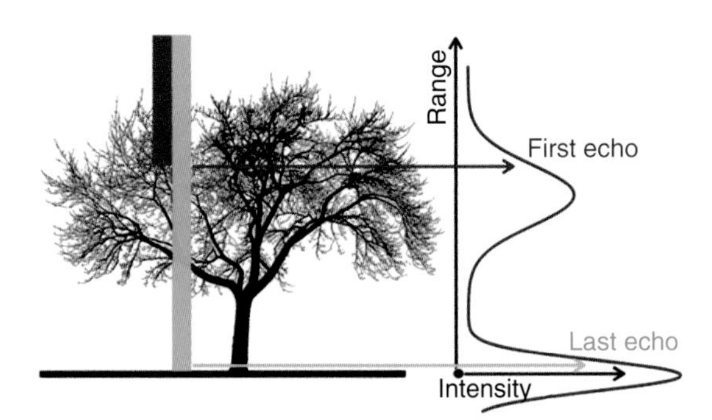

Plate 13.3 Illustration of waveform and triggering of first and last echo for a tree canopy.

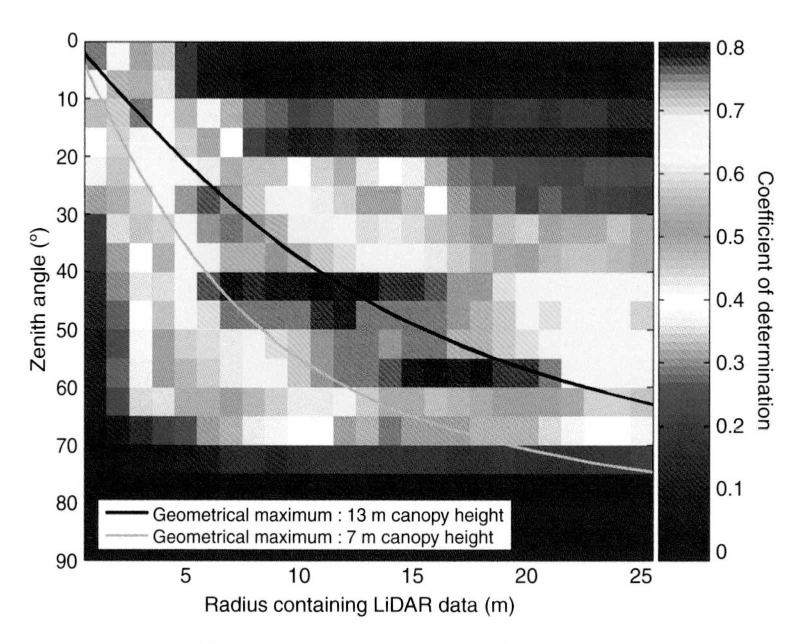

Plate 13.4 Matrix of coefficients of determination for regression of gap fraction from hemispherical photographs and ALS-derived fCover.

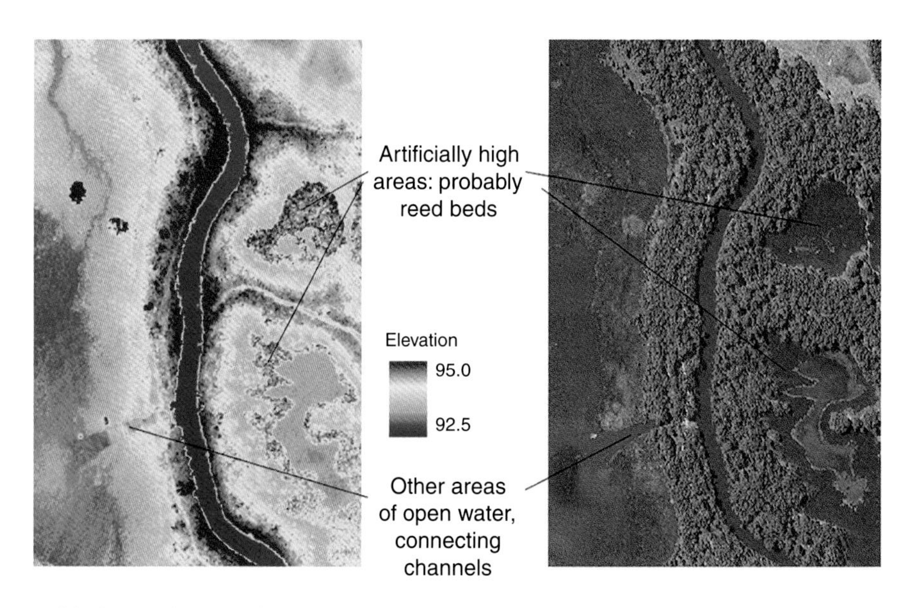

Plate 14.7 Reed beds in and around floodplain wetlands have been interpreted as ground elevations.

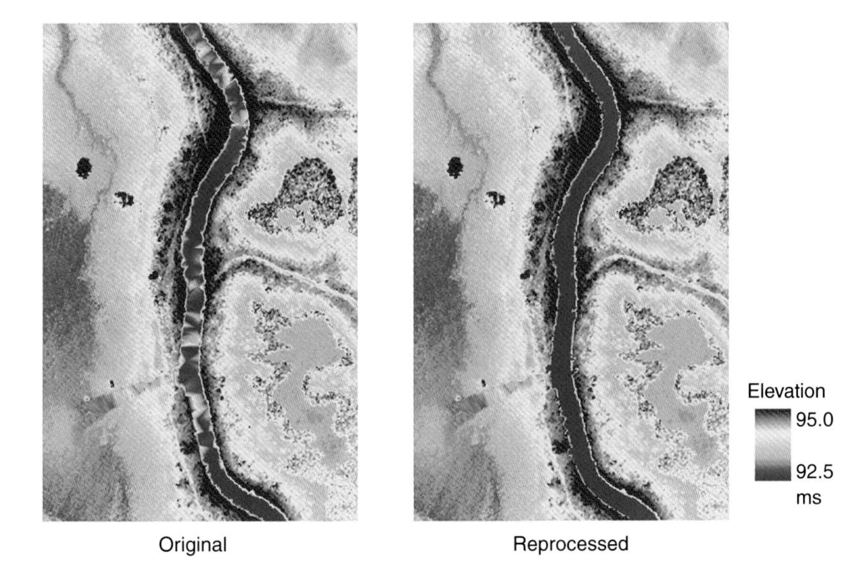

Plate 14.9 Elevations before (left) and after (right) the addition of river breaklines.

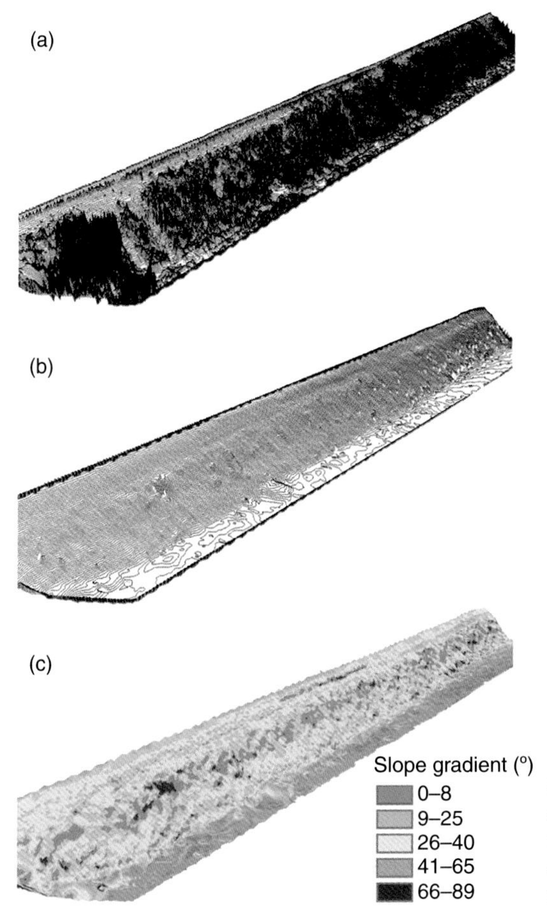

Plate 15.1 Utilisation of algorithms more commonly associated with aerial LiDAR allow vegetation on surfaces (a) to be removed (b) from TLS data and subsequent slope analysis to be performed (c), providing an effective tool for analysing slopes under vegetation.

Plate 15.3 Orthoimage of the Red Nabbs headland overlaid with a shapefile locating the spatial extent and position of changes recorded over the monitoring period. The headland was relatively stable throughout the monitored period with concentrations of change on vertical steps in the cliff face (Inset a) and towards the seaward edge of the site where the largest failure occurred (Inset b).

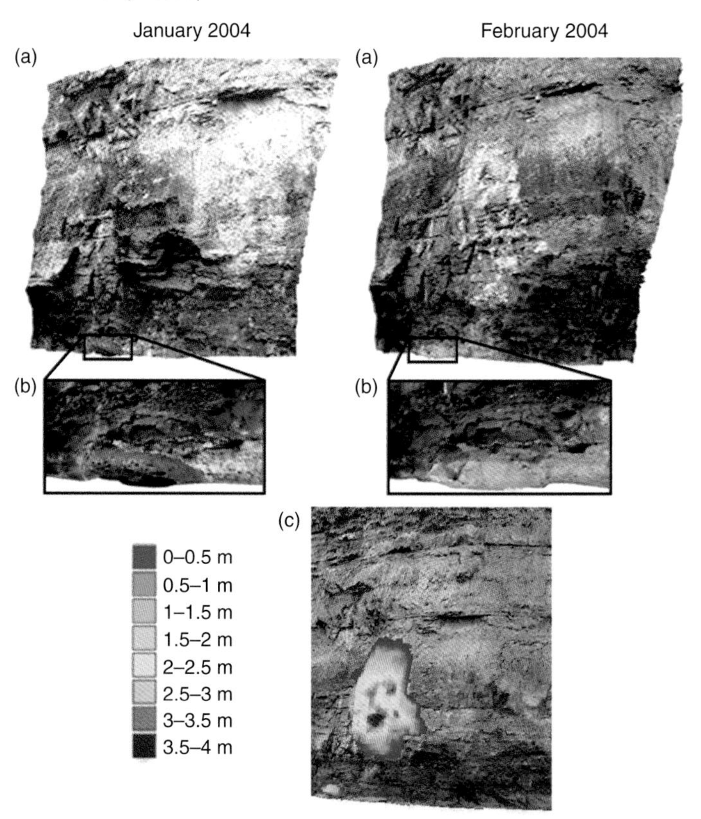

Plate 15.4 Sections of the surveys collected during January and February 2004 (a) illustrate a significant failure from undercut cliff material. An additional, smaller failure was revealed (b) through high-resolution differencing analysis (c). The LSS approach therefore indicates that the larger loss failed through a tangential sliding mechanism, truncating the adjacent cliff toe below in the process.

Point filtering

Because LiDAR observations consist of discrete returns from the ground and landcover, the raw data must be interpolated to represent the continuous ground surface. Prior to DEM generation, the point data are typically filtered to remove non-ground points due to such things as buildings, vegetation and other occluding objects (Slatton *et al.*, 2007). Many different filters have been proposed to remove non-ground points from LiDAR data, several of which are reviewed in Sithole and Vosselman (2004). In this work, the last-return points were filtered using the adaptive, multi-scale algorithm in Kampa and Slatton (2004) to remove non-ground points without altering the natural beach morphology. There are also commercial software packages that can be used with varying degrees of success to remove non-ground points, a summary of which is given in Fernandez *et al.* (2007).

Once the ground points were obtained through filtering and height biases between flights removed, the points were interpolated to 1 m × 1 m bare-earth elevation grids using ordinary kriging with a linear variogram model (Cressie, 1991). 1 m spacing can be achieved with high accuracy because the LiDAR system's point density is >1 point/m^2. A nugget effect of 0.07 m was added to capture measurement uncertainty for the LiDAR height error after adjustment. Figure 10.1 displays the filtered 1 m DEM for the June 2006 acquisition colour-coded by elevation and a zoomed in view near the pier showing the unfiltered result with buildings and vegetation and the filtered result with the majority of buildings and vegetation removed. In general, the filtering process will not eliminate all non-ground points. Vegetation that is both short and dense can be particularly difficult to remove completely since it greatly reduces the number of LiDAR returns from the ground. Fortunately, over beaches the non-ground objects that one typically must filter, such as lifeguard towers, people and trash bins, are quite easily removed. The interpolated elevation values underneath the pier in Figure 10.1 are excluded from the surface analysis since it is not possible to obtain LiDAR returns from underneath a solid object like a pier.

Shoreline delineation

There is no universally accepted method for delineating shorelines on beaches. Because of the dynamic nature of these systems, several proxies are used, depending on project needs and available data. These proxies include physical indicators such as a debris line or high-water line (HWL) or reference contours such as a tidal datum or zero-elevation (Stockdon *et al.*, 2002; Boak & Turner, 2005; Moore *et al.*, 2006). Traditionally, the HWL was used as a shoreline indicator as depicted on historical maps, and it is still commonly used today. The HWL is defined as the intersection of land with the water surface at high tide and is most often delineated by manual surveying or aerial photogrammetry. Additionally, many LiDAR systems record the return pulse intensity which can potentially be used to segment the HWL in LiDAR data (Starek *et al.*, 2007b). A problem with the HWL measurement is that it is a relative measurement dependent on date and time surveyed because the HWL varies as tide, waves and beach slope changes. The advantage of LiDAR data for shoreline delineation is that the user can accurately select a geo-referenced contour as an indicator for shoreline. Unlike the HWL, this creates a reliable common reference for comparing shoreline change from surveys conducted at different times (Stockdon *et al.*, 2002; Robertson *et al.*, 2004).

Because the LiDAR data were collected near low tide, the Mean Higher High Water (MHHW) tidal datum was selected as the proxy for shoreline in this study. MHHW is the average of the higher high water height of each tidal day observed over the tidal epoch. To obtain this contour, the offset between the airborne laser heights referenced to NAVD88 and the MHHW line referenced to the nearest tidal gauge's zero bench mark must be determined. The nearest NOAA tidal gauge, station 8720587, located near St Augustine Beach, was used for this work. The MHHW is 2.485 m above the station's zero level.

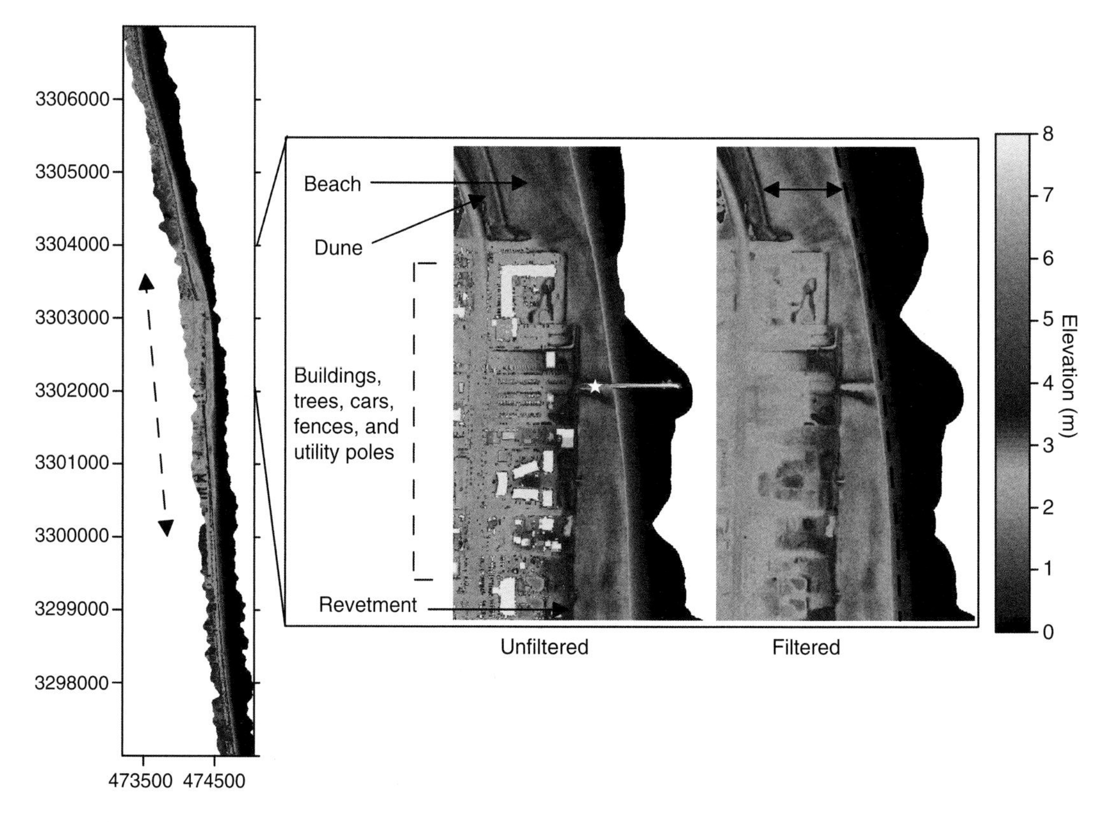

Fig. 10.1 (Left) The 1 m filtered DEM of the study region from June 2006, colour coded by elevation. The *x*, *y* axes are in UTM Zone 17 metres, and the dashed line indicates the general direction of the flights. (Right) Zoomed in view of pier vicinity before and after filtering. The double-arrow on the right plot spans the beach from dune toe to shoreline (dashed line). The interpolated surface underneath the pier (indicated by the star) was excluded from the analysis since no LiDAR returns occur there. The filtering serves to remove LiDAR returns from non-surface objects on the beach, such as lifeguard towers, trash bins and people. See Plate 10.1 for colour version of this image.

The NAVD88 datum is offset 1.870 m above the station's zero level. Because the LiDAR elevations are referenced to NAVD88, the 1.870 m offset was subtracted from MHHW to determine the corresponding MHHW elevation relative to the LiDAR elevations ($2.485 - 1.870 \approx 0.6$ m). Thus, the 0.6 m elevation line was contoured in each filtered DEM to delineate shoreline. Although not as common, methods other than contouring have been proposed to delineate the shoreline from LiDAR data or DEM (e.g. Stockdon *et al.*, 2002; Liu *et al.*, 2007).

Data mining

Because our goal is to determine which beach morphologies are most indicative of shoreline change, we must now construct a method to systematically extract several morphological features from the LiDAR data. The natural coordinate system traditionally used for studying shoreline change consists of a local 2D Cartesian system oriented with alongshore and cross-shore axes (Dean & Dalrymple, 2002). A computer program was developed (Starek *et al.*, 2007a,b) to sample the LiDAR-derived DEMs by extracting elevation values along cross-shore profile lines oriented roughly orthogonal to the shoreline. This provides the *x,y*-coordinates and elevation values along each profile at a user-defined spacing. To orient the profiles orthogonal to shoreline, the algorithm fits a piecewise trend line through the shoreline contour to determine the angle and direction of a landward baseline from which to extract the cross-shore profiles. Baselines can be aligned to match historical survey lines if desired, but it is important that the same baselines are used when comparing profile measurements across data sets. Once baselines are selected, the algorithm can then extract profiles at various intervals along the shore and calculate the intersection of each profile with the shoreline contour and other surface features (such as the fore dune).

For this analysis, the profiles were extracted every 5 m in the alongshore direction and they extend in the cross-shore from the dune toe line to the MHHW shoreline (see Figure 10.2). This provided a total of 2010 profiles for each data set.

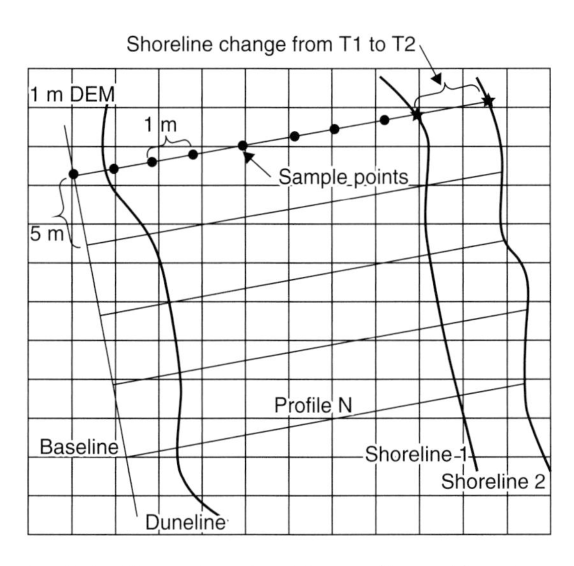

Fig. 10.2 Illustration of profile sampling and how temporal shoreline change was calculated from profiles extracted from lidar data acquired at time T1 and a later time T2. Inter-profile spacing was set at 5 m (not drawn to scale).

A 1D moving average filter of length 5 m is applied to each cross-shore profile to reduce high-frequency artefacts that may sometimes be present due to LiDAR scan line artefacts or isolated non-ground points that the filter failed to remove.

The temporal change in shoreline between the data acquisition periods listed in Table 10.1 was quantified using pairs of profiles by computing the distance between the shoreline position *x,y*-coordinates (Figure 10.2). Profiles were then segmented into binary classes, C_1 for 'erosion' or C_2 for 'accretion', depending on whether the shoreline had moved landward or seaward in that period. For instance, if a hypothetical Profile 1 in March 2005 experienced negative shoreline change for the March 2005 to May 2005 epoch, it was classified as an 'erosion' profile. Features extracted at a given time can therefore be interpreted as 'leading indicators'. For some epochs, only erosion or accretion occurred. In those cases, class C_1 still corresponded to the most negative shoreline change (high erosion or low accretion), and C_2 corresponded to the most positive shoreline change (low erosion or high accretion). In such cases, the median value in shoreline change was used as the decision rule to segment the profiles into C_1 or C_2.

Feature extraction

After the profiles are segmented into classes, the cross-shore morphologic features f from each profile l can be extracted. As an example, we obtain $f_1(l_m, C_1)$ for feature 1 and class 1, where m indexes the profiles. Results are accumulated over all features, profiles and both classes for each epoch. This approach allows us to mine the data by extracting several different features per profile and then to examine interclass separation and class-conditional probabilities $p(f_j | C_i)$ for $i = \{1,2\}$, where j indexes the feature set. A total of nine features were chosen based on their historical or suspected potential as indicators of shoreline change. Note that the features presented are only a subset of the many that could be extracted from the LiDAR data. If one chose additional features to investigate, their value could be assessed in the same manner as presented here.

Beach slope (m/m) was estimated using a linear fit through the elevations in each profile and represents the sub-aerial beach slope from dune toe to MHHW shoreline. For more steeply sloping beaches, the slope is sometimes measured only from the berm line seaward rather than dune line. *Near-shoreline slope* (m/m) measures beach steepness from the MHHW shoreline to a few metres landward. It was created to approximate the slope near the intertidal-wave run up zone. *Beach width* (m) is the width of the sub-aerial beach from dune toe to MHHW shoreline. It should be noted that this is instantaneous beach width at the time of a given LiDAR acquisition and should not be confused with the subsequent change in beach width that determines which class a given profile belongs to (C_1 or C_2). *Volume-per-width* (m^3/m) is the sub-aerial volume per unit width above 0m NAVD88 computed using trapezoidal integration. *Mean elevation gradient* (m/m) is the mean of the gradient estimated every 3m along a profile using a central-difference approximation. *Mean elevation curvature* (m/m^2) is the mean value of the gradient of the gradient. *Standard deviation of height* (m) is the standard deviation of the elevation values for the profile, which provides an estimate of surface roughness for that profile. *Deviation-from-trend* (m) is the orthogonal distance the shoreline deviates from the natural trend of the beach in the horizontal over the scale of a few kilometres and is a measure created to capture the deviation of the highly erosive pier region. The beach trend was estimated by a linear fit through the shoreline contour points, excluding the roughly two kilometre armoured section of St Augustine Beach that includes the pier. *Orientation* is the angle in degrees that the local shoreline makes relative to azimuth (i.e. clockwise from north). The local shoreline orientation was estimated using piecewise linear fitting.

As an example, Figure 10.3 displays the high alongshore variation in beach width, mean gradient and mean curvature extracted from the August 2003 profiles. The DEM presented on the left provides an understanding of alongshore scale. Referring to the beach width plot, notice the rise, sharp decrease and then subsequent increase in width over the range of 3000–5000m alongshore. This is an effect of the nourishment project on either side of the pier earlier that year. The reaction of gradient and curvature in the vicinity of the pier is due to the short width imposed by the revetment wall and large berm created from the beach nourishment. The short width results in less sample points and the large berm results in greater magnitudes in gradient and curvature. Because these measures are mean values, this combination of effects results in higher signed values for the profiles in this region. In contrast, notice the high mean gradients in the range of 1000–2000m alongshore, but the curvature does not react in this area. From the DEM, it was visually confirmed that this region has a large berm with short width but has a higher slope and is less uniform in the cross-shore than in the nourishment zone. Therefore, the non-uniformity and greater slope in this region is the likely cause for the disparity in behaviour between the two metrics.

As explained previously, features are extracted from the values along a profile line as opposed to using a moving 2D window or simple data clustering. This approach was selected for consistency with the standard practices used in coastal erosion studies, which utilize cross-shore and alongshore measurements. The major difference is that LiDAR provides several orders of magnitude increase in profile sampling density compared to traditional manual surveying methods. Figure 10.4 presents shoreline change for the October 2005 to January 2006 epoch. Shoreline change was calculated based on the LiDAR DEMs for profiles extracted every 5m and for profiles extracted every 300m (~1000ft), which is the standard spacing for manual transects for the State of Florida shoreline surveys (Foster *et al.*, 2000). As observed in Figure 10.4, general trends are captured in both survey resolutions, but there is significant loss of information relative to local features when using a 300m spacing. This small scale information is critical

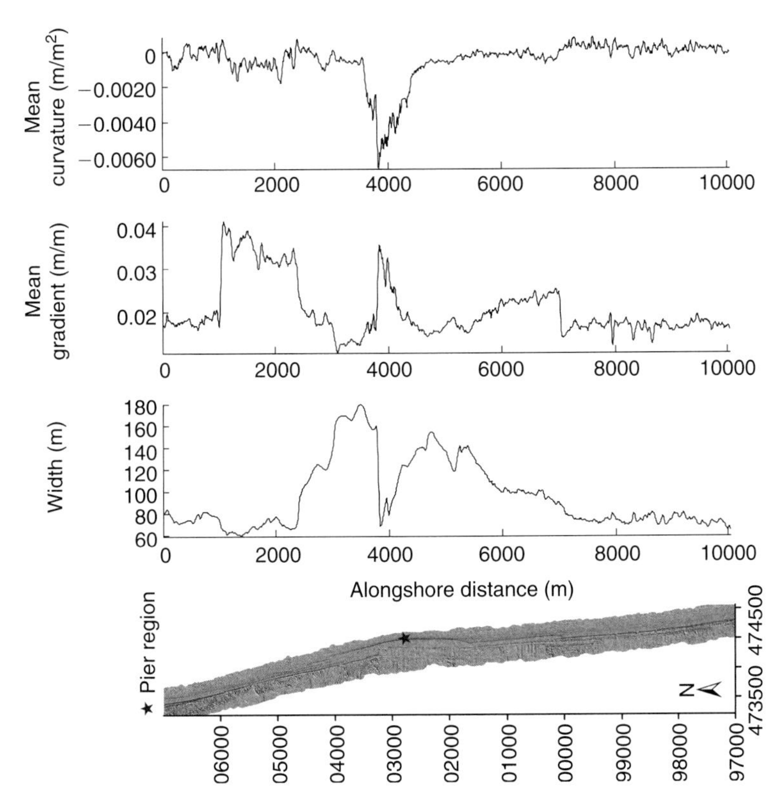

Fig. 10.3 Three features extracted from the August 2003 profiles; *y* axes are the relative alongshore distance (m). The DEM on left provides an understanding of the alongshore scale relative to the study area with *x*, *y* axes in UTM Zone 17 metres. The reduction in beach width near the alongshore distance of 4000 m coincides with the beginning of the revetment wall in the area of St Augustine pier.

when assessing local 'hot spots' of anomalous erosion, particularly near expensive development and civil infrastructure.

Shoreline change

Table 10.2 displays shoreline change statistics quantified for each data epoch using 1 m cross-shore profile sampling. Notice the extensive erosion in Epoch 1 (August 2003 to March 2005), approximately two years after the 2003 nourishment. The rapid loss of shoreline is mainly due to the anomalous Florida hurricane season of 2004. The 2005 nourishment to reclaim the lost beach can be observed in the large accretion values for Epoch 3 (May to October 2005). In the feature analysis, the around 2 km nourishment zone is excluded for impacted epochs.

Analysis

As explained previously, several cross-shore features were extracted from each profile and partitioned into 'more erosion tending' or 'more accretion tending' classes (C_1 or C_2) depending on the shoreline change experienced during the epoch following a particular LiDAR acquisition. To measure a feature's importance, we estimate class-conditional pdfs (likelihoods) for each feature–class pair, $P(f_j|C_i)$, and then rank

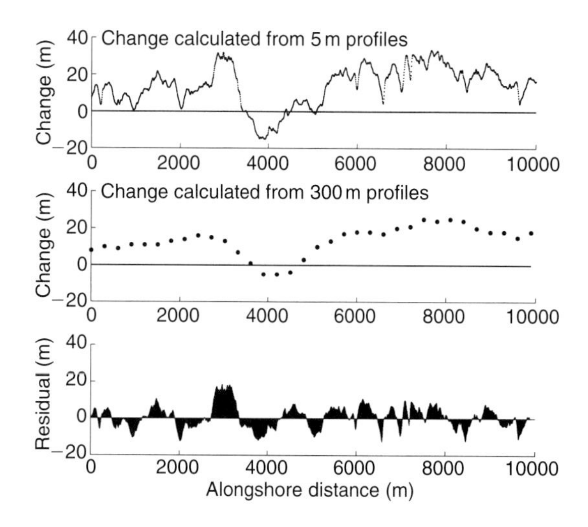

Fig. 10.4 Shoreline change plots between October 2005 and January 2006 showing the added information of the 5 m alongshore sampling relative to profiles located at intervals of 300 m, as is common for manual surveys. (Top) Shoreline change computed from LiDAR profiles every 5 m; (Middle) Same data with profiles taken every 300 m to simulate manual survey; (Bottom) Differences resulting from 5 m sampling versus linearly interpolated 300 m sampling.

Table 10.2 Shoreline change statistics for 10 km study area. Negative values refer to erosion and positive refer to accretion. In Epoch 1 the entire shoreline eroded, so the maximum value refers to the smallest negative change in shoreline.

Time period	Mean (m)	Min (m)	Max (m)	Std dev (+/− m)
Epoch 1	−60.96	−135.01	−10.40	26.49
Epoch 2	−0.65	−17.60	12.00	5.72
Epoch 3	33.53	−5.40	155.00	42.49
Epoch 4	0.63	−54.60	47.19	17.40
Epoch 5	13.99	−15.60	35.00	10.95
Epoch 6	−4.99	−39.20	23.20	16.20

interclass separation using pdf divergence measures (Starek *et al.*, 2007a,b). Generally, the more separation between class likelihoods for a given feature f_j, the more likely it is that classification will be successful. Consider a two-class decision problem with likelihoods $P(f_j|C_1)$ and $P(f_j|C_2)$. Classification using f_j is likely to be better than for other features if f_j yields the largest Euclidean distance between the class means. The relative performance will be even better if f_j yields the largest Mahalanobis distance, which is the Euclidean distance normalized by the standard deviations (Duda *et al.*, 2001).

Entropy is an information-theoretic measure of uncertainty for a random variable. It is conceptually similar to standard deviation, but encapsulates all information in the pdf rather than just the second-order statistical moment. The simplest divergence measure, the well-known Kullback–Leibler divergence $D_{kl}(P(x), Q(x))$, quantifies the relative entropy between two pdfs $P(x)$ and $Q(x)$ over the same random variable (Cover & Thomas, 2006). A large value for the relative entropy implies that a feature provides good separability between classes when we let $P(x) \rightarrow P(f_j|C_1)$ and $Q(x) \rightarrow P(f_j|C_2)$. In our case, features yielding greater divergence are more indicative of the 'erosion' or 'accretion' patterns observed along the beach. Because divergences are based on functional differences between pdfs, they serve as highly robust measures of class separability even when geometric measures of separation, such as Euclidean distance or Mahalanobis distance, fail due to similarity in themeans. While it is possible that feature transformation methods, such as principal components, could lead to features that produce even greater class separation, such methods were not used here because our intent was to assess the value of features relative to each other and retain clear physical meaning in the features for greater benefit to coastal researchers.

Because D_{kl} is not strictly a true metric in the mathematical sense, two metric forms of divergence are used to estimate feature separation between classes: Jensen–Shannon divergence (JSD) and a normalized form of Jensen–Shannon divergence (NJSD). NJSD was selected to determine if normalization by each feature's entropy had an effect on the separation between shore change classes. For an in-depth discussion on the feature ranking approach and the mathematical formulae the reader can refer to Starek *et al.* (2007a).

For comparison purposes, the squared correlation coefficient (R^2) was also computed to evaluate the second-order statistical relationships between the features and shoreline change. In general, a feature that is well correlated with measured shoreline change (i.e. change viewed as a random variable rather than a binary class label) would be expected to also exhibit a significant divergence in the class-conditional pdfs. Instances in which the feature ranking from the divergences is different from those suggested by correlation indicate cases where using the full pdf instead of merely a second-order description is important. For more details on information-theoretic measures the reader is referred to the authoritative treatment in the book by Cover and Thomas (2006).

The data in feature space can be multi-modal and non-Gaussian, as is often the case with LiDAR data. Therefore, the non-parametric Parzen windowing method was employed to estimate the pdfs from the data (Duda *et al.*, 2001). The size of the Parzen window in feature space, however, can strongly affect the accuracy of the estimated pdfs. The Parzen window should be small enough to capture the details of a pdf but not so small as to create artefacts due to a limited number of samples in the window. Windows that are too large yield pdf estimates that are overly smooth, and windows that are too small overfit the samples yielding pdf estimates that are spiky in appearance. Unfortunately, there is no universal procedure for determining the optimal size. In the general case, one can start with a large Gaussian window and progressively decrease it to allow more resolution in the pdf estimates. The final window size can be selected as that which is just large enough to avoid pdf estimates that exhibit sharp local gaps where the probability is close to zero. For this work, it was empirically determined using this procedure that a window size of 1/40th the range of values for each feature provided the best compromise between smoothing and over-fitting. It is the LiDAR technology that enables robust pdf estimation for terrain features because its high sampling rate reduces the sensitivity of the Parzen pdf estimate to window size.

Results

The overall rankings for each measure were computed by averaging the values across all epochs. Divergence metrics, JSD and NJSD, yielded the same average rankings and ranked deviation-from-trend (DT) highest followed by slope (S), width (W) and then volume-per-width (V). The high performance of DT is significant in that the pier region's deviation from the natural trend is a contributing factor to it being an erosion 'hot spot' (Foster *et al.*, 2000). Correlation, R^2, ranks DT highest followed by V, W, then curvature (C). These results suggest that the inclusion of the entire pdf is important for determining the relative importance of the S and C features. Overall, orientation ranked lowest and performed poorly even when the nourishment region was excluded. A likely explanation is that orientation influences on breaking wave angles and shoreline change are scale dependent, manifesting only at very large alongshore distances compared to the alongshore distances considered in this work.

To assess the classification potential of these cross-shore features for characterizing segments of beach more prone to erosion or accretion we implement a two-class naïve Bayes classifier. The Bayes classifier determines the posterior probability of a sample belonging to a class C given feature values $f \rightarrow P(C|f_1,...,f_n)$. In this example, C is either class 'erosion' or class 'accretion' defined previously. The set $(f_1,...,f_n)$ are the values of the cross-shore features for the specific profile we are attempting to classify. For a naïve Bayes classifier, one assumes conditional independence among the features (Duda *et al.*, 2001). This greatly simplifies the problem by estimating the likelihoods for one feature at a time rather than the joint (nine dimensional in our case) pdf. The assumption of independence will suffice in this case because we are concerned with each feature's individual relationship to shoreline variation and overall ranking relative to other features.

Each classifier is trained by estimating the class conditional pdfs, $P(f_n|C)$ using the extracted feature values. In this example, we test the classifier on each epoch having trained it using the data from the other epochs. For instance, if we are testing on Epoch 1, we train the classifiers using all other epochs and compare results to the actual change measured for Epoch 1. Only the top two ranked features for divergence and correlation are used to classify. There are 2010 profiles for each period to test except for Epoch 6, which has 852 profiles. Results for the classifiers are shown in Table 10.3 where the numerical values are the success rates (number of correctly classified profiles divided by the total number of profiles).

As observed in Table 10.3, the top two ranked features selected by the divergence metrics, JSD/NJSD, outperformed the correlation coefficient with an average success rate of 73% supporting the utility of the divergence method. Success rates of greater than 80% were achieved for some periods and 83% for Epoch 1. Overall, the results are very promising; however, they must be understood in the context for which they are intended. These predictions are based solely on probabilities of class occurrence for a given morphology 'learned' from the laser data with no direct inclusion of time, bathymetry or the governing physics. Therefore, these classification results are data dependent and not intended as a stand-alone tool for predicting change. Rather, the success of the results demonstrates that certain morphologies can be systematically extracted from LiDAR data sets and ranked to provide useful information about beach change dynamics, supporting the notion of morphologic change indicators and their potential power for beach characterization. With longer time series of LiDAR observations, feature extraction and pattern analysis methods should enable coastal researchers to better quantify and characterize coastal change.

Table 10.3 Illustrative results from a naïve Bayes classifier using just the top two ranked features according to pdf divergence and according to correlation (R^2). The average classification based on features ranked using divergence outperforms that based on features ranked using correlation by approximately 7%. In all six epochs, the divergence method outperforms the correlation method. DT = deviation from trend, S = slope and V = volume per width.

	Success Rate (# correct/total)		
Data set	Divergence method (DT, S)	R^2 method (DT, V)	Difference
Epoch 1	0.83	0.80	0.03
Epoch 2	0.64	0.63	0.01
Epoch 3	0.62	0.47	0.15
Epoch 4	0.80	0.79	0.01
Epoch 5	0.66	0.62	0.04
Epoch 6	0.81	0.67	0.14
Avg. success rate	**0.73**	**0.66**	**0.07**
Max. success rate difference			**0.15**

CASE STUDY 2: CALIBRATION OF A SPREADING MODEL

So called 'beach fill' models are used by coastal engineers to simulate the spatial spreading of sand over time deposited during beach nourishment. Here, we use the LiDAR data to calibrate one such type of model, the 'point fill', for the 2003 nourishment of St Augustine Beach. The point fill approach conceptualizes the nourishment as a hypothetical fill where all initial sand is deposited at a single point. This can be modelled mathematically using a Dirac delta function and solved in relation to the one-dimensional diffusion equation. The analytical solution yields a Gaussian model based on the longshore diffusivity (Dean & Dalrymple, 2002):

$$y(x,t) = M \Big/ \sqrt{4\pi Gt} * e^{-\left(x/\sqrt{4\pi Gt}\right)^2} \tag{10.1}$$

where M is initial planform area (m^2), G is longshore diffusivity parameter (m^2/s) with values of $0.002 < G < 0.014$ being typical for the east coast of Florida, t is time in seconds and is an arbitrary value initially to fit the curve, x is alongshore distance (m) and y is shoreline displacement (m).

This type of model is referred to as a one-line planform model because it describes the time history of the shoreline position relative to an alongshore baseline (Dean & Dalrymple, 2002). Furthermore, Equation (10.1) disperses, not erodes, the sediment over time. Although the model is not likely to be representative of a real beach fill, for practical purposes it is convenient because it requires no time series wave data or bathymetry and can be extended to approximate more realistic beach fills. It only requires initial values for diffusivity G and planform area M, where M can be estimated from the LiDAR data.

To calibrate the model, the August 2003 LiDAR data set, acquired shortly after completion of the 2003 beach nourishment, was used. A baseline is oriented parallel to the shoreline and extends the entire region of nourishment. This baseline forms our alongshore basis x, where $x = 0$ is the location of maximum shoreline displacement (Figure 10.5a). The initial shoreline displacement, y, is then calculated as the orthogonal distance from this baseline to the August 2003 shoreline. To fit a Gaussian to the data using Equation (10.1), the model is log-transformed and put in linear form. A least-squares approximation can then be used to estimate the unknown parameters.

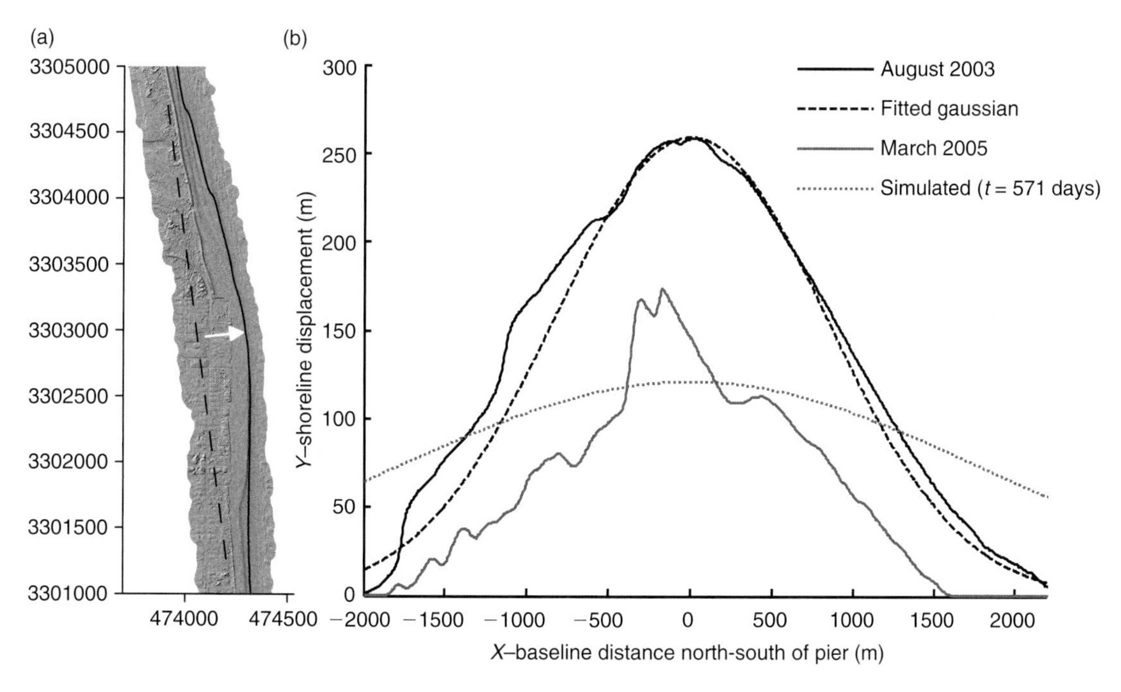

Fig. 10.5 (a) Top view of the August 2003 DEM in the region of nourishment showing the initial shoreline position. The alongshore distance (x) and orthogonal shoreline displacement (y) are measured relative to the baseline (dashed line). Left, right axes are in UTM Zone 17 metres. (b) The August 2003 shoreline displacement and initial (fitted) Gaussian, and the March 2005 displacement and simulated displacement after $t = 571$ days (time from August 2003 to March 2005 survey).

This provides a best-fit Gaussian model to the initial shoreline displacement as follows:

Log-transform to linearize

$$\ln(y) = \ln\left(M/\sqrt{4\pi Gt}\right) + \ln e^{-\left(x/\sqrt{4\pi Gt}\right)^2} \quad (10.2)$$

$$\ln(y) = \ln\left(M/\sqrt{4\pi Gt}\right) - \left(x/\sqrt{4\pi Gt}\right)^2 \quad (10.3)$$

let $C = M/\sqrt{4\pi Gt}, B = -1/(4\pi Gt), x = (x)^2$

At $x = 0$, C = max displacement of shoreline:

$$y(x=0,t) = M/\sqrt{4\pi Gt} * e^{-\left(0/\sqrt{4\pi Gt}\right)^2} = M/\sqrt{4\pi Gt} = C$$

Now rewrite in least-squares form because C is known:

$$\ln(y) - \ln(C) = Bx + V \quad (10.4)$$

where V is residual.

B can then be determined using least squares and back calculated to get planform area M. Figure 10.5b shows the fitted Gaussian to the August 2003 shoreline using M estimated from the data. As observed in Figure 10.5b, the Gaussian model is a reasonable approximation for the initial shape of the fill. To estimate the diffusivity parameter G we could refit the model using the March 2005 or May 2005 data and then use the change in y and elapsed time to solve for G; however, our goal is to simulate and compare to the measured change. Therefore G was set to $0.010\,m^2/s$ based on empirical values for the region from Dean (2003). Figure 10.5b shows a simulation run for t = 571 days (time between August 2003 and March 2005 survey) to compare with the actual shoreline displacement measured on March 2005.

Notice the smoothness and flattening of the Gaussian fill over time relative to the measured change in Figure 10.5b. Recall that this model is based on the diffusion relationship and therefore, all sediment is evenly spread away (alongshore) from the peak over time with no actual loss, as the total planform area must remain constant. Although the simulation results differ significantly in terms of spreading and smoothness relative to the measured change, the model is reasonable at capturing the general behaviour around the focus of deposition at $x = 0$. This ideal case is intended to provide an example of an analytic process model calibrated using LiDAR data. In general, a fill evolution model of this nature is unrealistic for most applications, but its simplicity does prove useful in certain situations where the nourishment can be modelled as a point source. The benefit of the high-resolution LiDAR data is that the parameters can be estimated directly using a regression fit with high degrees of freedom.

CASE STUDY 3: HURRICANE IMPACTS

Quantifying beach modification by hurricanes is critical not only for the accurate assessment of damage, but for improved prediction and mitigation of the impact on coastal communities. In 1999, Vero Beach, a resort community on the Atlantic coast of Florida south of Cape Canaveral, was mapped twice by an older topographic LiDAR system operating at 10,000 pulses per second, once on October 3 and again on November 18, to evaluate beach modification by Hurricane Irene. Irene passed through the Florida Keys and up the East Coast on October 15 and 16. The flight parameters consisted of a flying height of 900 m, flying speed of 60 m/s and approximate point spacing on the ground of 3 m. Small sections of the pre and post LiDAR DEMs for Vero Beach are shown in Figure 10.6 (Shrestha *et al.*, 2005). The alongshore profiles are plotted for pre-hurricane and post-hurricane. A systematic reduction in beach elevation of roughly 0.91 m (1.5 ft) can be seen along the profile. A modulation of the phase and frequency of the quasi-periodic beach topography is also evident. In Figure 10.7, changes in beach volume are computed directly over irregular polygons using the LiDAR-derived DEMs using three different integration methods. The 'cut volume' denotes

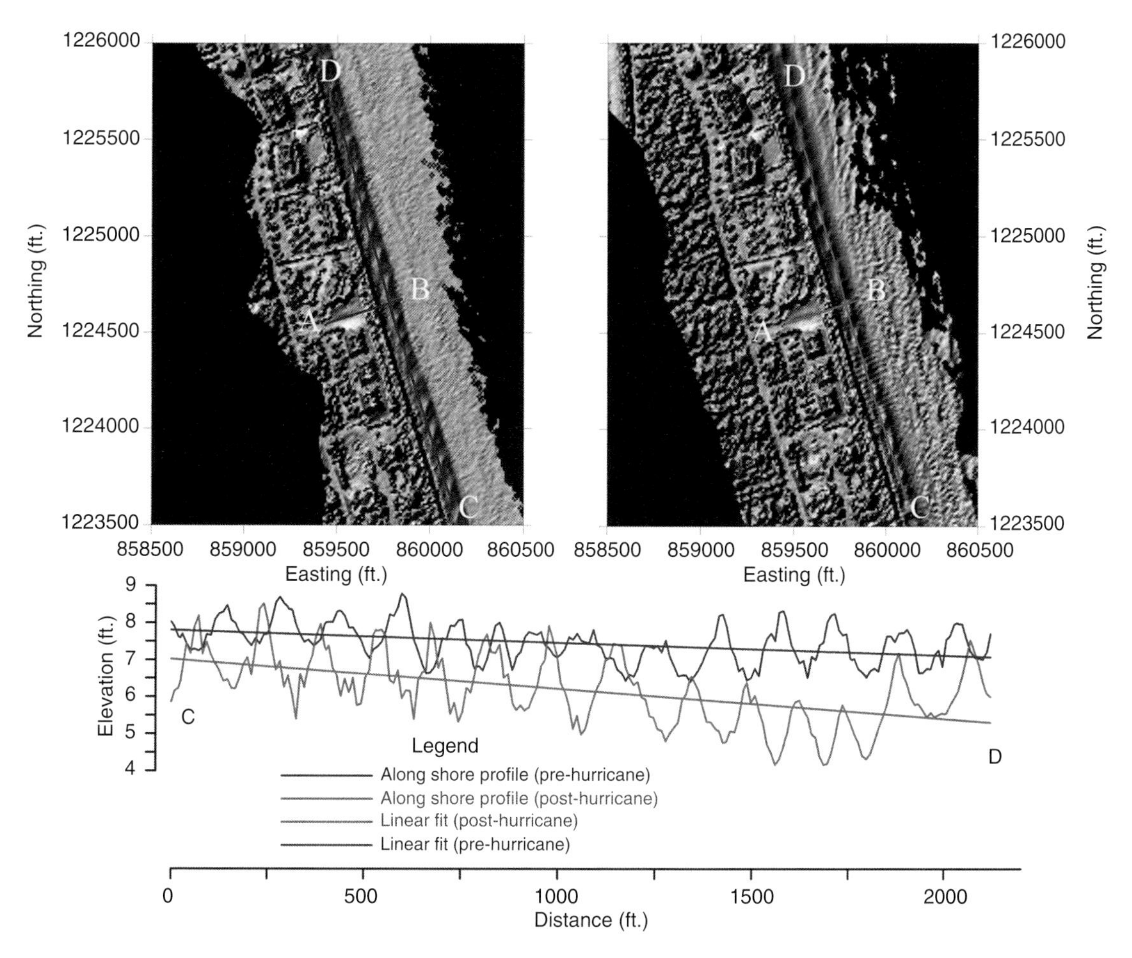

Fig. 10.6 Alongshore profiles (D-C) extracted from LiDAR data over the Vero Beach area before and after Hurricane Irene. The left DEM is pre-hurricane and the right DEM is post-hurricane. In addition to a net reduction in elevation along the profile, modulation of the quasi-periodic topography is evident. Elevations in NAVD88 (ft). Reprinted from *ISPRS Journal of Photogrammetry and Remote Sensing*, 59, Shrestha RL, Carter WE, Sartori M, Luzum BJ and Slatton KC, Airborne laser Swath mapping: quantifying changes in sandy beaches over time scales of weeks to years, 222–232, 2005, with permission from Elsevier. See Plate 10.6 for a colour version of these images.

erosion volume due to Hurricane Irene. Note that the plots and volume computations in Figures 10.6 and 10.7 (used with permission from Shrestha *et al.*, 2005) are presented in (ft) because the State of Florida employs a State Plane (ft) coordinate system for coastal mapping and surveying. Elevations are referenced to NAVD88 (ft). Having the high resolution images of beach elevation that LiDAR provides allows a more interdisciplinary approach to solving a wide range of problems because the data can be used by many different groups for distinct, yet complementary objectives. For example, the results in Figures 10.6 and 10.7 demonstrate that accurate calculations of parameters of critical importance to engineers working on storm damage mitigation are no longer restricted to sparse profile-based measurements. Yet, the same data set could be used by transportation engineers to assess vulnerability of evacuation routes near the shore.

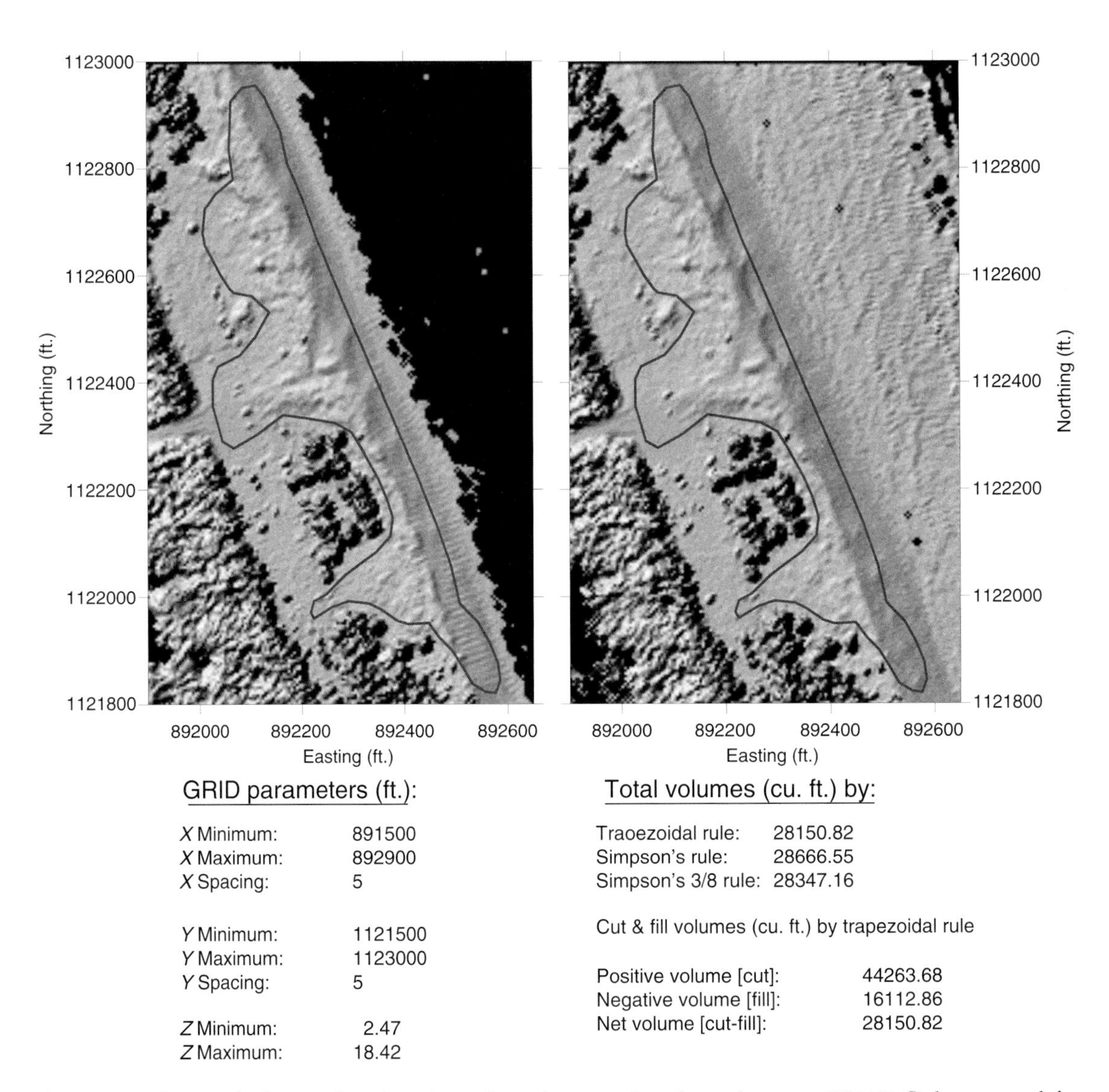

Fig. 10.7 Volume calculations based on irregular polygons taken from the same LiDAR flights as used for Figure 10.6. With dense LiDAR coverage, volume calculations are no longer restricted to profile-based information. The left DEM is pre-hurricane and the right DEM is post-hurricane. Cut volume is volume eroded due to the storm. Reprinted from *ISPRS Journal of Photogrammetry and Remote Sensing*, 59, Shrestha RL, Carter WE, Sartori M, Luzum BJ and Slatton KC, Airborne laser Swath mapping: quantifying changes in sandy beaches over time scales of weeks to years, 222–232, 2005, with permission from Elsevier.

CONCLUSION

Prior to the advent of airborne topographic LiDAR, our understanding of the three-dimensional evolution of beaches and their response to storms was limited by the data collection methods employed to measure beach change. The accuracy and high point density of LiDAR allows rapid sampling of the surface at sub-metre resolution, translating into orders of magnitude higher spatial resolutions and more frequent surveys than are possible with standard manual beach survey methods. This new information improves estimates of traditional parameters, such as volume and shoreline change, and it allows estimates to be made quickly after storm and nourishment events (Case Studies 2 and 3). But perhaps its greatest potential lies in the fact that the high sampling rates allow, for the first time, robust estimation of subtle morphological features and their pdfs (Case Study 1). The potential benefits from this new capability are just beginning to be explored and will undoubtedly advance the state of the art in coastal engineering and geomorphology. As longer time series of LiDAR observations become available, coastal researchers will be able to infer the dynamics of underlying physical processes at spatial and temporal scales not possible just a few years ago, which may lead to entirely new models for predicting coastal erosion and storm response.

REFERENCES

Axelsson P. 1999. Processing of laser scanner data – algorithms and applications. *ISPRS Journal of Photogrammetry and Remote Sensing* **54**: 138–147.

Baltsavias EP. 1999. Airborne laser scanning: basic relations and formulas. *ISPRS Journal of Photogrammetry and Remote Sensing* **54**: 199–214.

Boak EH, Turner IL. 2005. Shoreline definition and detection: a review. *Journal of Coastal Research* **21**: 688–703.

Brock JC, Krabill WB, Sallenger AH. 2004 Barrier island morphodynamic classification based on lidar metrics for north Assateague Island, Maryland. *Journal of Coastal Research* **20**: 498–509.

Cover TM, Thomas JA. 2006. *Elements of Information Theory*, 2nd edition. Wiley-Interscience: New Jersey.

Cressie NA. 1991. *Statistics for Spatial Data*. John Wiley and Sons: New York.

Crossett KM *et al.* 2008. *Population trends along the coastal united states: 1980–2008*. NOAA Coastal Trends Report Series: USA.

Dean RG. 2003. *Beach Nourishment: Theory and Practice*. World Scientific Publishing: River Edge, New Jersey.

Dean RG, Dalrymple RA. 2002. *Coastal Processes with Engineering Applications*. Cambridge University Press: Cambridge.

Duda RO, Hart PE, Stork DG. 2001. *Pattern Classification*. 2nd edition. Wiley: New York.

Fernadez JC, Singhania A, Caceres J, Slatton KC, Starek MJ, Kumar R. 2007. *An overview of lidar point cloud processing software*. University of Florida GEM Center: Report_2007_12_001; available at www.ncalm.ufl.edu.

Foster ER, Spurgeon DL, Cheng J. 2000. *Shoreline change rates – St. Johns County:* Florida. DEP Report No. BCS-00–03.

Gutelius G, Carter WE, Shrestha RL, Medvedev E, Gutierrez R, Gibeaut JG. 1998. Engineering applications of airborne scanning lasers: reports from the field. *Photogrammetric Engineering and Remote Sensing* **64**: 246–253.

Kampa K, Slatton KC. 2004. An Adaptive multiscale filter for segmenting vegetation in ALSM data. *Proceedings of the IEEE International Geoscience and Remote Sensing Symposium, IGARSS* 04, 3837–3840.

Liu H, Sherman D, Gu S. 2007. Automated extraction of shorelines from airborne light detection and ranging data and accuracy assessment based on monte carlo simulation. *Journal of Coastal Research* **23**: 1359–1369.

Luzum B, Slatton KC, Shrestha RL. 2004. Identification and analysis of airborne laser swath mapping data in a novel feature space. *IEEE Geoscience and Remote Sensing Letters* **1**: 268–271.

Meyer TH, Roman DR, Zilkoski DB 2004. What does height really mean? Part I: Introduction. *Surveying and Land Information Science* **64**: 223–233.

Moore LJ, Ruggiero P, List JH. 2006. Comparing mean high water and high water line shorelines: should proxy-datum offsets be incorporated into shoreline change analysis?. *Journal of Coastal Research* **22**: 894–905.

Neuenschwander A, Crawford M *et al.* 2000. Extraction of digital elevation models for airborne laser terrain mapping data. *Proceedings of the IEEE International Geoscience and Remote Sensing Symposium, IGARSS* 00, 2305–2307.

Robertson W, Zhang K, Whitman D. 2007. Hurricane-induced beach change derived from airborne laser measurements near Panama City, Florida. *Marine Geology* **237**: 191–205.

Robertson W, Whitman D, Zhang K, Leatherman SP. 2004. Mapping shoreline position using airborne laser altimetry. *Journal of Coastal Research* **20**: 884–892.

Sallenger AH *et al.* 2003. Evaluation of airborne topographic lidar for quantifying beach changes. *Journal of Coastal Research* **19**: 125–133.

Saye SE, Van Der Wal D, Pye K, Blott SJ. 2005. Beach–dune morphological relationships and erosion/accretion: an investigation at five sites in England and Wales using lidar data. *Geomorphology* **72**: 128–155.

Shrestha RL, Carter WE, Slatton KC, Dietrich W. 2007. *Research quality airborne laser swath mapping: the defining factors.* Ver.1.2, White paper available at www.ncalm.ufl.edu, National Center for Airborne Laser Mapping, Gainesville, FL.

Shrestha RL, Carter WE, Sartori M, Luzum BJ, Slatton KC. 2005. Airborne laser swath mapping: quantifying changes in sandy beaches over time scales of weeks to years. *ISPRS Journal of Photogrammetry and Remote Sensing* **59**: 222–232.

Sithole G, Vosselman G. 2004. Experimental comparison of filter algorithms for bare-earth extraction from airborne laser scanning point clouds. *ISPRS Journal of Photogrammetry and Remote Sensing* **59**: 85–101.

Slatton KC, Carter WE, Shrestha RL, Dietrich W. 2007. Airborne laser swath mapping: achieving the resolution and accuracy required for geosurficial research. *Geophysical Research Letters* **34**: L23S10, doi:10.1029/2007GL031939.

Starek MJ, Vemula RK, Slatton KC, Shrestha RL, Carter WE. 2007a. Automatic feature extraction from airborne lidar measurements to identify cross-shore morphologies indicative of beach erosion. *Proceedings of the IEEE International Geoscience and Remote Sensing Symposium, IGARSS* 07, 2511–2514.

Starek MJ, Vemula RK, Slatton KC, Shrestha RL. 2007b. Shoreline based feature extraction and optimal feature selection for segmenting airborne lidar intensity images. *Proceedings of the IEEE International Conference on Image Processing, ICIP* **4**: 369–372.

Stockdon HF, Sallenger AH, List JH, Holman RA. 2002. Estimation of shoreline position and change using airborne topographic lidar data. *Journal of Coastal Research* **18**: 502–513.

Tebbens S, Burroughs SM, Nelson E. 2002. Wavelet analysis of shoreline change on the Outer Banks of North Carolina: an example of complexity in the marine sciences. *Proceedings of the of National Academy of Sciences Colloquium: Self-organized Complexity in the Physical, Biological, and Social Sciences* **99**: 2554–2560.

White SA, Wang Y. 2003. Utilizing DEMs derived from lidar data to analyze morphologic change in the North Carolina coastline. *Remote Sensing of the Environment* **85**: 39–47.

Wollard JW, Aslaken M, Longnecker J, Ryerson A. 2003. Shoreline mapping from airborne LiDAR in Shilshole bay, Washington. The Hydrographic Society of America, US Hydrographic Conference 03.

Zhang K, Whitman D, Leatherman S, Robertson W. 2004. Quantification of beach changes caused by Hurricane Floyd along Florida's Atlantic coast using airborne laser surveys. *Journal of Coastal Research* **21**: 122–134.

11 LiDAR in the Environmental Sciences: Geological Applications

DAVID HODGETTS

School of Earth, Atmospheric and Environmental Sciences, The University of Manchester, Manchester, UK

INTRODUCTION

Geology is an inherently three dimensional subject, though until fairly recently geologists have had to rely on 2D representations of 3D problems in the form of maps and cross sections. Even in the field the geologist has relied on describing 3D data in a 2D form, then attempting to re-construct the data in 3D later. Recent years have seen the rapid adoption of digital quantitative mapping and surveying techniques in the geological sciences in order to do this. These techniques include the use of Global Positioning Systems (GPS), Ground Penetrating Radar (GPR), Total Station surveying and digital photogrammetry among others (Hodgetts *et al.*, 2004; Lee *et al.*, 2007; McCaffrey *et al.*, 2005; Pringle *et al.*, 2004). A more recent development is the use of LiDAR in the mapping of geological exposures to create Digital Outcrop Models or DOMs (Bellian *et al.*, 2005; Buckley *et al.*, 2008). These techniques allow data to be collected in an integrated three dimensional framework from the outset.

Laser Scanning for the Environmental Sciences,
1st edition. Edited by G.L. Heritage and A.R.G. Large.
© 2009 Blackwell Publishing, ISBN 978-1-4051-5717-9

One of the main drivers, but by no means the only one, in the development of the use of LiDAR and quantitative outcrop geology, has been the collection of outcrop analogue data for the oil industry (Bryant *et al.*, 2000). With the current trend in increasing oil price and declining resources, the need for improved understanding and modelling of hydrocarbon reservoirs is essential. Since the mid to late 1980s, quantitative three-dimensional information in the form of seismic data has been available for sub-surface hydrocarbon reservoirs in the oil and gas industry. Even today these seismic datasets only show a low resolution view of the subsurface (5 m resolution at best, typically 20–30 m). High resolution data can be obtained in the form of boreholes, with a vertical resolution in the centimetre scale, but these are expensive, and one-dimensional in nature (typically a 5 inch bore). Data from geological outcrops on land are used to fill in these gaps in resolution and areal coverage, and improve understanding of rock type distribution in the subsurface. Typically though, these outcrop datasets are presented in a 2D form of logs, cross-sections and maps, and in many cases are described qualitatively rather than quantitatively. This problem was first addressed in the mid to late 1990s when geologists used Differential Global Positioning Systems (dGPS) to place their outcrop data in an accurate 3D framework (Bryant *et al.*, 2000). Around this time, laser-based point surveying was also used to map data on vertical cliff faces where GPS did not work well (Hodgetts *et al.*, 2004). Development of robotic total-stations allowed a low resolution digital terrain model to be surveyed, which then led to the use of LiDAR (Light Detection and Range) systems to generate 3D models of the outcrops under investigation.

QUANTITATIVE DIGITAL OUTCROP GEOLOGY

The field of quantitative digital outcrop geology integrates a variety of digital and traditional outcrop data collection techniques into a coherent dataset in a 3D framework (Bellian *et al.*, 2005; Bryant *et al.*, 2000; Hodgetts *et al.*, 2004, McCaffrey *et al.*, 2005; Pringle *et al.*, 2004; Pringle *et al.*, 2006; Redfern *et al.*, 2007). The data may be interrogated in a quantitative manner and visualised in three dimensions. The development of LiDAR systems, particularly the terrestrial systems, has allowed for rapid improvements in this field, with high speed data acquisition and the ability to produce photo-realistic models of outcrops with relative ease. These Digital Outcrop Models or DOMs (Bellian *et al.*, 2005) provide a basis for geological interpretation and training. When using quantitative data collection approaches, such as LiDAR and DGPS, structural and sedimentological data are treated with equal importance. The resulting datasets are integrated into the same 3D framework, which means effects of structure on sedimentology (e.g. effects on facies distribution in syn-rift setting) and sedimentology on structure (e.g. fault geometry and distribution) can be investigated. LiDAR datasets have a high degree of coverage, and are indiscriminate about the data collected from the scan area, therefore the LiDAR data can be returned to at a later stage to look for other features which may have initially been missed in the field, or even interrogated at a later date for information which was not the focus of the original project.

The following sections cover the LiDAR workflow in geology, covering collection of data, processing, geological interpretation and visualisation. An example of a LiDAR study from the Gulf of Suez is given, then a few more potential applications are presented.

LIDAR WORKFLOW IN GEOLOGY

Most groups using LiDAR in geological applications use a similar workflow when collecting and processing data. Most differences in workflow are related to the model of scanner used rather than fundamental differences in techniques. The workflow generally falls into four major steps: (1) Survey planning, (2) Data collection (3) Post-processing and (4) Interpretation. Due to the well designed and tested nature of modern scanners, the data collection phase is the simplest of all. LiDAR software for the geological sciences is immature in development compared to the hardware. As a result of the lack of good geology specific LiDAR software, post-processing, and particularly the interpretation of the data, are the slowest parts of the workflow.

Survey planning

The aim of most surveys is to create a complete coverage of an exposure with a minimum of data shadowing. This requires a scan of each locality from two or more locations. Digital Terrain Models (DTM) and satellite imagery (e.g., Landsat and Quickbird) may be used to aid survey planning. Using the DTM data coverage, shadowing may be estimated from a planned survey position, and relative spacing of scan positions calculated before going into the field. These scan positions may be programmed into a handheld GPS, enabling them to be found easily in the field. It is, however, only once in the field that the actual scan position can then be chosen, which must give the best coverage of the exposure taking into account local topographic variations not seen on the DTM data.

Data collection

Data collection can be divided into the following steps (in this case based of the Riegl LMSZ420i system):

- Differential GPS measurement of the scanners position if global georeferencing is required.
- 360° panorama scan of the area (depending on make of scanner). This scan is used to provide a basis from which to select areas to scan in detail. Much of this scan may be of areas of no geological interest, though are still of use in the point cloud alignment.
- Image acquisition: Digital photographs are taken using the 14 mm lens. This lens has the same vertical field of view as the scanner (80°) and covers the 360° in seven images. Care must be taken to ensure good exposure by selecting the most appropriate metering mode and exposure compensation where necessary. This may be particularly problematic as the photographs are taken in all directions and range from directly into and directly away from the sun. Further post processing may be carried out on the images to compensate for the exposure variations, but the better the results from the field the better the final results will be.
- Colour extraction from images: The panorama scan as recorded by the scanner has *x,y,z* coordinates and intensity for each point. A colour value (RGB) is extracted from the photographs by mapping the *x,y,z* value back to the calibrated image. This is a simple and rapid process supported by the software running the scanner. Though fine scan areas can be selected from the intensity data, RGB coloured images are easier to use.
- Fine scans: The coloured panorama scan is used to identify the areas required to be scanned in detail. The average range of the area to be scanned is determined from the scanner software, and data spacing at that distance is defined. The data spacing chosen is dependant upon what level of information is needed. A good guide is to use a spacing of half the size of the smallest object of interest (e.g. to resolve a 10 cm thick bed, a data spacing of 5.0 cm is the maximum spacing needed to ensure that bed is imaged). As the scanner works on angular increments, the data spacing on the outcrop will increase with increasing range. This variation of spacing with distance has to be taken into account when defining areas to fine scan. In the experience of the author, between one and seven fine scans will be taken per scan position or station, with three to four being typical.
- Tilt mount photographs and off scanner photographs: The tilt mount used on the scanner allows longer focal length lenses to be used for the photographs. The camera has to be tilted with longer focal length lenses as the cameras field of view is less than that of the scanner. Off scanner photographs may be taken while the above processes are going on. These off scanner images are used to fill in data gaps, and may be taken from vantage points where it is not possible to get the scanner.

Time spent at each scan position depends on range of scanning and degree of resolution required. In the Nukhul study (see later section) approximately three scan positions per days could be collected for the longer range datasets (scanned at 500–800 m), and up to eight positions for areas where the range was less than 500 m.

Post-processing

This involves the basic checking of data, removal of bad data points (caused by dust, rain or false returns), and then checking the camera calibrations (particularly the mount calibration) to ensure a good match between the photographs and the scans.

Merging of data

Each scan position, when collected, has its own coordinate system, with the scanner itself at (0,0,0). This is known as the Scanners Own Coordinate System (SOCS). One scan is chosen as the base scan, and then other scans are linked to this via a point cloud alignment process done in software called Polyworks®. This gives a very accurate sub-centimetre alignment of the scans, which then build up into a merged model in a

Project Coordinate System (PRCS). Once the data is merged into the combined dataset, the PRCS needs to be transformed into a Global Coordinate Systems (GLCS). This may be done from DGPS data from scan positions, or from DGPS measurement of reflectors within a single scan. A minimum of three GPS positions are required in order to georeference the scan data correctly. Accuracy of georeferencing will depend on accuracy of the DGPS and how far apart the measurements are.

INTERPRETATION AND MEASUREMENTS POSSIBLE FROM LIDAR DATA

Figure 11.1 shows a LiDAR dataset with a variety of geological interpretations and data attached to it. The basic interpretation techniques are outlined here.

Basic interpretations

The most basic interpretation available to the geologist is the ability to draw polylines along the outcrop to delimit faults, bedding planes and any other bounding surfaces visible. Polyline interpretations should have an identifier attached to them to define what kind of feature is being interpreted, as well as a confidence level to show how sure the interpreter is of the position of the feature being interpreted.

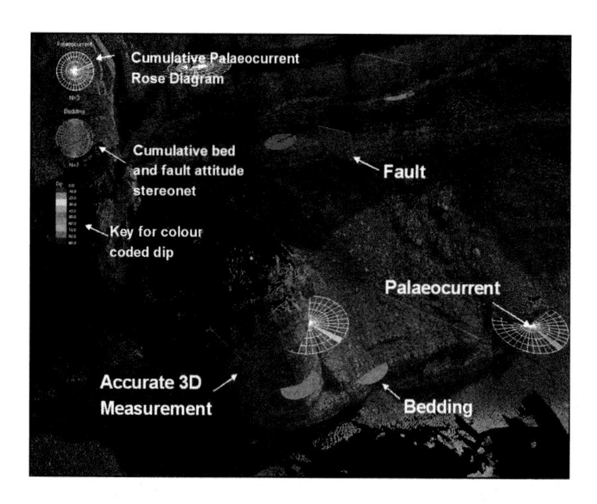

Fig. 11.1 LiDAR dataset with a variety of geological interpretations and data from both field observation and measurement from the scan data. A three-point measure from a visible surface in the LiDAR data can be used to calculate the attitude of the surface. This is applicable to the measurement of faults and bedding planes.

Auto-tracking of LiDAR data

Auto-tracking is the ability to automatically pick or track geological features without the need for manual interpretation. Auto-tracking is common in seismic datasets (Avseth *et al.*, 2005; Bacon *et al.*, 2003; Dorn, 1998), though in comparison LiDAR data is much harder to deal with due to the subtle nature of the features and the lesser degree of ordering in the data (seismic data is on a regular grid). Viseur *et al.* (2007) have demonstrated a method of detecting stratigraphic or tectonic features from user-defined seed points located on a digital outcrop model based on LiDAR data, using a 20 km long by 400 m high, continuous cliff face exposure, of the Vercors Urgonian Lower Cretaceous carbonate platform located in Gresse-en-Vercors (South of France). These techniques are currently in their infancy, but further development will yield improved tools for this purpose.

EXTRACTION OF GEOSTATISTICS

In petroleum geology the use of LiDAR data is increasingly common in order to extract geostatistics from objects such as faults, fractures, channels, mudstone barriers and baffles and bed thickness, as well as information on the distribution and lateral variation of parameters such as facies distribution (Coburn *et al.*, 2006; Yarus & Chambers, 1995). These geostatistics are used to populate reservoirs models (Grammer *et al.*, 2004; Hodgetts *et al.*, 2004; Pringle *et al.*, 2004; Redfern *et al.*, 2007). In outcrop, for example, it is possible to measure apparent channel width and thickness. If the palaeocurrent direction is known then true channel width may be calculated, and if the dip of the beds is known then true channel thickness may also be derived. Relationships between channel width and thickness may be investigated in different depositional environments, and any relationships observed in outcrop may be applied to the subsurface.

Structural measurements from LiDAR data

In order to measure the dip and azimuth of a bed, two apparent dips need to be visible. Three points are marked onto the LiDAR data demarking the bed and forming a triangle, the dip and azimuth of the bed can then be calculated from this triangle. An example of structural measurements in the form of bedding plane and fault attitude can be seen in Figure 11.2. Care must be taken to choose appropriate areas to measure the data, and the closer the triangle is to an equilateral triangle the better the measurement will be. Derived data has been compared with direct measurement in the field and has been found to be consistent. This approach allows bed attitude measurement to be made in cases where it would be difficult in the field due to lack of access, and as the measurements are made over a relatively large area compared to a measurement with a traditional compass, clinometer effects of sedimentary structures on the measurement are eliminated.

Fault attitude is measured in the same manner as for bed attitude using three observed points, though care must be taken to ensure this is done on a planar section of the fault, and areas of ramp float geometry must be treated with care. McCaffrey *et al.* (2005) show structural modelling results from a digital survey from Cullercoats (north east England), where a series of small scale faults in Permian sandstone in the hanging wall of the Ninety Fathom Fault were surveyed. Fault traces were interpreted from a LiDAR dataset and 3D surfaces fitted to the fault traces.

Integrating sedimentological data

Parameters such as grain size, shape and sorting are not derivable from LiDAR data, indeed there are many sedimentological features which cannot be extracted from LiDAR data as they are at or below

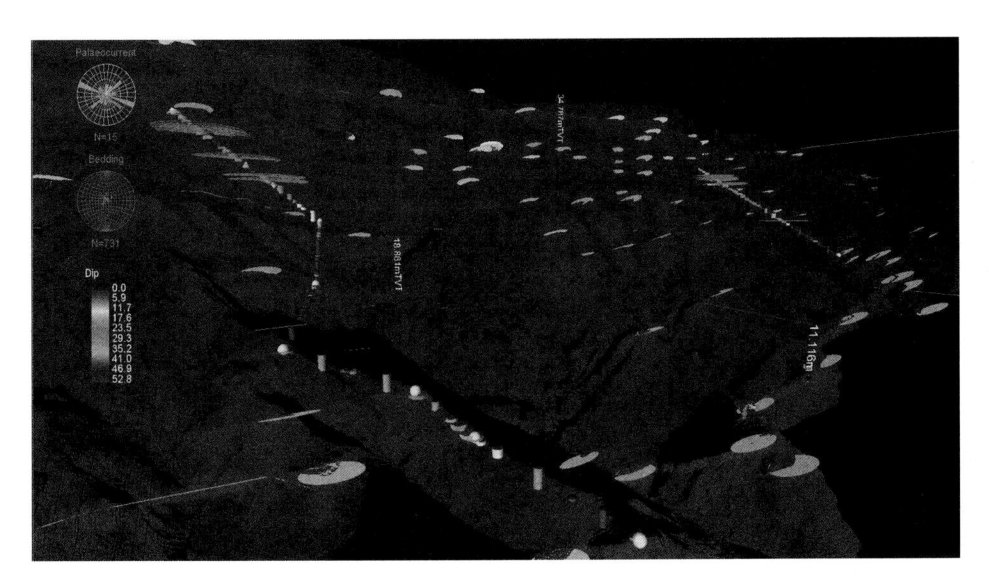

Fig. 11.2 Integration of sedimentological data with the LiDAR dataset. An area from the Wadi Nukhul dataset showing sedimentary logs attached to the scan data to act as ground control for interpretation. Also shown are structural measurements, palaeocurrents and accurate 3D measurements.

the resolution of current scanners. There are a few cases where palaeocurrent information may be derived from large scale cross-beds or scours, but in most cases these data are available from direct outcrop measurement only. Facies variations may be visible in the LiDAR data, but without ground control the exact nature of the individual facies will not be known. The inclusion of traditional field data and ground control is essential to the continued use and acceptance of LiDAR data as a geological tool (Figure 11.2). Sedimentary log data may be attached directly to the outcrop via GPS or DGPS data, with any errors between GPS and LiDAR data resolution being accounted for by snapping log deviation points to the correct position on the scan data, as the LiDAR data allows you to see where bed boundaries and marker beds are. These logs then provide the ground truth for interpreting the LiDAR data, and may even provide the training datasets for more automated classification approaches.

Cross-sections and squash plots

In seismic interpretation, squash plots allow lateral variations of thickness and changing stratal geometries on a large scale to be observed by reducing the horizontal, or increasing the vertical, scale. In 3D LiDAR datasets there is an extra level of complexity as the data in not in a single plane as in seismic (e.g. inline, cross line or arbitrary/random lines in seismic data) (Bacon *et al.*, 2003). This problem is addressed by defining a projection plane onto which all data points within a cut-off distance of that plane are projected (Figure 11.3). The orientation of

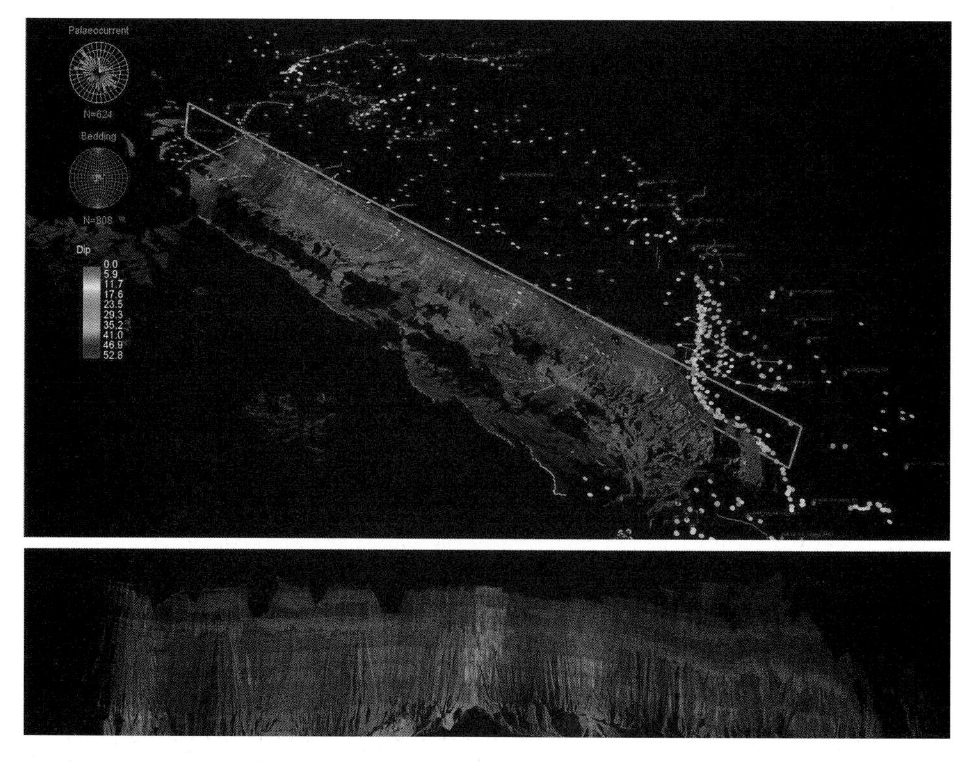

Fig. 11.3 Projection plane view of the southwest face of the Nukhul syncline outcrop (see Figures 11.5–11.7 with a four times vertical exaggeration. Note the presence of a subtle anticline structure visible in the projection plane. This feature is not easily visible in the field, and is only clearly visible in vertically exaggerated data. The feature is also visible in the contour map in Figure 11.5. See Plate 11.3 for a colour version of this image.

the plane is user-defined, and the projection direction is perpendicular to the plane. This means in strike or close to strike sections the plane may be oriented perpendicular to the dip direction, and the data projected down structural dip. The projected points are added to a bitmap image with defined *x* and y pixel dimensions. The resulting image derived from this method in the strike section case is a true stratigraphic thickness plot.

ATTRIBUTE ANALYSIS OF LIDAR DATA

The LiDAR data is not an unstructured cloud of point data, the data is organised into columns and rows due to the way the scanner maps the outcrop. This organisation of the point cloud makes creation of new attributes easy (Figure 11.4). In the standard dataset each data point has reflection intensity from the scanner, and red, green

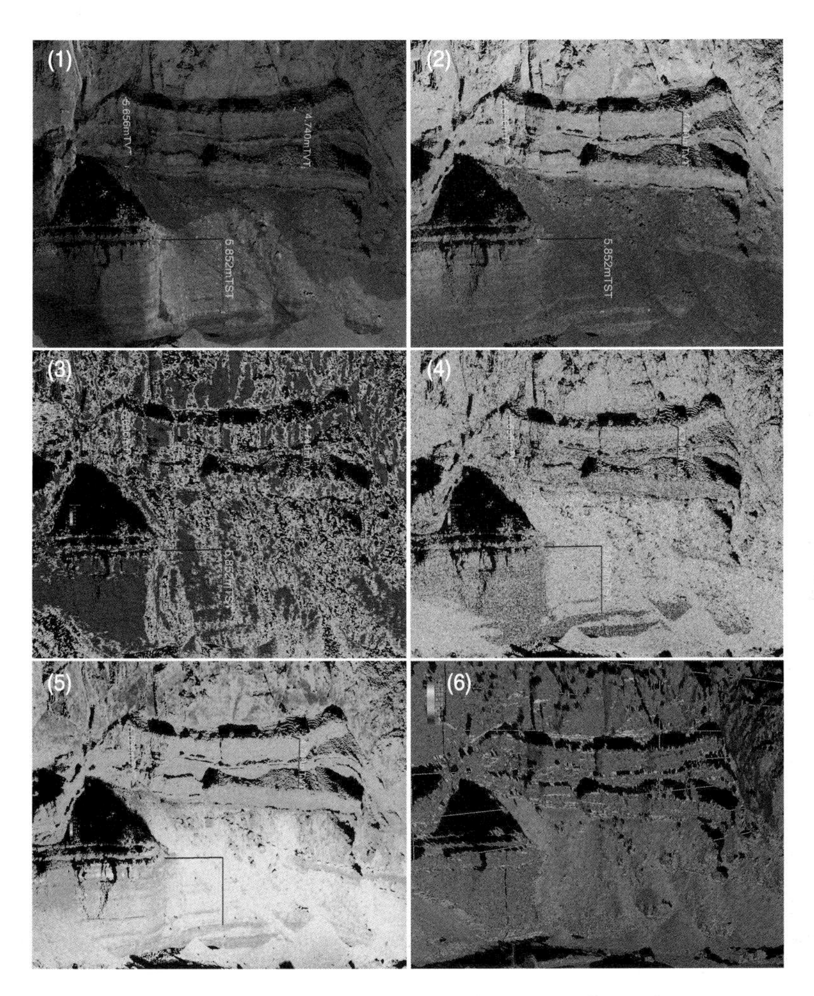

Fig. 11.4 Attribute analysis of LiDAR data. The scan data may be processed in order to extract more information. Parameters are (1) RGB colour, (2) Reflection Intensity, (3) Correlation between adjacent scan lines based on range from scanner (4) Surface roughness, (5) False colour intensity and (6) Dip of outcrop surface. Attribute analysis aids identification and analysis of subtle features hence aiding interpretation. See Plate 11.4 for a colour version of these images.

and blue (RGB) colour channels from the digital images. Each of these channels may be subjected to processing similar to image processing to increase contrast, enhance edges and so on, which may make interpretation easier. Another form of processing is based on comparing consecutive vertical scan lines to look for correlation from one line to the next, in a similar manner to continuity or coherency parameter in seismic data (Bacon *et al.*, 2003; Bahorich & Farmer 1995). This highlights variation across the outcrop, for example, termination of bedding planes at a fault, and so helps with structural interpretation. A variety of image processing filters may also be applied, such as laplacian or Gaussian filters, in order to aid edge detection and highlight subtle features.

Multi-component classification

By combining two or more co-registered attributes a new attribute may be derived, which may enhance trends or features not or poorly visible in the individual attributes. Bellian *et al.* (2005) used this approach combining intensity and angle of slope to derive a simple facies classification in a debrite and turbidite complex.

POINT CLOUDS VERSUS TEXTURED MESHES IN GEOLOGICAL APPLICATIONS

The LiDAR system produces a point cloud, which is a series of unconnected points in 3D space. These points may then be joined together to form a triangulated mesh. This meshed surface or triangulated irregular network (TIN) may then be texture mapped with image data to produce a photo-realistic model (Bellian *et al.*, 2005). At present, both point cloud data and texture mapped meshes are used, each having advantages and disadvantages. The point of meshing and texture mapping a point cloud dataset is to use the extra information in the photographs to fill in the gaps between the data points. This only works if the area a pixel covers on the outcrop is smaller than the data spacing of the points in the point cloud. The down side of texture mapping comes from the increase in the amount of data to be visualised – therefore slowing visualisation down. When drawing a point cloud each data point is drawn once. With triangulated data, however, each triangle has three points to it, and many points from the point cloud will exist in more than one triangle. In order to address this problem the mesh is decimated (Bellian *et al.*, 2005; Buckley *et al.*, 2008). This is achieved by analysing vertex in the mesh and checking to see if it were to be removed and that part of the surface re-meshed, how much difference it will make to the surface geometry. If the change is negligible then the point can be removed. A flat surface does not need many points to define it so more points would be removed, a complex surface, however, would change a lot and needs more points to keep its shape and therefore less are removed. Obviously in doing this the data is being degraded, the amount the data is degraded depends on the geometry and the parameters used in the decimation algorithm. Point cloud data, in comparison, can be visualised much faster and does not need to be decimated. This means the interpreter is always working on real, not interpolated, data. The interpretation though is limited by the data spacing. In practice, experience has shown that for making structural measurements working of undecimated point cloud data is the best approach. Each point selected from the model is a real measured point, not a result of interpolation between points which may give rise to measuring errors. For mapping boundaries and surfaces the texture mapped mesh is a better option as the extra information from the image makes feature identification easier. The problems of point clouds versus textured meshes are essentially ones of processing and graphics power, as computer power continues to increase these problems will tend to diminish.

GEOLOGICAL MODELLING

Once the LiDAR data has been interpreted and analysed the process of geological modelling begins. A variety of modelling packages are available (e.g. Petrel, RMS, and GoCad among others) for building 3D geological models. The interpretations from the LiDAR data are exported into the modelling systems where surfaces are modelled and 3D geocellular volumes are built based on the structural data. Facies data from the georeferenced sedimentary logs are used to build stochastic (statistically based) or deterministic models. Model building is an iterative process, where building the model may actually show errors in the data or reveal important data gaps. The modeller may then return to the LiDAR data to check interpretations, fill in data gaps where possible, or even just use the virtual outcrop model to improve their understanding of the system. This aspect of improved understanding of a geological system is particularly important where the outcrop model is being used as an analogue dataset for a sub-surface petroleum reservoir. An example of a large scale project using terrestrial LiDAR data is included here to show how this data may be used in a real application.

EXAMPLE GEOLOGICAL MODELLING APPLICATION: WADI NUKHUL, GULF OF SUEZ

The Rift Analogues Project (TRAP) aims to investigate the structural evolution and syn-rift stratigraphic development in extensional basins, utilising state of the art digital outcrop mapping and reservoir modelling approaches. The project aims to develop quantitative, high resolution sedimentological and sequence stratigraphic models of various rift settings in order to create reservoir models generally applicable to rift basins worldwide. The field area is located in the Hamman Faraun fault block, Gulf of Suez (Figure 11.5) and comprises very well exposed cliff faces of up to 120 m high with limited access, making it an ideal candidate for a LiDAR based study.

The Nukhul Formation is widely exposed in both the Hammam Faraun and El Qaa fault blocks and was deposited at the initial stages of rifting (rift initiation deposits) when the structural style was dominated by distributed extension on numerous small displacement active faults and associated fault-propagation folds (Gawthorpe *et al.*, 1997; Sharp *et al.*, 2000; Gawthorpe *et al.*, 2003). The Nukhul Formation stratigraphy comprises both tidal and shallow marine depositional systems. Detailed work in Wadi Baba (to the south of Wadi Nukhul) and several intra-block fault zones in the Hammam Faraun fault block (Gawthorpe *et al.*, 1997; Sharp *et al.*, 2000; Carr *et al.*, 2003, Jackson *et al.*, 2005; Jackson *et al.*, 2006) has developed a gross facies model and identified high frequency sequences, in the Nukhul Formation, which clearly reveal local variation of facies stacking and dispersal patterns related to growth folds and propagation/linking fault segments. Stratal geometries associated with these fault propagation folds are markedly different to those normally associated with syn-rift stratigraphy adjacent to normal fault zones, and have important implications for stratigraphic trapping, reservoir geometry and connectivity adjacent to normal faults. Documentation of the stratigraphic response to growth folds at fault tips, via the use of virtual outcrop models, is directly relevant to understanding along strike variability in sequence and reservoir development in transfer zones, and the evolution of transfer zones during lateral propagation and hard linkage of fault tips.

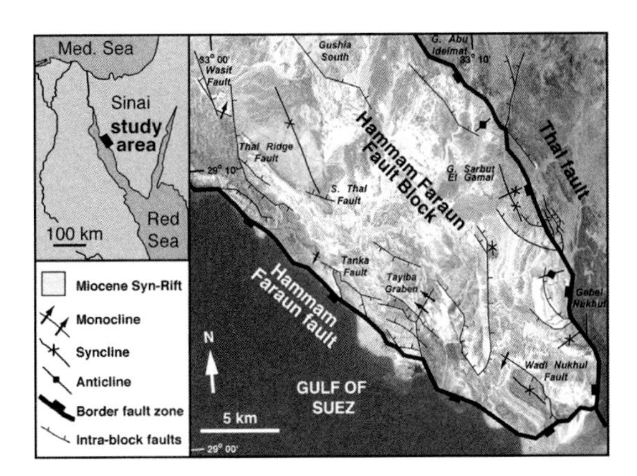

Fig. 11.5 Location of the Wadi Nukhul study area in the Sinai Peninsula, Gulf of Suez. The Nukhul syncline lies in the south east part of the Hamman Faraun fault block, where syn-rift tidal sediments are exposed.

Hardware

The LiDAR data was collected using a Riegl LMSZ420i terrestrial laser scanner (www.riegl.com). This laser scanner has a data collection rate of up to 12,000 points per second and a 360° × 80° field of view using a near-infrared, eye safe laser. Maximum useful range of the scanner is approximately 800 m, though returns are regularly seen from exposures as far as 1200 m. Attached to the scanner via a precision camera mount was a calibrated Nikon D100 camera, which was used to provide colour for the point cloud, and texture maps for the triangulated point cloud data. The camera had 14, 28, 35, 50, 85, and 180 mm lenses to ensure best field of view and minimise the number of photographs needed for the subject under survey. A second Nikon D2X 12 mega pixel digital camera was used for off-scanner imagery to ensure best photographic coverage, taking pictures from positions where it was not possible to take the scanner. The scanner was driven via a Panasonic CF29 Toughbook laptop PC with 1 gigabyte of ram. Georeferencing of the scan data was done using a Trimble Pro XR differential GPS, giving real-time sub-metre accuracy without the need for a dedicated base station. Merging of data was performed using Innovmetrics Polyworks® software.

The dataset

The Nukhul datasets was collected over a period equal to four weeks over several field seasons. The total study comprises 85 linked and georeferenced scan positions, more than 5000 digital photographs (both on and off scanner), 829 strike/dip measurements (from both scan data and outcrop) 924 palaeocurrents (from field observation) and over 30 outcrop logs which have been digitised and attached back to the scan data in their correct positions. In addition, Quickbird satellite imagery has been used for obtaining more regional information. A field team of a minimum of three people was required for transportation of all the necessary equipment (scanner, tripod, batteries, camera, DGPS, laptop, food and water for the operators) as most localities did not have direct vehicular access.

Results

The resulting dataset is shown in Figure 11.6. The field area has an average data spacing of 5.0 cm on the sections of outcrop of interest. Cliff exposure range from a few metres in height to 140 m or more, with most of these areas inaccessible by foot; therefore the only method of obtaining accurate data is by LiDAR. The structural data measured from the scan data have been used to delimit the structure of the Nukhul syncline, along with polyline interpretations of the key stratal surfaces in the cliff exposure (Figure 11.7). Using bed attitude measurements and the polylines a resulting hanging wall geometry surface has been derived with a much higher degree of control than was previously possible. The integration of LiDAR and traditional field measurements allowed the modelling and constraint of the synclinal structure on a scale not possible with field data alone. In many cases data derived from the LiDAR dataset is far more accurate than achievable by direct measurement in the field (due to weathering and sedimentary structures on the surfaces to be measured). Particularly high resolution structural variations (on a scale of less than a few centimetres) are still best measured, if possible, in the field.

LIDAR IN GEOCONSERVATION

Bates *et al.* (2008) describe the use of LiDAR in geosconservation at a dinosaur track way site in the Barcelona province, Catalonia. The trackway site comprises steeply dipping rock of 60 in an abandoned quarry (Figure 11.8). The nature of the exposure makes access to the individual track ways problematic, and they have suffered significant weathering since their exposure by open-air lignite mining in the 1980s (Schulp & Brokx, 1999;

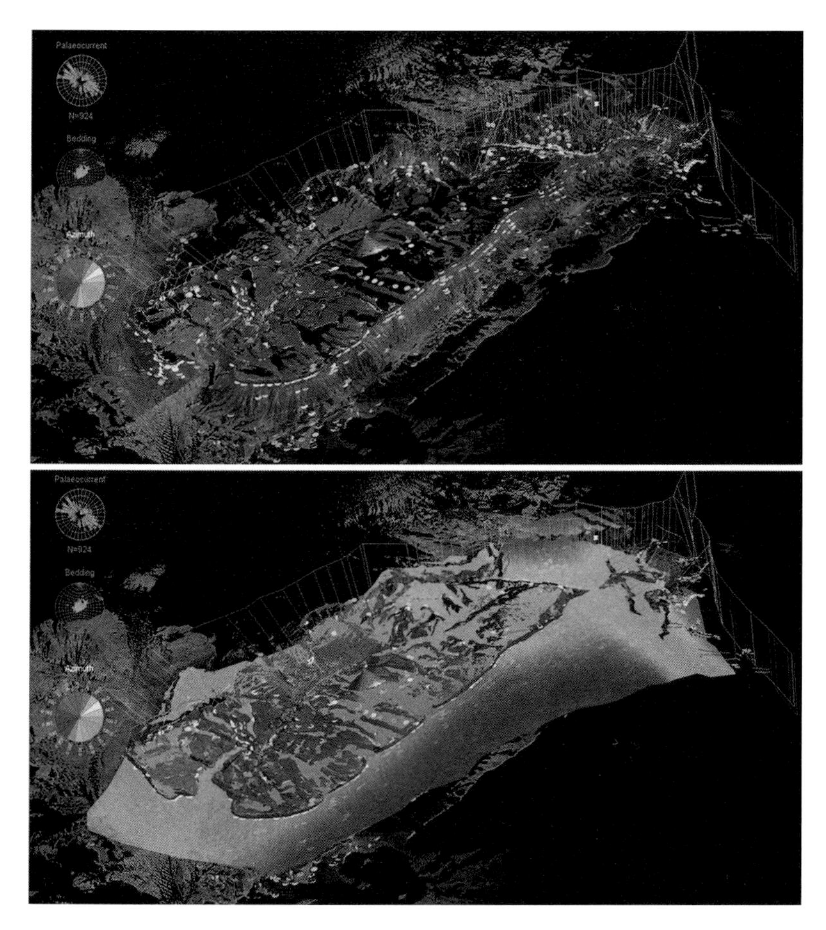

Fig. 11.6 The complete Wadi Nukhul dataset comprising 85 scan stations, 924 palaeocurrents, 829 structural measurements and some 30 logs. The lowermost image shows the dataset with one of the hanging wall structural marker surfaces generated from the polyline and structural interpretations. Model is viewed from the north east. See Plate 11.6 for a colour version of these images.

Vila *et al.*, 2005; Bates *et al.*, 2008). As such, much important information is being lost at a rapid rate. The LiDAR data was used to create a DOM of the site as a record of these footprints in their current state, as well as allowing quantitative 3D analysis of the track way data. Breithaupt *et al.* (2004) also used terrestrial LiDAR imaging and digital photogrammetry separately to record and map small sections of outcrop in Wyoming and Colorado (USA) containing abundant dinosaur tracks and skeletal remains.

LIDAR DATA AS A FIELD TOOL

Software developed at the University of Manchester allows the merged and georeferenced scan data to be taken back out into the field and used as a very high resolution base map. By linking a hand held GPS into a tablet PC with daylight viewable screen, real-time navigation through the scan data while in the field is possible. The advantage of this is that field observations of features which are not available from

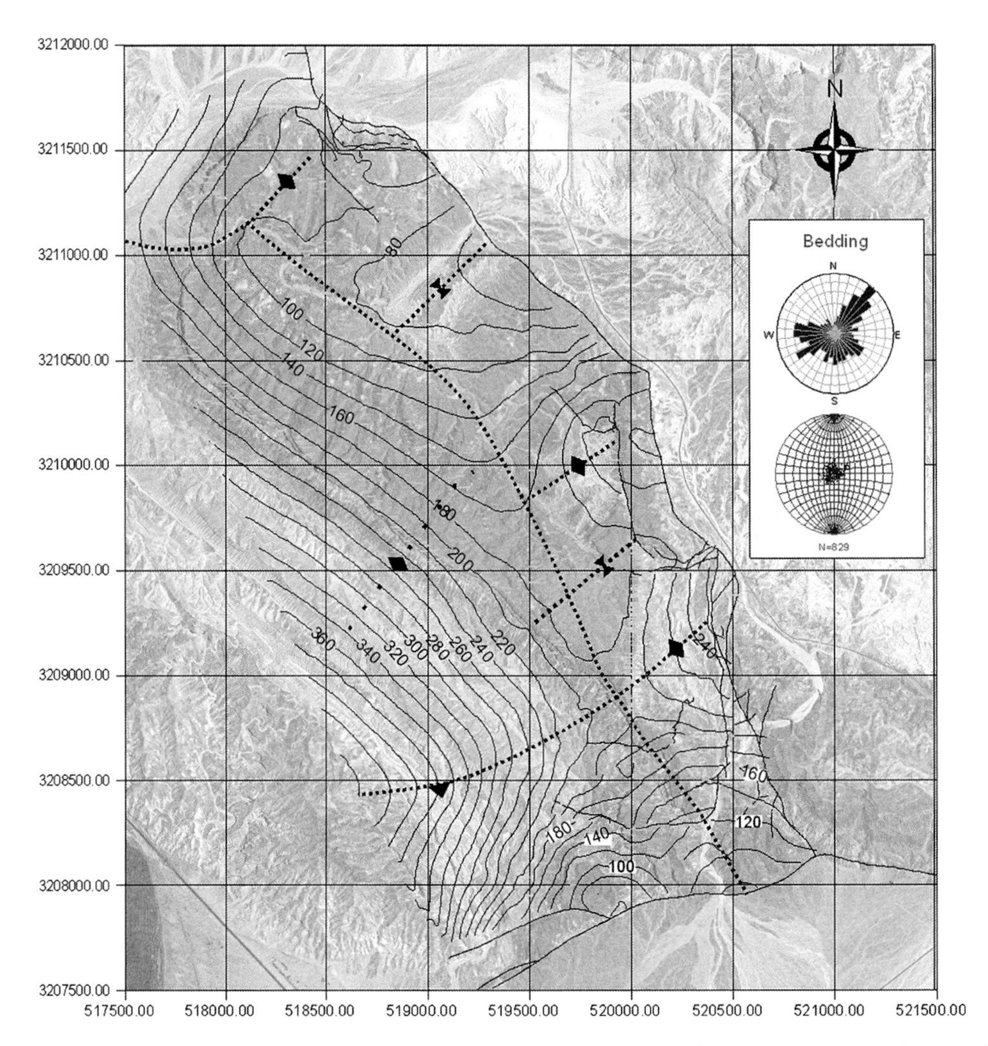

Fig. 11.7 The detailed structural surface model for a hanging wall marker surface (see Figure 11.6) from within the Nukhul syncline study area. Positions of major and minor anticlines, synclines and faults are shown. Though this map could have been constructed without the LiDAR data, the degree of detail and accuracy would have been much less, and some of the more subtle features may have been missed all together.

the scan data (such as grain size and facies) may be attached directly to the scan data while in the field, thus ensuring the accurate georeferencing of all information. Another advantage of having the scan data in the field is the ability to look at the digital data from a variety of viewpoints and distances while standing at a fixed position in the field. In many cases large scale features are obscured when close to an outcrop, being able to zoom out on the datasets while looking at a small scale feature in outcrop can provide valuable insights into both structural and sedimentological variations. A natural progression of this approach is the use of augmented reality. Augmented reality allows the combination of real word data with computer based information,

Fig. 11.8 Riegl LMSZ420i at the dinosaur track way outcrop. Note the steep nature of the beds making direct access problematic. Erosion of the quarry face means loss of the information stored in the trackways. LiDAR provides a means of conserving this information.

typically overlaying information onto a user's field of view via a transparent headset. (Bellian *et al.*, 2005) have described this kind of approach, using terrestrial scanning LiDAR, RTK GPS and Augmented reality to collect and interpret Permo-Triassic carbonate strata.

Virtual outcrop models as a training tool

The teaching of geology has always included large amounts of fieldwork. Fieldwork, however, is expensive, particularly with large groups of students. Virtual outcrop models are far from a replacement for real fieldwork because, as already mentioned, there are many important geological observations which are not available in LiDAR data. Virtual outcrop models are however a useful addition to fieldwork as pre-trip training, or post trip summary tool, therefore making time in the field more effective. Finally, many outcrop localities are logistically impossible to reach with large groups, the use of virtual outcrop models here thus provide the only feasible means of viewing these outcrops in three dimensions.

CONCLUSIONS

In conclusion, the benefits of using an integrated data collection approach with LiDAR in geology are outlined below:

- High speed data collection – LiDAR can scan at up to 12,000 data points per second (depending on which machine is used), with each data point having *x*, *y* and *z* positional information, reflection intensity and colour provided by a calibrated high resolution digital camera effectively providing, therefore, a three dimensional photograph.
- Structural geology and sedimentology are treated with equal importance. The datasets are integrated in the same framework and means effects of structure on sedimentology (e.g. effects on facies distribution in syn-rift setting) and sedimentology on structure (e.g. fault geometry and distribution) can be investigated.
- Datasets once collected may be interrogated at a later date for information which was not the focus of the original project. LiDAR datasets have a high degree of coverage – therefore the LiDAR data can be returned to at a later stage to look for other features which may have initially been missed in the field.
- Easy collection of accurate spatial data (e.g. apparent widths of geo-objects such as channels, scours etc.) for object statistics for reservoir characterisation.
- Virtual outcrop analogues may be constructed for use in visualisation suites – an essential teaching/training tool for example, where fieldtrip is not possible, or act as a precursor to a fieldtrip.

In addition to the LiDAR data, other information needs to be collected which cannot be extracted from LiDAR data alone:

- Outcrop logging/sampling provides data which LiDAR cannot capture (facies, palaeocurrent data, grain size variations etc.) The LiDAR data provides the 3D framework into which the log data is integrated.
- Differential GPS/TotalStation mapping allows focused data collection on surfaces/features of special interest (e.g. key stratal surfaces), data which may not be easily visible (e.g. amalgamation surfaces), or point data collection (samples,

palaeocurrent data, grain size and facies) away from the sedimentary logged sections. Again all data is fully georeferenced.

The current disadvantages of terrestrial LiDAR acquisition in geology are the expense of the equipment and the weight of the equipment for carrying in the field. The software issues are rapidly being addressed as LiDAR data acquisition becomes more common in geology. Overall then, the application of LiDAR to geology represents a great improvement in the ability to truly work in 3D from outcrop data, with improved interpretation, analysis and modelling possible. Understanding of lateral facies changes are improved due to the good coverage of outcrop LiDAR data, and subtle large scale features are easier to identify in 3D datasets. The LiDAR dataset provides an important additional tool to the field geologist, making field time more productive and allowing further data collection and analysis to be undertaken while at the workstation. The LiDAR datasets also provide an invaluable training tool which may be applied in both academic and industrial environments.

ACKNOWLEDGMENTS

The author would like to thank the Rift Analogues Projects (TRAP) and its sponsors StatoilHydro and ConocoPhillips for their support and permission to use TRAP data in this publication. The author would also like to thank Dr Mimi Hill, and an anonymous reviewer, for their valuable comments on this manuscript. Franklin Rarity, Paul Wilson, Ivan Fabuel Perez, Xavier van Lanen and Karl Bates are thanked for providing images and comments.

REFERENCES

Avseth P, Mukerji T, Mavko G. 2005. *Quantitative Seismic Interpretation*. Cambridge: Cambridge University Press.

Bacon M, Simm R, Redshaw T. 2003. *3D Seismic Interpretation*. Cambridge: Cambridge University Press.

Bahorich M, Farmer S. 1995. 3D seismic discontinuity of faults and stratigraphic features: The coherence cube. *The Leading Edge* **14**: 1053–1058.

Bates, KT, Rarity F, Manning PL, Hodgetts D, Vila B, Oms O, Galobart A, Gawthorpe RL. 2008. High-resolution LiDAR and photogrammetric survey of the Fumanya dinosaur tracksites (Catalonia): Implications for the conservation and interpretation of geological heritage sites. *Journal of the Geological Society* **165**: 625–638.

Bellian, JA, Kerans C, Jennette DC. 2005. Digital outcrop models: Applications of terrestrial scanning LiDAR technology in stratigraphic modeling. *Journal of Sedimentary Research* **75**: 166–176.

Breithaupt BH, Matthews NA, Noble TA. 2004. An integrated approach to three-dimensional data collection of Dinosaur tracksites in the Rocky Mountain West. *Ichnos* **11**: 11–26.

Bryant I, Carr D, Cirilli P, Drinkwater N, McCormick D, Tilke P, Thurmond J. 2000. Use of 3D digital analogues as templates in reservoir modelling. *Petroleum Geoscience* **6**: 195–201.

Buckley SJ, Howell JA, Enge HD, Kurz TH. 2008. Terrestrial laser scanning in geology: data acquisition, processing and accuracy considerations. *Journal of the Geological Society* **165**: 625–638.

Carr ID, Gawthorpe RL, Jackson CAL, Sharp IR, Sadek A. 2003. Sedimentology and sequence stratigraphy of early Syn-Rift tidal sediments: The Nukhul Formation, Suez Rift, Egypt. *Journal of Sedimentary Research* **73**: 407–420.

Coburn, TC, Yarus, JM, Chambers RL. 2006. Stochastic Modeling and Geostatistics. Principles, Methods, and Case Studies, Volume II. In: *AAPG Computer Applications in Geology* **5**.

Dorn GA. 1998. Modern 3D Seismic Interpretation. *The Leading Edge* **17**: 1262–1273.

Gawthorpe RL, Jackson CAL, Young MJ, Sharp IR, Moustafa AR, Leppard, CW. 2003. Normal fault growth, displacement localisation and the evolution of normal fault populations: the Hamman Faraun fault block, Suez Rift, Egypt. *Journal of Structural Geology* **25**: 1347–1348.

Gawthorpe RL, Sharp I, Underhill JR, Gupta S. 1997. Linked sequence stratigraphic and structural evolution of propagating normal faults. *Geology* **25**: 795–798.

Grammer GM, Harris PM, Eberli GP. 2004. Integration of Outcrop & Modern Analogs in Reservoir Modeling. In: *AAPG Memoir* **80**, 394.

Hodgetts D, Drinkwater NJ, Hodgson DM, Kavanagh J, Flint S, Keogh K, Howell J. 2004. Three-dimensional geological models from outcrop data using digital data collection techniques: an example from the Tanqua Karoo depocentre, South Africa. In: *Geological Prior Information: Informing Science and Engineering. Geological Society Special Publication*, Curtis A, Woods R (eds), Geological Society of London **239**: 57–75.

Jackson CAL, Gawthorpe RL, Carr ID, Sharp IR. 2005. Normal faulting as a control on the stratigraphic development of shallow marine syn-rift sequences: the Nukhul and Lower Rudeis Formations, Hammam Faraun fault block, Suez Rift, Egypt. *Sedimentology* **52**: 313–338.

Jackson CAL, Gawthorpe RL, Sharp IR. 2006. Style and sequence of deformation during extensional fault-propagation folding: examples from the Hammam Faraun and El-Qaa fault blocks, Suez Rift, Egypt. *Journal of Structural Geology* **28**: 519–535.

Lee K, Tomasso M, Ambrose WA, Bouroullec R. 2007. Integration of GPR with stratigraphic and LiDAR data to investigate behind-the-outcrop 3D geometry of a tidal channel reservoir analog, upper Ferron Sandstone, Utah. *The Leading Edge.*

McCaffrey KJW, Jones RR, Holdsworth RE, Wilson RW, Clegg P, Imber J, Holliman N, Trinks I. 2005. Unlocking the spatial dimension: digital technologies and the future of geoscience fieldwork. *Journal of the Geological Society* **162**: 927–938.

Pringle JK, Howell JA, Hodgetts D, Westerman AR, Hodgson DM. 2006. Virtual outcrop models of petroleum reservoir analogues: A review of the current state-of-the-art. *First Break* **24**: 3–13.

Pringle JK, Westerman AR, Clark JD, Drinkwater NJ, Gardiner AR. 2004. 3D high-resolution digital models of outcrop analogue study sites to constrain reservoir model uncertainty: an example from Alport Castles, Derbyshire, UK. *Petroleum Geoscience* **10**: 343.

Redfern J, Hodgetts D, Fabuel-Perez I. 2007. Digital analysis brings renaissance for petroleum geology outcrop studies in North Africa. *First Break* **25**: 81–87.

Schulp A, Brokx WA. 1999. Maastrichtian Sauropod Footprints from the Fumanya site, Berguedà, Spain. *Ichnos* **6**: 239–250.

Sharp IR, Gawthorpe RL, Underhill JR, Gupta S. 2000. Fault-propagation folding in extensional settings: Examples of structural style and synrift sedimentary response from the Suez rift, Sinai, Egypt. *Geological Society of America Bulletin* **112**: 1877–1899.

Vila B, Oms O, Galobart À. 2005. Manus-only titanosaurid trackway from Fumanya (Maastrichtian, Pyrenees): further evidence for an underprint origin. *Lethaia* **38**: 211–218.

Viseur S, Richet R, Borgomano J, Adams E. 2007. Semi-Automated Detections of Geological Features from DOM – The Gresse-en-Vercors Cliff. In: *69th EAGE Conference and Exhibition incorporating SPE EUROPEC 2007*, ExCel London.

Yarus J M, Chambers RL. 1995. Stochastic modeling and geostatistics principles, methods, and case studies. In: *AAPG Computer Applications in Geology* **3**, American Association of Petroleum Geologists.

12 Using LiDAR in Archaeological Contexts: The English Heritage Experience and Lessons Learned

SIMON CRUTCHLEY

English Heritage, Kemble Drive, Swindon, Wilts, UK

INTRODUCTION

The Aerial Survey and Investigation team of English Heritage first became involved with LiDAR in 2001. This chapter summarises the results gained so far, the lessons learned and the usefulness of LiDAR to archaeological survey. It begins by looking at the general development of LiDAR and its use in archaeological contexts before moving on to examine some of the practical issues related to using LiDAR. Using the projects with which English Heritage has been involved, these cover two main questions. One relates to the data itself and what can and cannot be recorded; the other to the practical application of the data by professionals in their day to day work, looking at how to optimise the use of the data for as many people as possible. Finally the chapter examines the use of LiDAR by others in the archaeological community together with future advances and enhancements of the technique.

Laser Scanning for the Environmental Sciences, 1st edition. Edited by G.L. Heritage and A.R.G. Large.
 ISBN 978-1-4051-5717-9

LONG-TERM CONTEXT

For over 100 years people have been taking aerial photographs and using them as a simple but effective remote sensing tool. In Britain, the use of aerial photographs in archaeological applications really began to become a serious tool in the 1920s. Crawford's publications (1923 and 1928) set the main standards that remained in use throughout the twentieth century, his most notable work was *Wessex from the Air* (Crawford & Keiller, 1928), showing how aerial photography and fieldwork could dramatically change our understanding of ancient landscapes (Barber, 2009).

The quality of the film and cameras has changed over time and in the last decade there has been an increased use of digital cameras. However, in the majority of cases, the taking of aerial photographs specifically for archaeological purposes still utilises the same procedure of using a hand-held camera located in a small aircraft. Since the 1970s there has been access to satellite imagery at improving resolutions and there have been advances in multi-spectral digital imagery from airborne platforms, but unlike ground based remote sensing, there had been no radical new techniques. This changed in the last eight years when airborne laser scanning (LiDAR) became more readily available for civilian researchers with some dramatic early results which led many to suggest this technique could revolutionise aerial survey.

Prior to the advent of LiDAR, the Aerial Survey and Investigation team at English Heritage had been using standard aerial photo interpretation techniques for well over 30 years, developing expertise in photo interpretation and mapping technologies. A major result of this has been the development of a systematic scheme of mapping from aerial photographs that can be used as the basis for research and management of the historic environment. This National Mapping Programme (NMP) aims to systematically map 'to a consistent standard by interpretation, mapping, classification and description ... all archaeological sites and landscapes in England which are visible on aerial photographs' (Bewley, 2001, 2002). At present, conventional aerial photographs, both vertical and oblique, are the main source consulted, but over the years the Aerial Survey and Investigation team at English Heritage (or its previous incarnation as the Air Photography Unit of the Royal Commission on the Historical Monuments of England) have carried out experiments with other media/sources. These have included some of the earliest examinations of the possible use of multi-band film in the early 1970s (Hampton, 1974) and an analysis and evaluation of Infra Red Linescan Imagery as used by the Royal Air Force (RAF) in the UK (Small, 1994). When English Heritage were first shown LiDAR in 2001, the potential was immediately obvious, and they therefore rapidly began a series of projects to evaluate its usefulness. With each project, more was learned about the possibilities of LiDAR and about the various problems and other issues involved with its capture and use.

LIDAR – HISTORY AND DEVELOPMENT

LiDAR (Light detection and ranging) survey is based on the principle of measuring distance through the time taken for a pulse of light to reach the target and return. Airborne LiDAR does this with a pulsed laser beam which is scanned from side to side as the aircraft flies over the survey area, measuring between 20,000 to 100,000 points per second to build an accurate, high resolution model of the ground and the features upon it. (For further details of the technological specifications see Bewley *et al.*, 2005; Holden *et al.*, 2002; Heritage & Large, this volume.) It was developed in the 1960s by the US military for submarine detection and has been in commercial use by various bodies, mainly for bathymetric measuring, since the 1970s. The Environment Agency (EA) in England and Wales has been working with LiDAR for at least 10 years, but until recently the standard resolution used was 2 m or less. The resolution of LiDAR refers to the resolution of features on the ground and is based on the number of hits on the surface within a 1 metre square for the point cloud data; 2 m resolution therefore equates to approximately one hit within every 2 m^2 block (Crutchley, 2009). This is perfectly adequate for measuring large-scale topographic changes for flood modelling etc, the purpose for which the flights were commissioned; however, it was generally considered that this was not good enough for spotting a wide range of archaeological features. Certainly this was the experience from previous analysis of satellite imagery.

Prior to 2000, the archaeological community in the UK had not considered LiDAR for archaeological use and indeed very few had even heard of the technique. This was not necessarily the case elsewhere in the world. One of the earliest examples of the use of LiDAR for archaeological research came as a result of the First Remote Sensing Conference in Archaeology in 1984 when Tom Sever of NASA joined Payson Sheets, an archaeologist from the University of Colorado, who had a National Science Foundation grant to excavate prehistoric villages in the Arenal Region of Costa Rica. Devastated by ten volcanic eruptions over the past 4000 years, these villages were preserved to some extent under layers of ash. Sever joined Sheets' research team to investigate the utility of remote sensing technology in a tropical environment. NASA initiated two series of flights using a specially equipped Learjet flying at an altitude of about 1000 ft. Together with colour and false-colour infrared photographs, seven spectral bands from Landsat's Thematic Mapper, thermal data and two bands of synthetic aperture radar data, they also collected LiDAR data (McKee and Sever, 1994). Whereas multi-spectral data, and even satellite imagery to a certain extent, have been used in the succeeding years for archaeological projects, there is little evidence for further use of LiDAR for examining heritage landscapes in the Americas or Europe until the last few years.

The situation in the UK changed in November 2000 when Nick Holden of the Environment Agency (EA) gave a demonstration of the system at a NATO funded seminar on using aerial survey techniques for archaeological purposes held in Leszno, Poland. He showed several examples of how LiDAR was used to create digital terrain models (DTMs) that the EA used for modelling floods etc. One of the sites he showed was of Newton Kyme in Wharfedale, North Yorkshire. This had long been known through cropmarks as the site of a Roman fort and settlement, but had been under the plough for decades and was assumed to have been completely levelled. Using false sunlight on the LiDAR DTM at an extremely low elevation it was possible to make out the ramparts of the fort, suggesting that there were still traces of the feature visible on the ground. These traces, however, would have been considerably less than a metre in height and had obviously been missed by previous surveys of the area. This suggested that LiDAR was capable of picking up traces of features that were apparently not readily recoverable by the usual archaeological survey techniques. It was also recognised that improving on the standard 2 m resolution of EA LiDAR data was likely to enhance its usefulness for studying archaeological features.

ENGLISH HERITAGE'S USE OF LIDAR FOR ARCHAEOLOGICAL SURVEY

Through growing experience with LiDAR there are two related issues that English Heritage have tried to address, both in the different projects with which we have been involved and with reference to future projects. One is the issue of the quality of the data in its various forms and what can be derived from it. This has been addressed through examining different sources of data at different resolutions and in different formats. The other issue relates to the actual use of this data. No matter in what format it is provided (with the exception of flat image files, the inadequacies of which are dealt with below) there is a requirement for relatively complex software and hardware to view the data in an easily usable form. English Heritage's investigation of this technology has been to assess **what** can be derived from LiDAR data, and also to establish **how** that data can be made available to those who need it in a readily usable form. How these two themes have been tackled will be demonstrated through the projects described below.

Stonehenge

To investigate the potential of LiDAR more fully, English Heritage commissioned the Environment Agency to fly the area of the Stonehenge World Heritage Site (WHS) (Figure 12.1) with LiDAR at a 1 m ground resolution, based on the previous assumption that 2 m would be too coarse to record many of the monuments. The survey was carried out in December 2001. The Stonehenge landscape is probably one of the most thoroughly investigated landscapes in the world, having been analysed archaeologically using a variety of survey techniques over a considerable period of time. Aerial photographic interpretation has been used on several occasions, most recently with the Stonehenge World Heritage Site Mapping Project (SWHSMP) (Crutchley, 2002). The chance to compare LiDAR data with the results of recent mapping was one of the key factors that led to the choice of Stonehenge WHS as a test area. Additionally, this fitted in well with future research requirements and management needs due to planned changes in the site's surroundings.

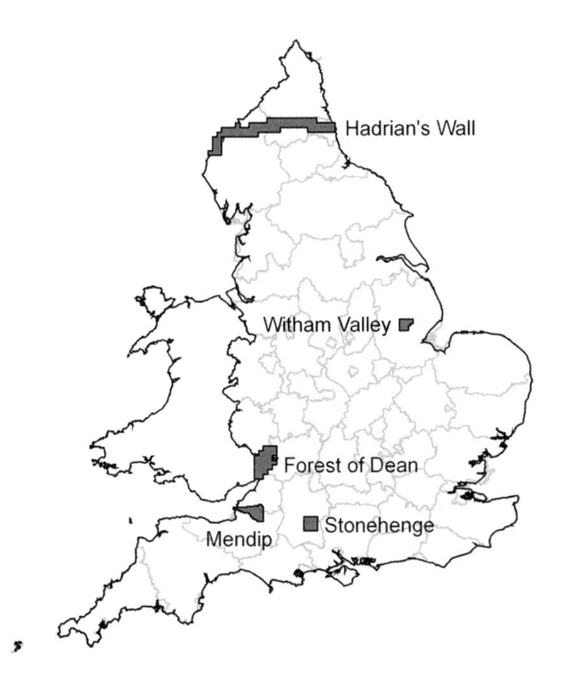

Fig. 12.1 Location diagram for project areas.

Given the intensity of previous investigations it was not expected that LiDAR would find very much that was new, but rather that this would be a test to see how many known sites could be recorded, with the added benefit of knowing which had not been completely levelled by ploughing. In the event, the results of the survey exceeded expectations. They are described in detail in Bewley *et al.* (2005) but can be summarised as follows. The LiDAR data not only recorded a small number of new sites and added significant data to previously known ones (Figure 12.2); but it also gave a much greater degree of accuracy to all those other known sites that it recorded.

Most of the archaeological features had been mapped by rectifying oblique photographs of varying quality against the Ordnance Survey 1:10,000 maps and therefore could only be assumed to be positionally accurate to within 5–10 m or occasionally even worse. Comparison of the LiDAR data with the OS 1 : 2500 scale maps showed that they had a comparable accuracy; that is to within 1–2 metres. Another key finding of the survey was that a number of features that had previously been assumed to have been completely levelled by ploughing still had a slight earthwork component.

Whilst the Stonehenge project had been very successful in terms of learning what could be derived from LiDAR data, it had also raised a number of issues with regard to actually using the data. At the time of the project, English Heritage did not have the capability to view the LiDAR data themselves. Instead, the comparison with previous NMP results was carried out using flat jpg images of hill shaded DTMs created by a third party (Colin Shell of the Department of Archaeology, Cambridge University); these were imported into AutoDesk software as background data and were then directly compared to the air photo transcriptions. They were originally provided as a set of pre-prepared georeferenced LiDAR images with relief shading from different directions, but as the comparison went on it was necessary to obtain several different views. Although a pragmatic approach for a pilot project, it was clear that this was not the most effective way of working; there would be significant benefit in the archaeological specialist having direct access to the LiDAR data, with the ability to manipulate it in real time.

Other findings reported on in Bewley *et al.* (2005) did not relate specifically to this 'standard' LiDAR data, but rather to the different uses that could be made of the DTMs derived from it and by reprocessing the data in other ways. The key element of these was the possibility of using algorithms to remove surface features such as trees, something which is examined in greater detail below.

Witham Valley

Another project gave English Heritage the chance to gain direct experience of manipulating LiDAR data and test the technique in a different landscape. To this end, English Heritage planned to carry out work in the Witham Valley as part of a broader project working in partnership with

Fig. 12.2 Newly identified banks defining fields of probable prehistoric date in an area south of Stonehenge. The features highlighted in white were newly discovered using the LiDAR data; those in black were recorded by the NMP survey using conventional photography. Underlying LiDAR data lit from the north. Source Environment Agency (Dec 2001).

Lincolnshire County Council and others. The project had developed from the original Witham Valley project that had its origins in the 1980s, when a series of rescue excavations uncovered sections of an Iron Age causeway and numerous votive offerings at Fiskerton on the north bank of the Witham (Catney & Start, 2003). The Witham valley landscape is essentially an arable one and so most of the archaeological sites are known through cropmarks and findspots. However, there are also significant pockets of earthwork survival, most notably of the medieval monastic sites. These are discussed by Everson and Stocker with particular reference to some of the causeways associated with them and their location in the landscape (Everson & Stocker, 2003).

The timetabling of the project meant that the LiDAR data used was that already available from the Environment Agency and again the results were compared with the NMP mapping from aerial photographs after that stage of the project had completed. The data provided by the EA was in the form of ASCII gridded files that could be read in ArcGIS, the software used by English Heritage for the data processing and visualisation. Data was provided in 2 km by 2 km squares, based on the OS grid, that were initially viewed as simple greyscale images with pixels graded from black to white according to the height recorded for each 2 m block, before being imported into ArcScene so that it could be visualised as a three dimensional surface.

As was expected, the visibility of features could be enhanced by using a low glancing light to emphasise the shadows of slight features; as is the case with slight earthworks and aerial photography. The more perpendicular the direction of lighting to the line of archaeological features, the better defined they are, so the advantage of being able to select the direction of illumination (e.g. having the artificial sun low, in the north) was a noticeable benefit. Another advantage of the LiDAR data in the landscape of the Witham Valley, where a lot of the features have been severely reduced by ploughing, was the ability to exaggerate the height ratio in the images, sometimes up to 20 times, to make features evident (Figure 12.3).

The results of the comparisons are described in greater detail elsewhere (Crutchley, 2006), but can be summarised as follows. LiDAR, with its ability to record slight differences in height, proved to be an excellent means of recording the subtleties in the geomorphology of the landscape, particularly the presence of palaeo-channels that can be an important indicator of the likely location of previous settlement and other activity. Relict-channels and related features were shown so clearly that it was decided it was unnecessary to map them in any further detail as they were sufficiently clear as a background image (Figure 12.4).

For the recording of the more usual type of archaeological features there were mixed results. Many of the sites recorded by the interpretation of aerial photography were visible only in the form of cropmarks or soilmarks and therefore did not show up in the three-dimensional model provided by the LiDAR data. One particularly successful result was in the area around Seney Place, Southrey, a moated retreat house owned by Bardney Abbey. Here the LiDAR data revealed extensive evidence of field systems and enclosures that had not been identified on the aerial photography. However, there was less success at the Bronze Age barrow cemetery at Barlings where a number of known barrows, clearly visible on aerial photographs could not be recognised on the LiDAR data.

Another site provided an excellent example of the value of using all available sources when undertaking this form of archaeological survey. Examination of features in the vicinity of Bardney revealed an interesting roughly 'playing card'

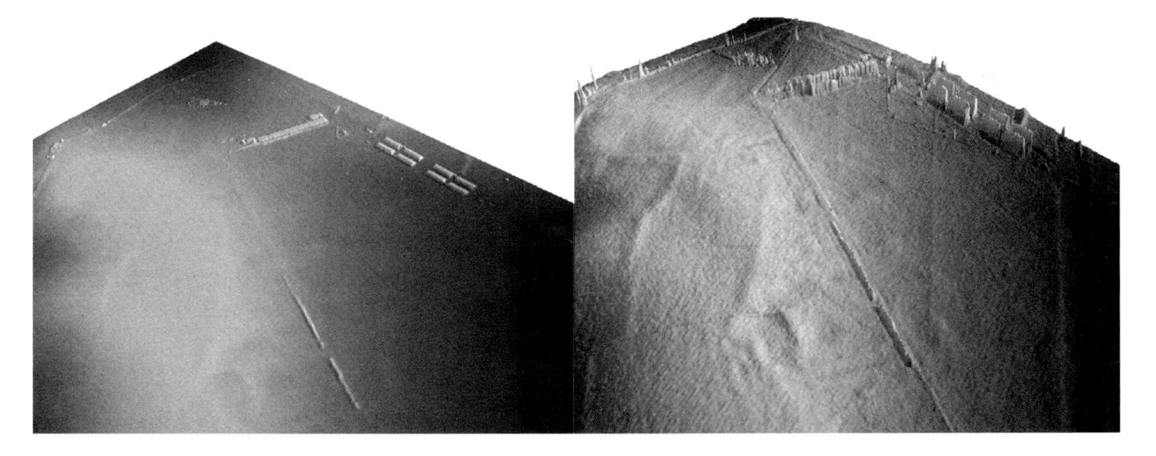

Fig. 12.3 An example of the advantage of possible height exaggeration on site visibility at Bardney Airfield, Lincolnshire. The image on the left shows the actual height of features; the image on the right has the height exaggerated ten times. Apart from the line of the runway being much clearer there are traces of the banks of possible prehistoric fields towards the left edge of the second image. LiDAR courtesy of Lincolnshire County Council: Source Environment Agency (March 2001).

Fig. 12.4 Tattershall Bridge adjacent to the Witham. On the RAF photograph (left) it is possible to see some geomorphological features e.g. relict watercourses, but the false sunlit image from the LiDAR data shows the features much more clearly. (RAF/CPE/UK/2009/2478). RAF photography NMR/LiDAR courtesy of Lincolnshire County Council: Source Environment Agency (March 2001).

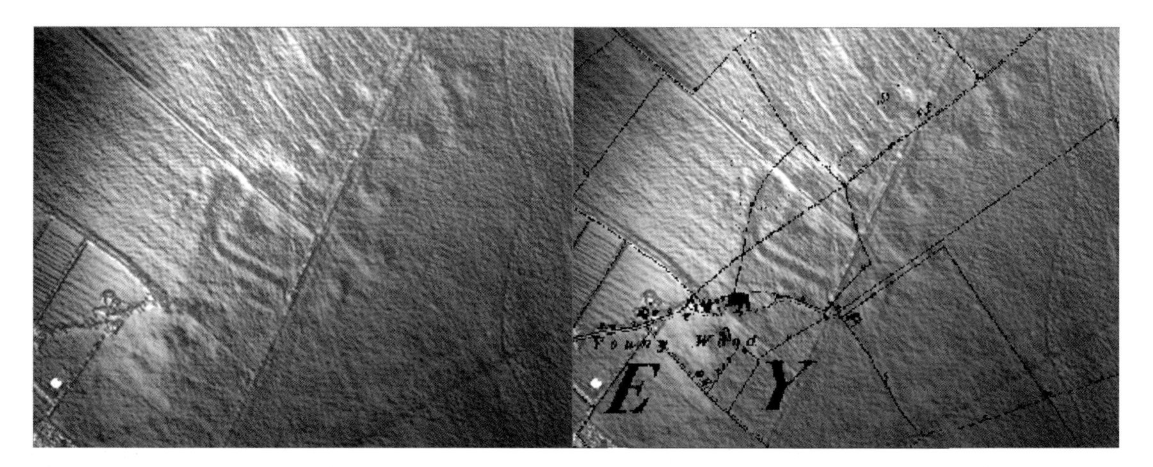

Fig. 12.5 The potential 'fort' seen on LiDAR data (left) and comparison with 1st Edition OS map (right). LiDAR courtesy of Lincolnshire County Council – Source Environment Agency (March 2001); base map (c) and database right Crown copyright and Landmark Information Group Ltd. (All rights reserved 2005).

shaped feature. Following the standard validation procedure, this structure was checked against the current OS base map and the 1st Edition OS map from the nineteenth century, and was visible on neither. Indeed the 1st Edition map even seemed to show a field boundary, that has since been removed, as appearing to curve around the feature as though respecting its presence (Figure 12.5).

Examination of the site in its landscape context (particularly with the height exaggerated ten times) revealed that it was on a slight ridge with a commanding view over the valley below (Figure 12.6).

This combination of its size and shape and its location gave very clear indications that this might be a previously unknown Roman fortlet or signal station. However, further examination of the other sources revealed a different story. It was noted when examining the current OS map, that the site lay on the edge of a former airfield and inspection of aerial photographs from during, and immediately after, World War II showed that this was an area of hard standing (i.e. a concreted area), leading to a possible hangar or storage building (Figure 12.7).

This in no way invalidates the value of LiDAR data, but simply confirms the necessity to use all available sources when carrying out a survey. The Witham project was the first opportunity for an English Heritage specialist to directly use the full functionality of 3D GIS to manipulate and display the LiDAR data. This proved especially useful when creating views that showed the features to their best advantage and also allowed sites to be viewed in their landscape setting. (Figure 12.8). More importantly it confirmed the need for each archaeologist working on the survey to have a work station enabled for looking at the LiDAR data in 3D; the other team members, working with exported tiles only, were limited in the level of interpretation that was possible on each site.

The project also proved helpful in assessing the usefulness of the standard EA data, and confirmed that 2 m resolution data was simply not sufficiently well defined to allow the identification of large numbers of features. It also raised questions about the very nature of the data supplied by EA: gridded ASCII files. Whilst work had been taking place on the Witham Valley, English

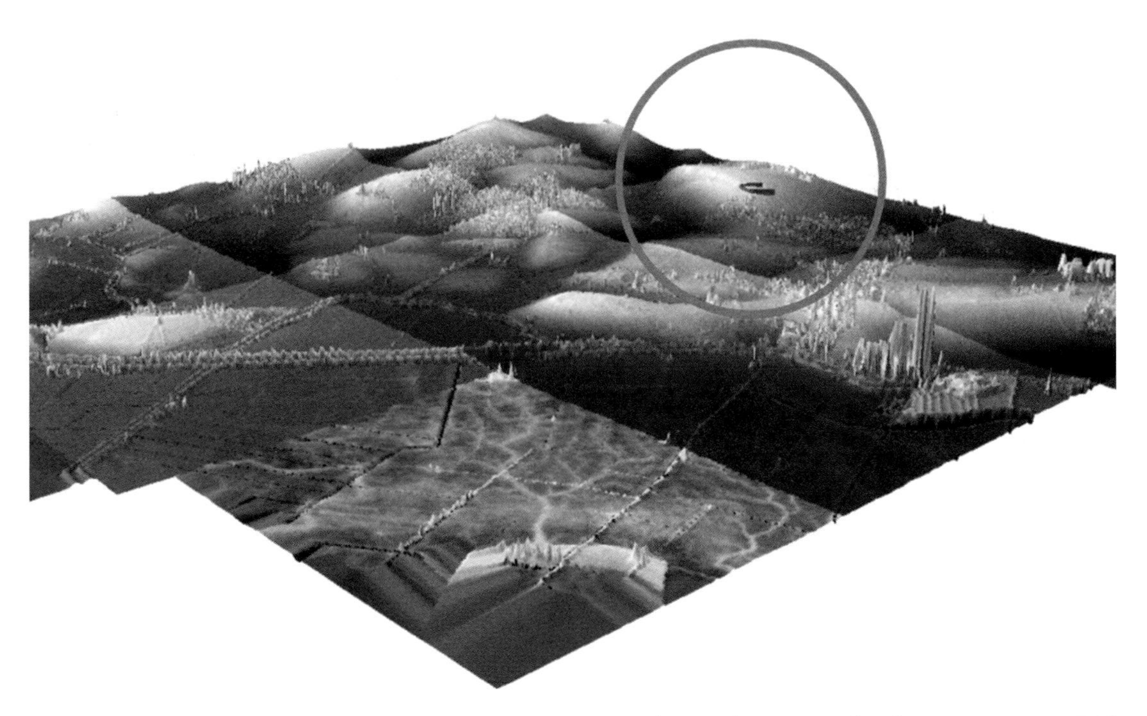

Fig. 12.6 Topographic setting of potential fort (top right) at Bardney with ten times height exaggeration. LiDAR courtesy of Lincolnshire County Council: Source Environment Agency (March 2001).

Fig. 12.7 A combination of LiDAR and photographic evidence showing the true nature of the site at Bardney airfield. LiDAR: Courtesy of Lincolnshire County Council – Source Environment Agency (March 2001); Photo: RAF 3G/TUD/UK 197 5449. English Heritage (NMR) RAF photography.

Heritage had also become further involved with Cambridge University on another project area discussed below.

Forest of Dean

For this project the specialists had direct access to the raw tabulated LiDAR data rather than gridded files and, whilst this created some issues due to the size of files, it showed that there was significantly more useful information available in the raw data. One issue that developed out of the Stonehenge survey was the use of algorithms for the removal of surface features such as trees. This was a technique that had not been considered at the time of the initial survey, but had been used in the paper published on the results of the survey (Bewley *et al.*, 2005). Almost since the first uses of LiDAR there have been basic semi-automated filters that are applied to the raw LiDAR data to produce surface models with modern features removed (e.g. the standard EA data is provided as both unfiltered and filtered). Unfortunately these appeared to be less useful for archaeological

Fig. 12.8 Stixwould Priory and the possible causeway leading from the river (centre). LiDAR courtesy of Lincolnshire County Council: Source Environment Agency (March 2001). See Plate 12.8 for a colour version of this image.

analysis. A 'bare-earth' model uses algorithms to remove all those features that it estimates to be above the natural ground surface by comparing the relative heights of recorded points. This paper is not the appropriate place to go into these in detail, but those wishing further details should see Sithole & Vosselman (2004). Because they were designed to produce a 'bare-earth' model for the purpose of calculating topologies and so on, these filtered terrain models were not concerned with the sort of small scale variations that archaeologists are usually interested in; what the algorithm sees as noise to be removed, the archaeologist sees as a feature to be interpreted. Equally worryingly, the resulting surface from using these early algorithms may have processing artefacts that can be confused with archaeological features (Figure 12.9).

What has changed recently is that mathematicians and specialists in remote sensing who have an interest in archaeology have begun to develop more specialised algorithms that will remove vegetation, but without the creation of too many artefacts. They have also looked at the raw data provided by LiDAR and how this can be used. Although LiDAR is described above as using a scanning, pulsed laser, until relatively recently the laser was treated as if it provided only a single collected point of data, the first pulse return. More recently, the raw data has been interrogated to provide more information. The composition of the laser beam is such that the nature of the

Fig. 12.9 Filtered data of the Stonehenge Cursus from the Environment Agency. Note the presence of artefacts appearing as a regular grid pattern across the flat area where trees have been 'removed' by the filter. N.B. these should not be confused with the traces of a real field system visible particularly within the banks of the cursus (centre). Underlying LiDAR data lit from the north east. Source Environment Agency (Dec 2001).

surface that it hits determines how much of the beam is reflected at one time. If the beam hits a solid surface such as a road or exposed stone all the laser energy is reflected back to the scanner simultaneously. If, however, it hits a surface that is effectively porous to the laser beam (e.g. forest canopy) it will be reflected back in degrees. There will be some reflection from the very top of the canopy, but as some energy penetrates further down there will be further reflections from lower in the canopy. If the beam is sufficiently powerful and there are enough gaps in the canopy some of the signal will reach the ground surface and be reflected back from there; the last return.

As part of the research into the potential of LiDAR, English Heritage teamed up with The Unit for Landscape Modelling (ULM), Cambridge University and the Forestry Commission to look at a test area in the Forest of Dean. This is a densely wooded region with a mix of long established deciduous woodland interspersed with more recent stands of conifers. With the exception of the extensive remains of industrial activity in the area, there are relatively few known archaeological remains from within the central forest and even those such as the massive hillfort at Welshbury are virtually invisible from the air (Figure 12.10).

Dense woodland is neither conducive to standard field survey using an EDM nor for GPS survey due to the probability of losing a signal under a dense canopy, so it was felt that any technique that might aid the recording of features in such woodland must be worth investigating further. The project used data flown by ULM and collected at a higher than average resolution allowing the creation of a 0.25 m grid. This was provided to staff at English Heritage as gridded files of both first and last return data, together with an image file of the data once it had been processed using the vegetation removal algorithm. The first return data simply recorded the canopy in much the same way as the standard aerial photograph (Figure 12.11), but the last return effectively removed the bulk of the tree cover revealing the features beneath (Figure 12.12).

This data was compared with data acquired from the NMP project previously carried out on the area. The last-pulse LiDAR data was seen to reveal the entire layout of the Welshbury fort together with the associated field system recorded by the Royal Commission on the Historical

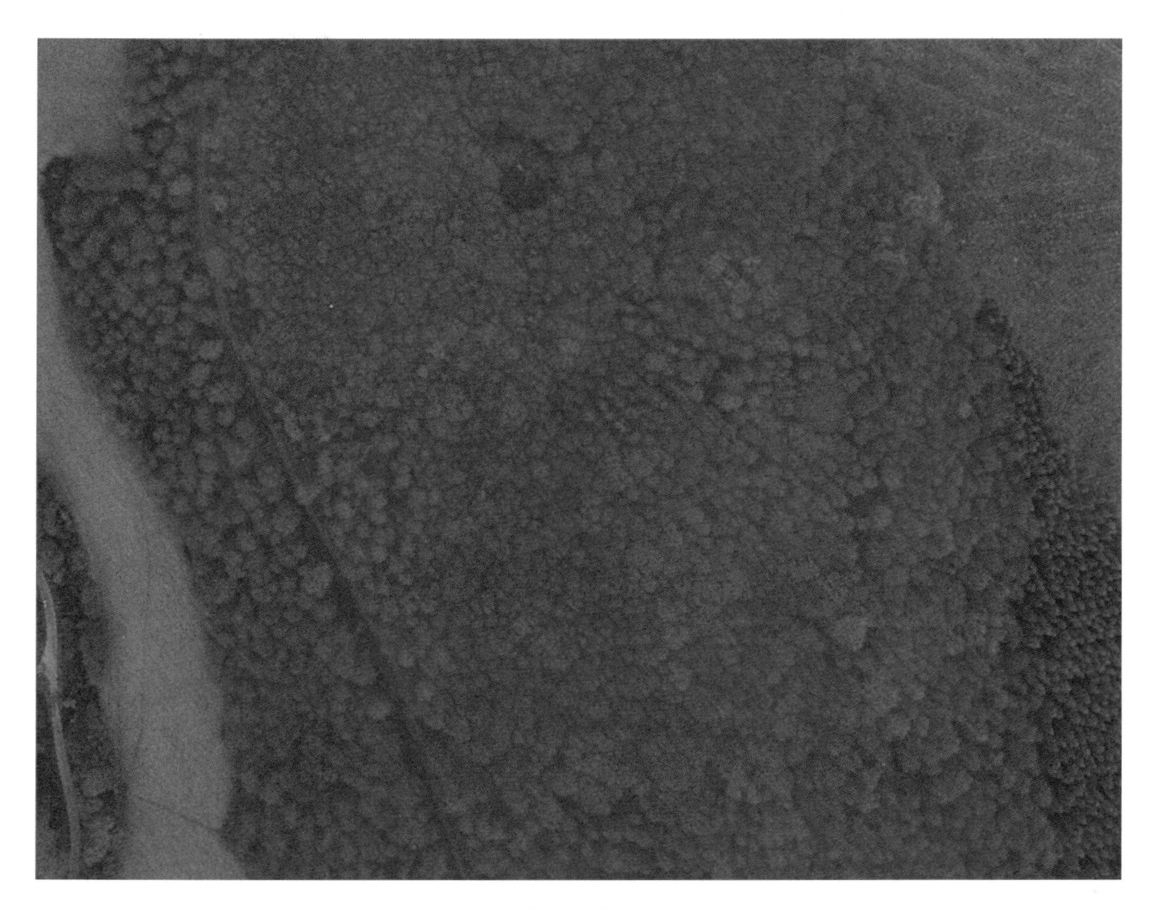

Fig. 12.10 Conventional aerial photograph of Welshbury Hillfort showing the dense tree canopy. Image courtesy of the Forestry Commission: Source Cambridge University Unit for Landscape Modelling (March 2004). See Plate 12.10 for a colour version of this image.

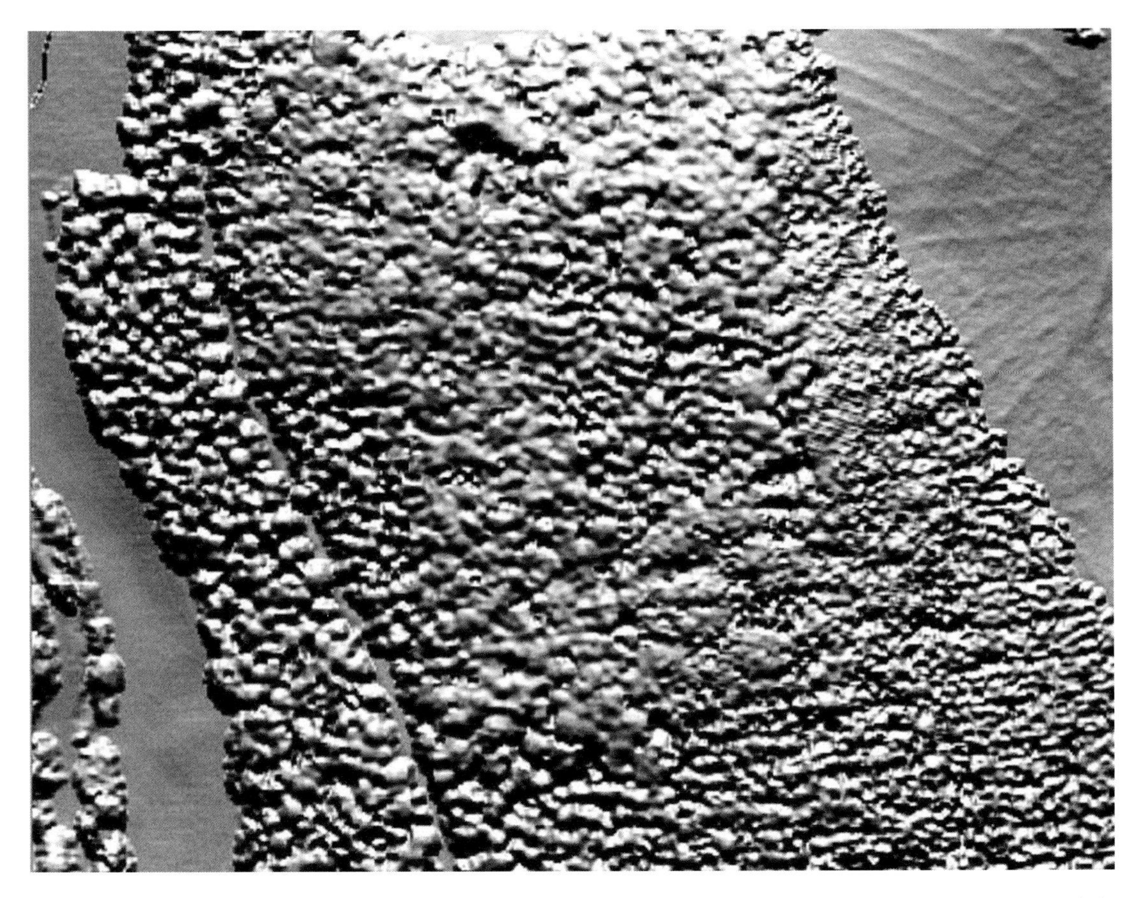

Fig. 12.11 First return data relief shaded – what is recorded is the top of the tree canopy. LiDAR courtesy of the Forestry Commission: Source Cambridge University Unit for Landscape Modelling (March 2004).

Monuments of England (RCHME) in a detailed field survey carried out in 1996. The data was also provided to Forest Research from the Forestry Commission, who carried out a walk through survey based on the data. The broad results of this project are reported in Devereux *et al.* (2005), but the key finding was that considerable detail was visible under the tree cover using the LiDAR, and it was able to record even some quite subtle features such as charcoal burning platforms.

Initial comparison between the 'raw' last return data and the processed algorithm, suggested that whilst the algorithm provided more detail than the raw data, the gains were not sufficiently high to warrant the additional resources required to obtain this when the raw data could be processed 'in house'. For this reason, when English Heritage became involved with the next LiDAR project (below) they opted against the use of the gridded data, a decision that has since been revised.

Hadrian's Wall

A further small project looked at the use of LiDAR in a landscape where archaeological remains survive as earthworks, mainly in rough pasture and

Fig. 12.12 Last return data to 'remove' the trees revealing the ground surface. The remaining 'trees' probably represent areas of particularly dense foliage or thick tree trunks/stumps. LiDAR courtesy of the Forestry Commission: Source Cambridge University Unit for Landscape Modelling (March 2004).

moorland. The Hadrian's Wall World Heritage Site is being mapped by the Aerial Survey and Investigation team as part of NMP, but LiDAR was flown only for a small area in the centre, around the fort at Brocolitia (Carrawburgh). To interpret and map the archaeological features from the LiDAR data, it is necessary to make it available in a suitable mapping package. The key factor is that whilst ArcMap and other basic GIS packages can read LiDAR data (processed as a grid) they can view it only in two dimensions as a rough greyscale image (Figure 12.13) that is not easy to interpret. It is only when the image has been processed within a package such as the ArcScene module, or something similar that allows rendering and movable shadows that aid analysis, that it can be properly interpreted (Figure 12.14). Those packages that allow manipulation of the light source and the exaggeration of the height element (e.g. Quick Terrain Modeler from Applied Imagery www.appliedimagery.com or Landserf www.landserf.org developed by City University, London) are particularly useful in highlighting slight features.

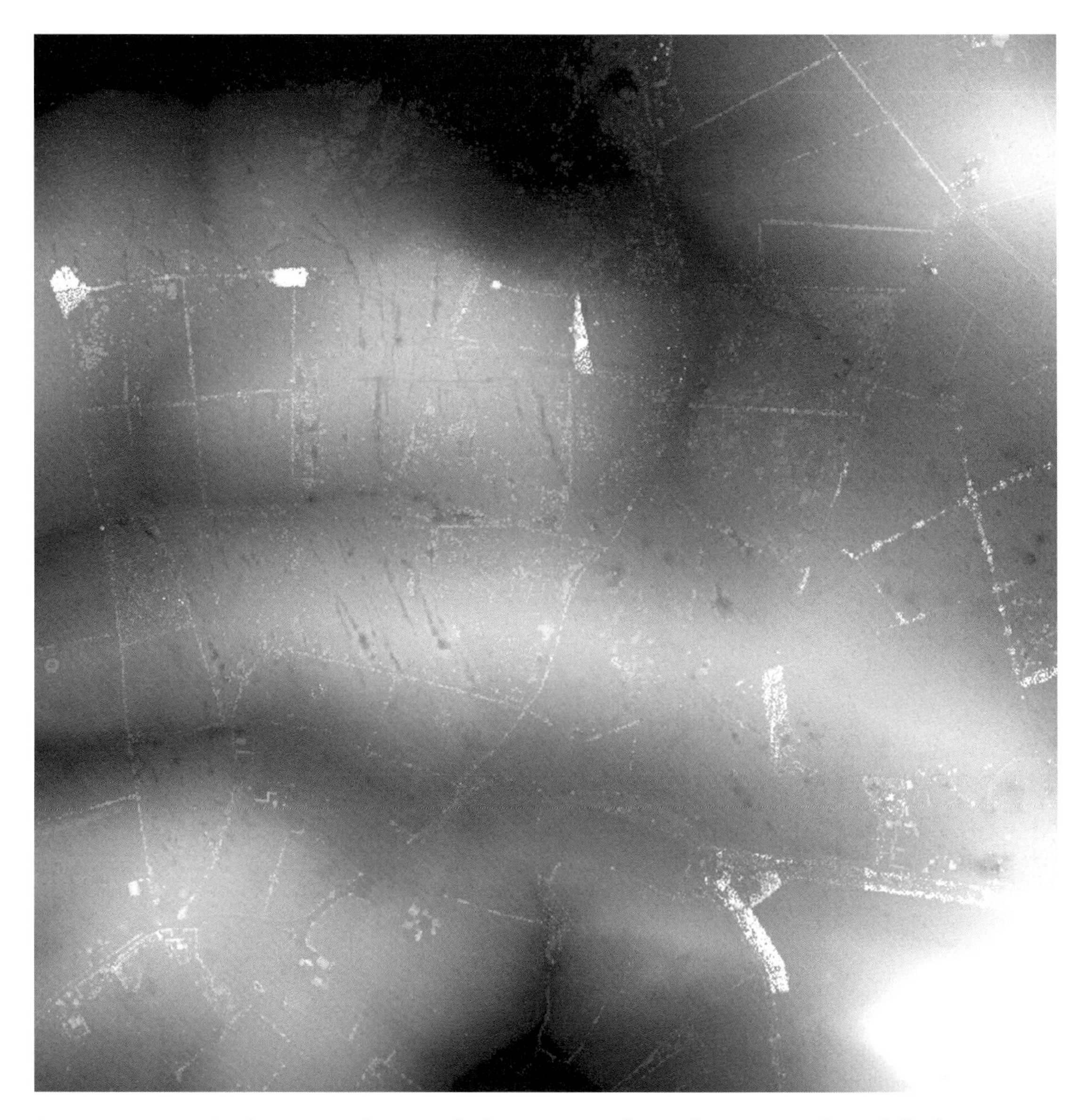

Fig. 12.13 Raw LiDAR data as viewed in a standard viewer. Note the rough appearance that is difficult to interpret. LiDAR courtesy of Mendip Hills AONB: Source Cambridge University Unit for Landscape Modelling (April 2006).

The problem is that whilst packages like ArcScene produce very good graphic images, they are designed for visualisation, not for mapping; hence the export of a registered image is not seen as a priority. The result is that it takes a series of somewhat convoluted processes to get an image output that can be read into a standard drawing or GIS package as the basis from which to map and interpret features. The Hadrian's Wall project provided the opportunity to experiment

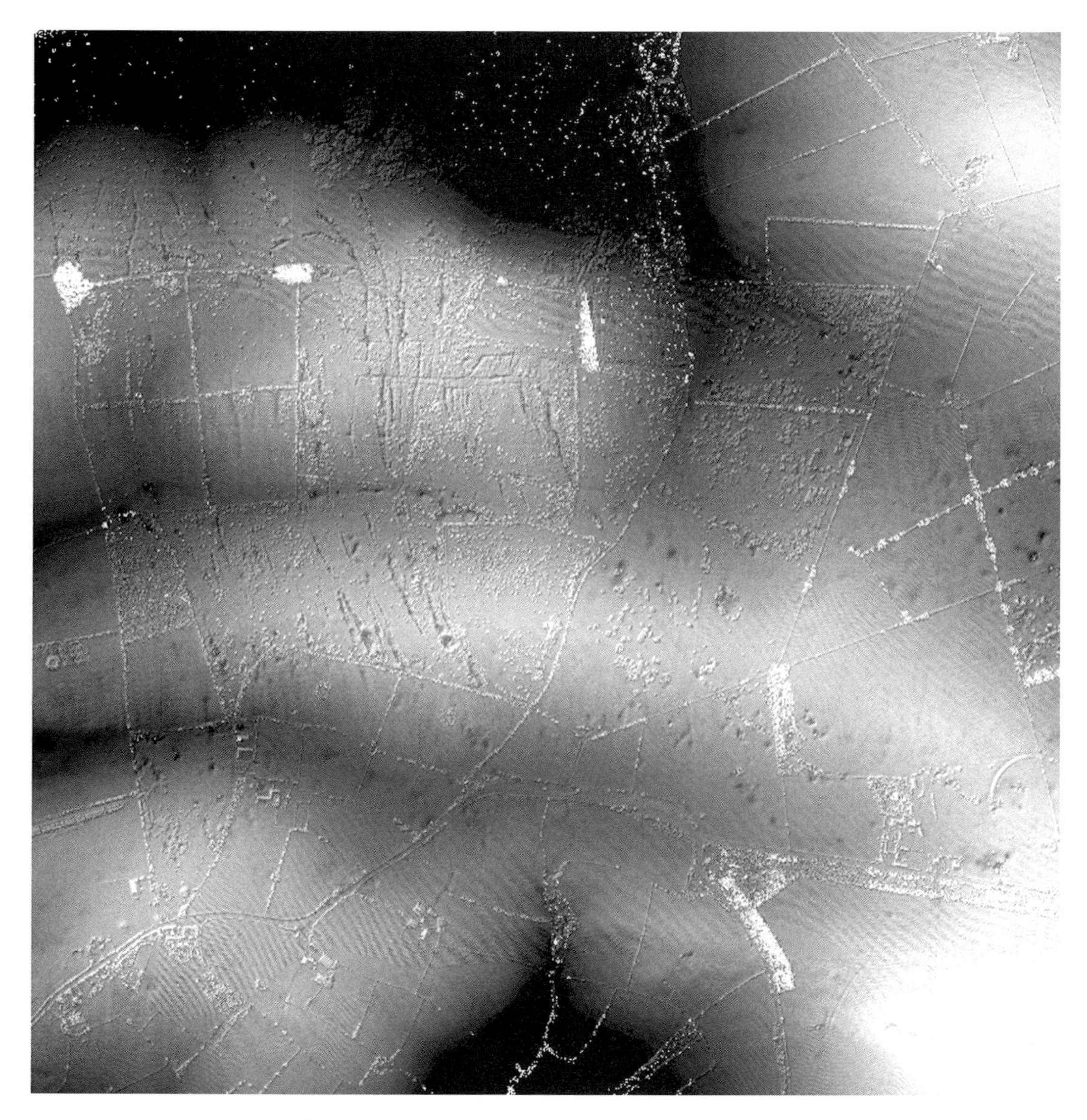

Fig. 12.14 LiDAR data viewed through the ArcScene module allowing rendering creating an image far easier to interpret. LiDAR courtesy of Mendip Hills AONB: Source Cambridge University Unit for Landscape Modelling (April 2006).

with these outputs, but it was the most recent project where the various aspects of the analysis of LiDAR data all came together, allowing the data to be used interactively as a primary source.

Mendip AONB

From the very beginning of working with LiDAR data, English Heritage have been looking to use it as a primary source rather than for comparison after conventional mapping has taken place.

This has recently been achieved for the Mendip Hills AONB project, looking at specially commissioned LiDAR data at the same time as examining the conventional aerial photographs. The data were collected by ULM Cambridge with a ground resolution of *circa* 1 m and so provide good visibility of archaeological features. The project also offered the opportunity to address the other key issue that has been raised by all the projects to date; the need to view the LiDAR data interactively without the need for complex and expensive hardware and software for every user. Working solely with 2D images (as in all previous projects), has proved to be an unsatisfactory method and it is clear that the people doing the interpretation need direct access to the data in a 3D interactive environment (Figure 12.15). The solution that has been investigated in the Mendip Hills project is the use of a combination of programs that allow the creation of files in a proprietary format that can then be viewed via a free downloadable viewer. The primary mapping continues to be carried out using AutoDesk Map, the standard mapping tool for all English Heritage NMP projects, but this is used alongside Quick Terrain Reader, a free downloadable reader of files produced in Quick Terrain Modeler (see above). This means that for each tile of data a flat image file with shaded relief can be provided to the interpreter for importing into AutoDesk together with a file that can be interrogated interactively through the viewer. This enables the interpreter

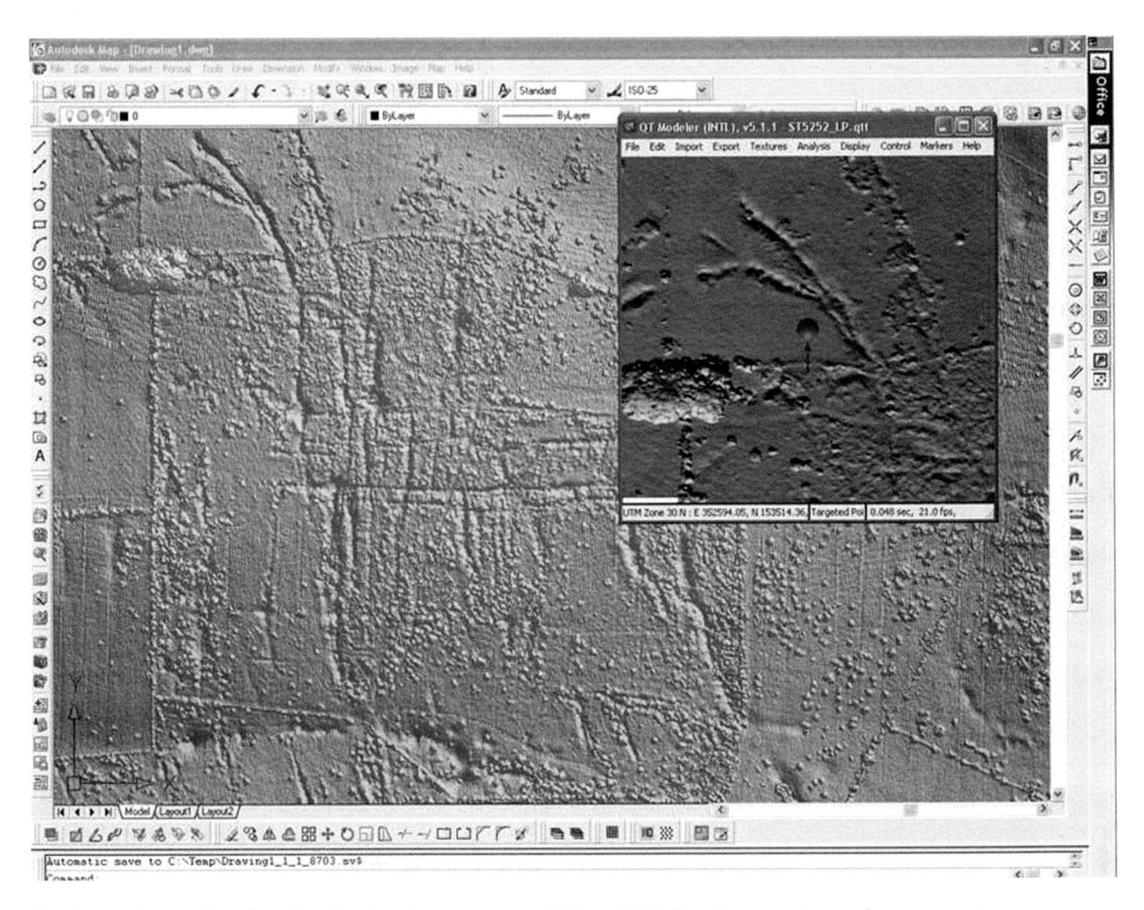

Fig. 12.15 Screenshot showing the simultaneous use of 2D and 3D data for mapping and interpretation.

to examine the data themselves, changing the elevation and azimuth so as to emphasise features of interest. Whilst it is not possible to plot features against the 3D image, it is possible to view the 3D and flat images simultaneously and map features onto the flat image based on what is visible in 3D. If it is not possible to see the features on the flat image even with the assistance of the 3D image, a revised flat image can be created using the parameters determined by the interpreter in 3D. The added advantage of this system is that it is possible to provide data to third parties (e.g. partner organisations or independent researchers) as 3D as well as 2D data, with the interpreted vector data overlaid on the DTM if necessary.

WIDER USE OF LIDAR FOR ARCHAEOLOGY

As well as the research being carried out by Cambridge University there has been a limited take up of the use of LiDAR for archaeological survey and other related work. In 2001, Defence Estates on Salisbury Plain Training Area carried out research to assess the suitability of using remote sensing techniques in the management of archaeological monuments; this was designed primarily to map changes of land use and the consequent threat to the archaeology, rather than predominantly looking to record the archaeological remains themselves (Barnes, 2003). A number of people have used the data previously collected by the EA as a source for examining the geomorphology of river valleys such as the Trent (Challis, 2005, 2006), Ribble and the Vale of York; these have been predominantly interested in using LiDAR to trace palaeochannels and other aspects of the geomorphology of the landscape, rather than looking at archaeological monuments. Even Time Team, the popular archaeology television programme produced by Channel 4 in the UK, used LiDAR data in one episode when they were looking for two crashed World War II bombers in Warton Marsh near Preston. However, again the 2 m interval data was only used for plotting the general surface terrain of the marsh and not to identify archaeological features. The LiDAR data commonly records the *x,y,z* values for first and last returns of each pulse, together with a measure of the intensity of the returned signal. The majority of work relating to LiDAR data has dealt with the locational information to create DSMs and DTMs that can be examined for the presence of features of archaeological interest. However, the intensity of the signal return can be used to analyse the reflectivity of the surface being hit by the laser; an ongoing project is investigating this as a potential means of assessing the moisture content of the surface being scanned and how this might be used to predict the likelihood of preservation of archaeological remains (Challis *et al.*, 2006, 2008).

Outside of the UK, LiDAR is becoming more commonly experimented with across Europe for archaeological purposes. A project carried out by a team including staff from Cambridge University, used LiDAR purely for archaeological research on the Loughcrew landscape in Ireland and produced dramatic results (Shell & Roughley, 2004). Here LiDAR data was used to examine known sites in the landscape and also to record new sites. The survey increased the data held by the body responsible for the local record of ancient monuments by nearly 900%, but this was prior to further analysis and the survey accepts that some of the features recorded may have relatively recent origins. Even so this is a spectacular example of the efficacy of LiDAR in the right environment. Further afield, Sittler carried out some research that dramatically revealed extensive medieval ridge and furrow cultivation preserved in later woodlands near Rastatt in south west Germany (Sittler, 2004).

The latest advance for LiDAR is full waveform digitisation of all the laser returns, a process that produces a complete 3D model from each pulse, rather than just the first and last returns. By combining the added detail from the whole pulse of the beam, such as the echo width and amplitude, it is possible to produce much more accurate models of the ground surface by eliminating ground cover that can give a false reading that appears to be the ground surface (Doneus &

Briese, 2006; Doneus *et al.*, 2008). The downside of this is that it produces even greater amounts of data, something that is already an issue and is discussed above.

THE FUTURE

Whilst the use of the 3D viewer is a pragmatic solution to the problem, it is not ideal and there are still serious issues with regard to mapping from the LiDAR data. For it to be most useful there needs to be a way of mapping directly from the 3D data whilst in an interactive mode. The issue has been that, until recently, the majority of those working with LiDAR data, and indeed laser scanned data in general, either wanted to use it to create 3D models for presentation or wanted to be able to extract data automatically from point clouds; there was little call for the manual extraction of data from processed LiDAR data. However, the more recent versions of AutoDesk Map (2007 onwards) include the facility to view raster surfaces with interactive hill shading defined by a user-controlled light source together with height exaggeration. Placing these facilities within the CAD environment means that it is now possible to combine the mapping elements of CAD with the 3D facilities that allow the enhancement of LiDAR data, which are so crucial in its interpretation. These techniques will be explored more fully in the forthcoming projects to be carried out by English Heritage and will be published elsewhere.

The second element that needs further examination is the use of algorithms for canopy penetration. Whilst the initial analysis of raw last return LiDAR data from the Forest of Dean appeared to promise good results, these have not been borne out by further examples. With some small exceptions the LiDAR data for Mendip has been unsuccessful at locating features in woodland, due in part to lack of canopy penetration. A high percentage of the woodland on Mendip consists of coniferous plantations, which have proved the least susceptible to canopy penetration in all cases, and this may be a major factor in the lack of success in this area. However, there are clearly some issues with regard to canopy penetration using just last return data, as there are examples where what appears to be one swath of LiDAR has achieved relatively good penetration whereas that immediately above or below it has not. Further work has also been carried out with the Forestry Commission and Cambridge University specifically looking at the benefits of algorithms versus raw last return data and the evidence is strongly in favour of algorithms (P Crow: pers comm). The issue remains that the process of using algorithms is quite specialised and requires relatively complex and expensive hardware and software to carry out. The more detailed analysis of full waveform data promises even greater accuracy and detail from LiDAR in woodland, but again requires even greater analysis and creates files of such magnitude they cannot easily be dealt with using standard computer equipment. The joint projects with the Forestry Commission and Cambridge University show the potential for collaboration in looking at LiDAR data, but there is a requirement for more of the same if the end users are to get the greatest benefit from the wealth of data that LiDAR is undoubtedly capable of providing.

CONCLUSION

LiDAR data is clearly an important new source of information for aerial survey and has the potential to provide major benefits to those carrying out such work. For areas where there are expected to be large numbers of extant earthwork features, LiDAR could form the basis of surveys that could then be supplemented by traditional aerial photographic analysis and ground survey. There is already a substantial, and rapidly increasing, amount of LiDAR data in existence, most of which has been flown for non-archaeological reasons. Where the data is of sufficiently high resolution there is no reason why this should not be utilised for archaeological survey. After all, archaeologists, particularly aerial surveyors, have been using data collected for other purposes for years.

ACKNOWLEDGEMENTS

Copyright for all images, with the exception of Figure 12.10, belongs to the author and English Heritage. All images with the exception of Figure 12.10 were created by the author, based on data provided by various other bodies. LiDAR data for Stonehenge was provided by the Environment Agency and initially processed by Colin Shell of Cambridge University, who has been a great help to the author in understanding how to manipulate LiDAR data. LiDAR data for the Witham Valley was provided by Lincolnshire County Council although the original source was the Environment Agency. The author is grateful to Lincolnshire County Council for the provision of the data and the research opportunities it provided. LiDAR data for the Forest of Dean was provided by the Forestry Commission. The data was collected by the Unit for Landscape Modelling (ULM), Cambridge University funded by the Forestry Commission. The author is most grateful to Peter Crow at the Forestry Commission and the staff of ULM for use of the LiDAR data and the reproduction of some of the findings of their research in the Forest of Dean. The author would also like to thank Peter Horne of the English Heritage Aerial Survey and Investigation team for his comments on early drafts of the paper.

REFERENCES

Barber M. 2009. *Mahta Hari's glass eye and other tales.* English Heritage.

Barnes I. 2003. Aerial remote-sensing techniques used in the management of archaeological monuments on the British Army's Salisbury Plain training area, Wiltshire, UK. *Archaeological Prospection* **10**: 83–90.

Bewley RH. 2001. Understanding England's Historic Landscapes: An Aerial Perspective. *Landscapes* **2.1**: 74–84.

Bewley RH. 2002. Aerial Survey: Learning from a Hundred Years of Experience? In: *Aerial Archaeology - Developing Future Practice' Proceedings of the NATO Advanced Research Workshop on Aerial Archaeology, 15–17 November 2000,* Bewley RH, Rączkowski W (eds), NATO Life Sciences, Series 337; 11–18.

Bewley RH, Crutchley SP, Shell C. 2005. New light on an ancient landscape: LiDAR survey in the Stonehenge World Heritage Site. *Antiquity* **79, No. 305**: 636–647.

Catney S, Start D. 2003. Time and tide: The archaeology of the Witham Valley. *Proceedings of the Witham Archaeological Seminar of December 2002,* WVARC.

Challis K. 2005. Airborne LiDAR: A tool for geoarchaeological prospection in Riverine Landscapes. In: *Archaeological Heritage Management in Riverine Landscapes,* Stoepker H (ed.), Rapporten Archeologische Monumentenzorg 126; 11–24.

Challis K. 2006. Airborne laser altimetry in alluviated landscapes. *Archaeological Prospection* **13**: 103–127.

Challis K, Howard AJ, Kincey M, Carey C. 2008. *Analysis of the Effectiveness of Airborne LiDAR Backscattered Laser Intensify for Predicting Organic Preservation Potential of Waterlogged Deposits.* ALSF Project Number 4782. University of Birmingham.

Challis K, Howard AJ, Moscrop D, Gearey B, Smith D, Carey C, Thompson A. 2006. Using airborne LiDAR intensity to predict the organic preservation of waterlogged deposits. In: *From Space to Place: 2nd International Conference on Remote Sensing in Archaeology,* Campana S, Forte M (eds), BAR International Series 1568; 93–98.

Crawford OGS. 1923. Air Survey and Archaeology. *Geographical Journal* **LXI**: 324–366.

Crawford OGS, Keiller A. 1928. *Wessex from the Air.* Clarendon Press: Oxford

Crutchley SP. 2002. *Stonehenge World Heritage Site Mapping Project: Management Report.* Unpublished report, English Heritage.

Crutchley SP. 2006. LiDAR in the Witham Valley, Lincolnshire: An assessment of new remote sensing techniques. *Archaeological Prospection* **13**: 251–257.

Crutchley SP. 2009. *The Light Fantastic – Using Airborne Laser Scanning in Archaeological Survey.* English Heritage.

Devereux BJ, Amable GS, Crow P, Cliff AD. 2005. The potential of airborne LiDAR for the detection of archaeological features under woodland canopies. *Antiquity* **79, No. 305**: 648–660.

Doneus M, Briese C. 2006. Full-waveform airborne laser scanning as a tool for archaeological reconnaissance. In: *From Space to Place: 2nd International Conference on Remote Sensing in Archaeology*, Campana S, Forte M (eds), BAR International Series **1568**: 99–105.

Doneus M, Briese C, Fera M, Janner M. 2008. Archaeological prospection of forested areas using full-waveform airborne laser scanning. *Journal of Archaelogical Science* **35**: 882–893.

Everson P, Stocker DA. 2003. Coming to Bardney... – The landscape context of the causeways and finds groups of the Witham Valley. In: *Time and Tide: The archaeology of the Witham Valley: Proceedings of the Witham Archaeological Seminar of December 2002*, Catney S, Start D (eds), Witham Valley Archaeological Research Committee; 6–15.

Hampton JN. 1974. An experiment in multispectral aerial photography for archaeological research. *Photogrammetric Record* **8** No. 43: 37–64.

Holden N, Horne P, Bewley R. 2002. High-resolution digital airborne mapping and archaeology. In *Aerial Archaeology – Developing Future Practice' Proceedings of the NATO Advanced Research Workshop on Aerial Archaeology, 15–17 November 2000*, Bewley RH, Rączkowski W (eds), NATO Life Sciences, Series **337**: 173–175.

McKee BR, Sever TL. 1994. Remote Sensing in the Arenal Region. In: *Archaeology, volcanism and remote sensing in the Arenal Region, Costa Rica*. Sheets PD, McKee BR (eds), University of Texas Press, Austin; 136–141.

Shell CA, Roughley CF. 2004. Exploring the Loughcrew Landscape: a New Approach with Airborne LiDAR. *Archaeology Ireland* **18**, No. 2, Issue No. 68: 20–23.

Sithole G, Vosselman G. 2004. Experimental comparison of filtering algorithms for bare-earth extraction from air-borne laser scanning point clouds. *ISPRS Journal of Photogrammetry and Remote Sensing* **59**: 85–101.

Sittler B. 2004. Revealing historical landscapes by using airborne laser scanning. In: *Laser-Scanners for Forest and Landscape Assessment: Proceedings of the ISPRS Working Group VIII/2, Volume XXXVI, Part 8/W2*, Thies M, Koch B, Spiecker H and Weinacker H (eds), 258–261.

Small FJ. 1994. *RAF Infra Red Linescan Imagery: An Analysis and Evaluation*. Unpublished report, English Heritage.

13 Airborne and Terrestrial Laser Scanning for Measuring Vegetation Canopy Structure

F.M. DANSON[1], F. MORSDORF[2] AND B. KOETZ[3]

[1]Centre for Environmental Systems Research, School of Environment and Life Sciences, University of Salford, Manchester, UK
[2]Department of Geography, University of Zurich, Zurich, Switzerland
[3]ESA – ESRIN, EO Science, Application and Future Technologies Department, Frascati, Italy

INTRODUCTION

Terrestrial vegetation covers 60% of the Earth's land surface, varying spatially and temporally in cover, composition and function with climatic gradients and disturbance patterns. The vegetation canopy is the interface between the land surface and the atmospheric boundary layer and controls radiative energy exchanges and the fluxes of gases including water vapour and carbon dioxide. At regional to global scales these processes are closely coupled to climate dynamics and there is growing evidence of the importance of terrestrial vegetation as both a source and a sink within the global carbon cycle. Terrestrial vegetation contains around 500 Gt of carbon stored in its biomass and exchanges around 60 Gt/year of carbon with the atmosphere; although currently a net carbon sink, it may become a net source of carbon within the next 50 years (Cox *et al.*, 2000) influencing atmospheric carbon dioxide concentration and global temperature changes. At local to regional scales vegetation canopy structure affects light interception and evapotranspiration and represents a key factor in determining net primary productivity. Vegetation canopy characteristics also affect ecological processes like fire vulnerability, habitat structure and successional processes. These characteristics are not static, however, so that a complete functional description of a vegetation community should take into account changes in canopy structure with phenology, plant stresses, ecosystem dynamics and disturbance.

Vegetation canopy structure describes the size, shape and three-dimensional distribution of canopy components including stems, branches, shoots and leaves (Weiss *et al.*, 2004). In very simple regular canopies it may be possible to reconstruct a detailed three dimensional architectural model using photogrammetry, manual 3D digitization or direct measurement. For more complex canopies, height and plant density may be measured relatively easily,

Laser Scanning for the Environmental Sciences,
1st edition. Edited by G.L. Heritage and A.R.G. Large.
 ISBN 978-1-4051-5717-9

but other canopy structural characteristics are normally determined by reference to some average value or distribution function. Airborne and terrestrial laser scanning promise to revolutionise the 3D measurement of complex plant canopies, like forests and woodland, providing data on leaf and stem arrangement with unprecedented accuracy and detail. A simple canopy volume containing only leaves may be described by three variables, the leaf inclination distribution function (LIDF), the leaf area density and a clumping factor. Leaf area index (LAI), defined as the one-sided leaf area per unit ground area, is the sum of the leaf area density integrated over the full depth of the canopy. Gap fraction, the probability of a beam of light intercepting a leaf in a given direction, is determined by the combination of LIDF, leaf density and clumping. Gap fraction in zenith direction of 0°, looking directly up from below the canopy, is the canopy cover.

Information on canopy structure, and LAI in particular, is required at a wide range of spatial scales to provide, for example, inputs to regional scale ecosystem productivity models and global models of climate, hydrology, biogeochemistry and ecology (Bonan, 1993). Local scale information is required for ecological management, disease and stress detection and a range of other management tasks. The key tool for spatial mapping of vegetation canopy characteristics is satellite remote sensing which provides a synoptic view of the Earth's surface and is now used for operational mapping of vegetation species distribution and quantitative canopy characteristics (Boyd & Danson, 2005). The application of satellite remote sensing normally relies on variation in the spectral reflectance of the surface. The spectral reflectance characteristics of vegetation vary with the biochemical and biophysical properties of the canopy components so that a wide range of variables may be mapped by inference using canopy 'spectral signatures'. Remotely sensed estimates of canopy structure provide two-dimensional maps of the variables of interest, normally derived through the use of empirical regression models and spectral vegetation indices, like the normalized difference vegetation index (NDVI) (Cohen *et al.*, 2003; Colombo *et al.*, 2003), or in some cases by radiative transfer modelling (Danson *et al.*, 2001; Kötz *et al.*, 2004; Schlerf & Atzberger, 2006). A limitation of these approaches however, is that they provide no information on the three-dimensional structure of the canopies. Airborne laser scanning (ALS) has recently provided an opportunity to derive three dimensional information about vegetation canopy structure for the first time. This technology, coupled with the very high resolution information that may be derived from terrestrial laser scanners (TLS), has the potential to provide new insights into spatial and temporal variation in vegetation canopy structure.

The next section of this chapter provides an overview of the traditional methods to measure vegetation canopy structure on the ground, including direct and indirect methods and destructive and non-destructive approaches. The following section considers the nature of the interactions of laser light with vegetation canopies, emphasizing the importance of multiple scattering within a single laser beam. The next section reviews the existing research on airborne and ground based laser scanning of vegetation structure, further illustrated with reference to airborne and terrestrial laser scanning experiments by the authors at a test site in eastern Switzerland. Recent research aimed at modelling the interaction of lasers with vegetation canopies is then considered, and the chapter concludes by commenting on research priorities for laser scanning of vegetation structure.

GROUND BASED METHODS TO MEASURE VEGETATION CANOPY STRUCTURE

Direct measurement of vegetation canopy structure normally involves destructive sampling at the plant level (Jonckheere *et al.*, 2004) and the extrapolation of the measurements to plot or canopy scale. Direct contact-based methods used to measure LAI include stratified clipping of leaf material and litter fall collection; indirect contact based methods include the application of previously determined allometric equations and the use of point quadrat theory. The inclined point quadrat method involves inserting thin needles into the canopy at known zenith angles, and counting the number of contacts with green leaf material. LAI may be determined from simple relationships with contact frequency, even when the canopy elements are not randomly distributed. Such measurements are however extremely time consuming for all but the simplest vegetation canopies.

Non-contact-based methods have been developed from the concept of contact frequency and measure gap fraction or gap size distribution. Canopy directional gap fraction is defined as one minus the probability that a beam of light will intercept a canopy element in a given direction. The relationship between LAI and contact frequency is linear but with gap fraction it is non-linear. For a canopy with randomly distributed infinitely small foliage elements the gap fraction *Po* is given by the Poisson model (Equation (13.1))

$$P_0(v) = \exp\left(-G(v)\frac{L}{Cosv}\right) \tag{13.1}$$

where $P_0(v)$ is the gap fraction at zenith angle v and $G(v)$ is the mean projection of unit leaf area in direction v and L is the LAI.

There is now a range of optical instruments to measure canopy gap fraction including line sensors and hemispherical sensors. Line sensors like the Accupar-80 (Decagon Devices) and SunScan (Delta-T Devices) use an array of detectors to determine the gap fraction in some direction, normally that of the Sun, whilst hemispherical sensors like the LAI-2000 (Licor Inc.) sample the canopy at multiple zenith and azimuth directions (Jonkheere *et al.*, 2004). The Demon instrument (CSIRO, Canberra, Australia) measures direct beam penetration through the canopy at a range of solar zenith angles and the TRAC (3rd Wave Engineering) instrument measures sunfleck distributions to determine gap fraction and gap size distributions. These instruments are relatively easy to use and have been applied in a large number of studies.As an example, Weiss *et al.* (2004) used gap fraction measurements derived from an LAI-2000 instrument in a wide range of different canopies including agricultural crops, plantation woodlands and semi-natural forests. They compared four different methods to estimate canopy LAI and LIDF, use of gap fraction at 57.5° (which is independent of LIDF), Millers formula, gap fraction model inversion using numerical optimization and gap fraction model inversion using a look-up table. Using the look-up table results as a surrogate for ground data, they found that the gap fraction at 57.5° was most strongly correlated to the estimated LAI and proved to be a simple but effective estimation method.

The main limitation of light interception methods is that they do not provide a permanent record of canopy structure and they do not allow separation of leaf and woody material. They measure plant area index (PAI), the cross-sectional area of all leaf and woody material in the canopy, and require correction to estimate LAI. A widely used alternative to estimate canopy gap fraction and LAI is the use of hemispherical photographs taken from below the canopy which do provide a permanent record. The recent availability of digital cameras equipped with hemispherical lenses has made this approach cheap and accessible. The strengths and weaknesses of hemispherical photography are reviewed by Jonkheere *et al.* (2004) who conclude that digital hemispherical photography may currently be regarded as the gold standard for gap fraction measurements.

Hemispherical photography may be used to directly estimate canopy gap fraction at a range of zenith angles and the methods used by Weiss *et al.* (2004) may then be applied to estimate LAI. Photographs are, of course, two-dimensional representation of three-dimensional scenes, and cannot therefore be used to directly estimate gap size distributions, canopy height or canopy size. They also generally require post-processing with manual intervention to classify the images into leaf, wood and sky elements (Jonkheere *et al.*, 2004). It is this limitation that points to the application of airborne and terrestrial laser scanning to provide a permanent three-dimensional record of vegetation structure, and the potential to extract more detailed and accurate information about the three dimensional distribution of leaf and woody material within vegetation canopies.

INTERACTION OF LASERS WITH VEGETATION CANOPIES

Laser scanning is normally based on measuring the time of flight of a pulse of light, which is scattered from an object back to the receiver at the measuring device. The laser pulse that is emitted is altered both in energy and shape by the scattering process depending on the properties of the scattering object. For simple objects such as buildings, the scale at which the structure of the object is changing (10–100 m) may be larger than the footprint size of commonly used airborne laser scanners (0.1–1 m). This simplifies the detection of echoes from such return signals compared to those reflected from vegetation, where the size of the scattering elements (e.g. leaves or needles) may be much smaller than the footprint diameter. In this case the radiation is scattered in many directions and the scattering elements may also transmit part of the laser energy. In order to further study the effects that the vegetation has on the backscattered laser scanner data, we first need to understand the nature of the signal returned from simpler targets.

In most applications of laser scanning the measured distance of an echo is considered as corresponding to a specific object at the location of the echo. This is not always correct, and a number of studies (Axelsson, 1999; Hofton *et al.*, 2000; Jutzi & Stilla, 2003; Wagner *et al.*, 2006) have shown that the interaction of complex scatterers and laser pulses is not fully understood. However, it is acknowledged that these interactions can have a substantial influence on the measured distance of an object to the laser emitter, and thus on the return statistics for vegetation canopies. In the following sections we explore how laser pulses and scatterers interact. This can be considered as a one-dimensional problem, with the relevant axis being either time or range, with both being equivalent due to the constant speed of light. The underlying relations describing the measured signal strength for both airborne and ground based laser scanners are the same as those for radar remote sensing (Wagner *et al.*, 2006). It is therefore convenient to use the radar equation, introduced below, which can be used for computing signal strengths for these systems.

Return power

The detectability of a target depends on the return energy contained in the reflected pulse. Several factors influence the return energy, and for simple targets this energy can be computed, if properties such as beam divergence, distance target-sensor, emitted pulse energy and target properties (geometry and reflectance) are known. Due to optical beam divergence the laser beam widens with distance from the device. Assuming a diffuse target that reflects uniformly into the hemisphere, Baltsavias (1999a) computed the energy of the return signal as:

$$P_r = \rho \frac{M^2 A_r}{\pi R^2} P_t \qquad (13.2)$$

where P_r is the power at the receiver, ρ the reflectivity of the target, M the atmospheric transmission, A_r the illuminated receiver area, R the distance between the laser and the target and P_t the laser pulse energy at the transmitter. Wagner *et al.* (2006) developed a special form of the radar equation to compute the return power of a laser pulse for a given geometric setting, in which they introduce the cross-section of the scatterer and neglect atmospheric transmittance:

$$P_r = \frac{P_t D_r^2}{4\pi R^4 \beta_t^2} \sigma \qquad (13.3)$$

where D_r is the aperture diameter of the receiver optics, β_t^2 is the beam divergence and σ is the backscatter cross-section, defined as:

$$\sigma = \frac{4\pi}{\Omega}\rho A_s \quad (13.4)$$

where Ω is the angle defining a backscattering cone due to surface roughness, ρ is the reflectivity of the scatterer and A_s is the illuminated area of the scattering element. Baltsavias (1999a) provided an example of the return energy for a standard ALS scenario, calculating the ratio of return energy to transmitted energy to be as low as 1.2×10^{-9} to one. Since the transmitter energy cannot be too high due to eye safety constraints, very sensitive detectors with a very good signal-to-noise ratio are required. One of the main implications of Equations (13.3) and (13.4) is that target visibility for laser scanner systems not only depends on target size, but also on target reflectivity; and in some cases, more on the latter than on the former. The implication for vegetation targets is that variations in the spectral response of leaves, woody material and understorey vegetation, or other substrate, and their size and distribution within the laser beam, will contribute to the complex return signal from such targets.

Wavelength

Since the detectability of a specific object is dependent on its reflectance, spectral reflectance at the specific wavelengths of the laser needs to be known. Wavelengths of ALS systems are typically in the near infrared, with the Optech ALTM systems using 1064 nm, and the TopoSys Falcon II a laser wavelength of 1560 nm. Terrestrial laser scanners use wavelengths between 500 and 1500nm with longer wavelengths less sensitive to atmospheric effects; shorter wavelengths are more problematic because of eye-safe considerations but can produce smaller footprints (Pfeifer & Briese, 2007) which may be an advantage when investigating vegetation characteristics and cover. The importance of laser wavelength for vegetation studiesis clear if the spectral reflectance of woody material is compared with that of leaf material.Figure 13.1 shows the typical reflectance spectrum, measured with a spectroradiometer, for needles and tree bark from a Scots pine (*Pinus sylvestris*) tree over the 400–2400 nm range. The 900nm wavelength of the Riegl LMZ210 terrestrial laser scanner is close to the maximum reflectance for vegetation with a large spectral difference to the woody material. Spectral differences at 1064nm are smaller because of higher reflectance of the bark. 1560nm is in the region of water absorption for vegetation and reflectance

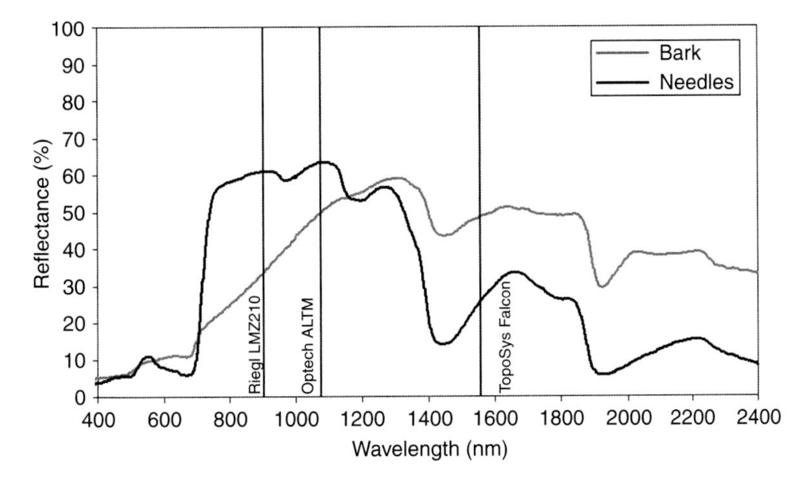

Fig. 13.1 Spectral reflectance of leaf and wood samples showing commonly used laser wavelengths.

differences at this wavelength are related to variations in leaf water content (Danson & Bowyer, 2004). The spectral reflectance contrast with woody material is lower however, so that discrimination between these two targets, based on the intensity of the returns from a laser device, may be problematic.

Waveform

Wagner *et al.* (2006) presented an equation describing the form of the return signal power with time, the so-called waveform. This equation may be used as the basis of a model to simulate the return signals of some typical scatterers to provide some insight into the nature of the waveform and its importance in studying vegetation canopy structure. Following Wagner *et al.* (2006), the power at the receiver over time t can be expressed as:

$$P_{r,i}(t) = \frac{D_r^2}{4\pi R_i^4 \beta_t^2} P_t(t) * \sigma_i(t) \qquad (13.5)$$

where σ_i is the differential cross section of the scatterer and * denotes the convolution operator. To illustrate this relation the return waveform of a synthetic multi-layer vegetation canopy is simulated. In Figure 13.2, a Gaussian shaped laser pulse is used for the simulation, since this is the shape most often used by commercial ALS systems, while the theoretically nearly rectangular shape used is less commonly used. Using the superposition of Gaussian functions, a theoretical vegetation cross-section can be obtained (Figure 13.2) consisting of upper canopy, understorey and the ground, which are represented by the peaks from left to right. Here, an emitted laser pulse of Gaussian shape with a duration of 10 ns and an equivalent length in the range domain of three metres was used. It is clear that some objects that are separated in the cross-section cannot be separated in the return waveform (at least not by simple discrete return detections methods), for example the low vegetation close to the ground peak (right-most peak). Another problem arises from the fact that the ground peak might be widened due to the effect of terrain slope or roughness. In such cases, the determination of terrain height might contain errors that might be as large as the footprint of the ALS beam diameter for a slope of 45 degrees.

Echo detection

Most ALS and TLS do not record the full return waveform, but detect, in real time, a discrete number of echoes in the return signal. For many systems, two discrete echoes are detected, which are called first and last echo (or return) in the literature. Some ALS allow the detection of up to five echoes, with each being stored. Echo detection is normally achieved by thresholding the return signal. When the first rising edge of the return signal exceeds a given intensity, a first echo is triggered. A last echo is triggered as the last rising edge of the return signal. An illustration of how this might work for a tree canopy and an ALS is given in Figure 13.3. A first echo will be triggered as soon as the vegetation density inside the canopy reaches a critical value. A part of the laser pulse moves on through the gaps to be reflected from the ground to trigger a last echo at the receiver. Depending on the final form and amplitude of the return signal, a system that uses a constant threshold for triggering may induce ranging errors. A solution is an adaptive threshold, which takes signal intensity into account to correctly trigger first and last echoes. Here the threshold is set to a fraction (e.g. 1/e or 0.5) of the peak amplitude of which the rising edge is to be detected.

Since the amplitude of the peak is not known *a priori*, echo triggering must be done from a buffered version of the full return waveform. Another critical property related to echo detection is the instrument dead-time, the time that is needed in-between two echoes for them to be recorded as separate echoes. This time may be translated into the vertical distance that two objects in the illuminated area of a laser beam must be separated to be recorded as distinct (e.g. first and last echo) returns. This distance is related to the duration of the emitted laser pulse, and is at

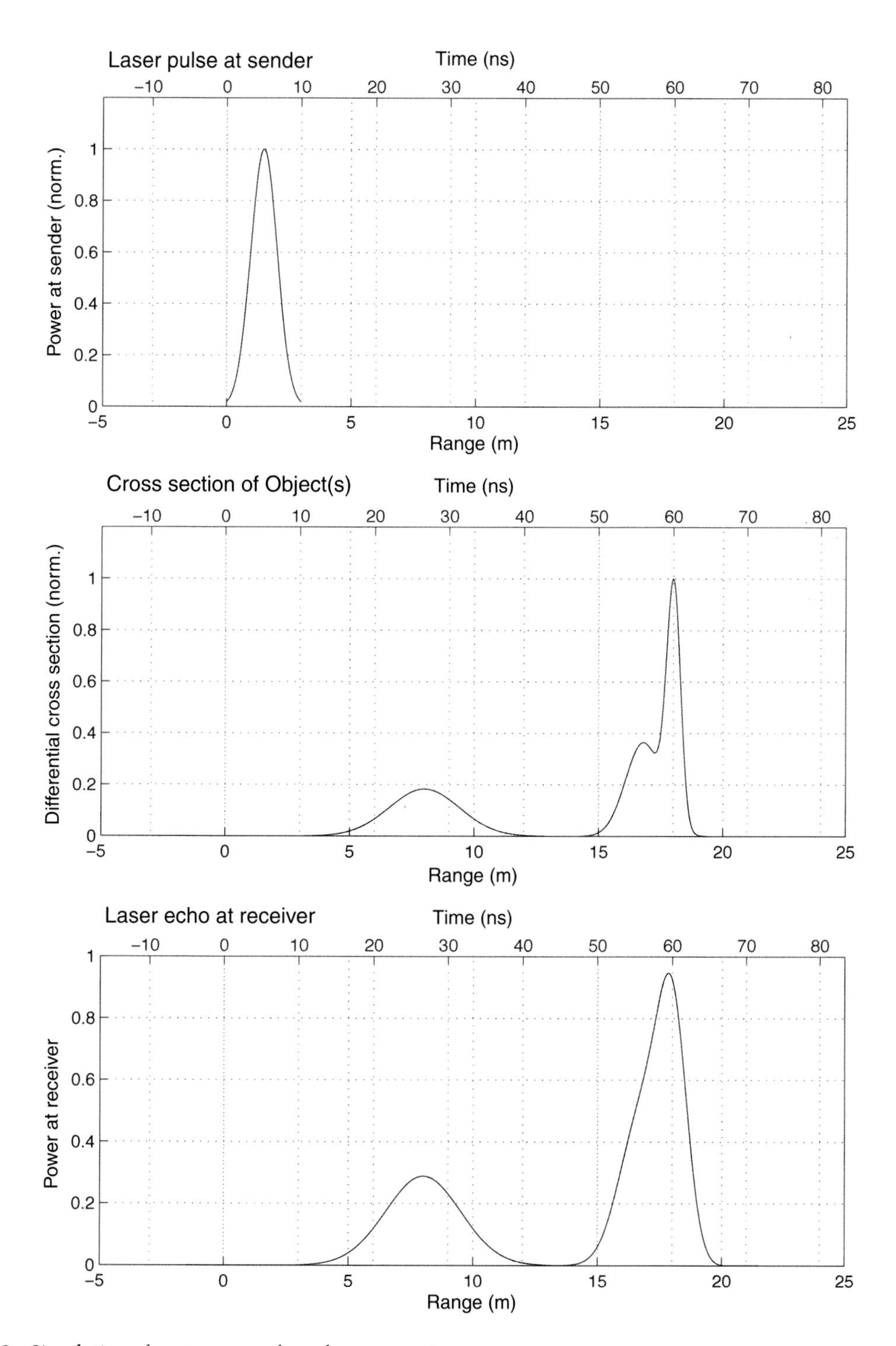

Fig. 13.2 Simulation of a return waveform for a vegetation canopy.

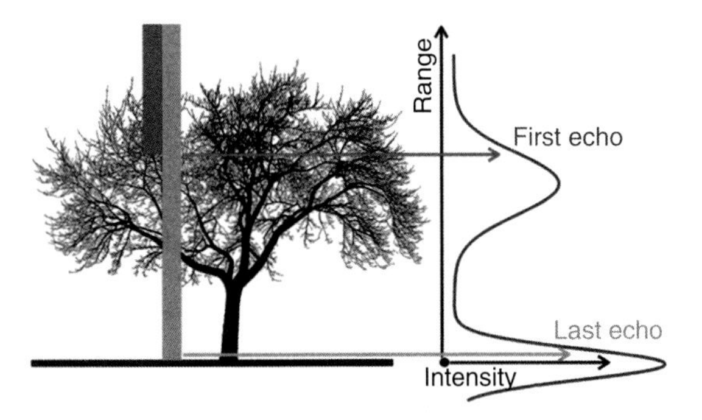

Fig. 13.3 Illustration of waveform and triggering of first and last echo for a tree canopy. See Plate 13.3 for a colour version of this image.

least half the pulse length of the emitted pulse. For an ALS using a laser pulse of 10 ns (or about three metres) in length, the minimum detectable spacing that could be measured by one laser shot would be 1.5 metres. For a TLS with a pulse duration of 0.1 ns the spacing between objects in the range direction would be approximately 15 mm. In practice, this distance can be much larger due to the widened returns of single scatterers, which overlap in the time domain and cannot be separated by traditional means of echo detection. For vegetation canopies this effect determines the minimum distance between two objects that is required for them to be recorded as two separate echoes. For ALS this is illustrated by the lack of returns between the ground and the lower part of a vegetation canopy.

Point density and footprint size

Point density for ALS is defined as the number of laser echoes per unit area. The point density can be considered as the equivalent to resolution for passive imaging sensors. The higher the point density, the more precise and accurate the ALS data products will be, with saturation towards very high point densities, when the Earth's surface will be oversampled. For instance, for building reconstruction a point density of one to two points per square metre might be sufficient, while for power line mapping a point density of 10 points per square metre may be needed (Maas & Vosselman, 1999).

For ALS systems using a rotating or oscillating mirror, the point spacing on the ground depends on the scan frequency and the maximum scan angle, which can be varied for most systems. The scan pattern on the ground is either *Z*-shaped or sinusoidal, depending on the mirror operation (Baltsavias, 1999). For ALS systems using a fibre array, the point density in the across track direction is controlled by flying height and angular fibre spacing across track; and in the along track direction by flying speed and line scan frequency. Point spacing can influence both the applicability of physically based methods to analyse LiDAR data, affecting the accuracy with which vegetation structural characteristics may be extracted (Morsdorf *et al.*, 2004).

Most commercially available ALS are so-called small footprint systems, where the beam diameter on the ground under normal acquisition conditions (e.g. altitudes up to 1000 m above ground level) is of the order of one metre. These systems enable the generation of high resolution terrain and surface models, which is their primary use. The footprint diameter (A) depends on beam divergence γ and flight altitude h (and in some cases the aperture (D) of the transmitter/receiver optics):

$$A = D + 2h\tan\left(\frac{\gamma}{2}\right) \qquad (13.6)$$

Since D can be neglected in most cases, and γ is generally very small, Equation (13.6) can be rewritten as:

$$A = h\gamma \qquad (13.7)$$

It is known that the size of the footprint alters the ability of the laser pulse to penetrate vegetation (e.g. Nilsson, 1996; Chasmer *et al.*, 2006). The smaller the footprint, the larger the chance of not receiving a last echo from the ground in denser vegetation. Thus, for systems recording first and last echo, the penetration of vegetation will in fact be better for systems using larger footprints (Schnadt & Katzenbeisser, 2004). For some systems (e.g. Optech) beam divergence can be set for different applications, for example to allow greater penetration into vegetation. Some systems, such as the Laser Vegetation Imaging Sensor (LVIS) (Blair *et al.*, 1999), have much larger footprint diameters (in the order of ten metres) allowing coverage of larger areas at lower cost but with lower spatial resolution.

Footprint size for terrestrial laser scanners is particularly important in characterizing vegetation canopies because the targets may be at ranges between a few metres and a few tens of metres. For a terrestrial laser scanner with a typical beam divergence of 3 mrad, the footprint size, or beam width, would be 30 mm at a range of 10 m, and 60 mm at a range of 20 m. It is therefore important to consider the range and beam width in TLS measurements of vegetation canopies in order to better understand how different canopy components (leaves, branches, stems) contribute to the return signal.

ALS incidence angle and flying altitude

Acquisition settings of modern small footprint ALS influence laser return statistics, as was shown by Chasmer *et al.* (2006), who studied the effect of pulse energy on canopy penetration and found that pulses with higher energy penetrate further into the canopy. The need to identify universal indicators for biophysical vegetation products was addressed by Hopkinson *et al.* (2006), who used a vertical standard deviation as predictor for canopy height and found it to be robust with respect to varying site conditions and acquisition settings. Furthermore, it is expected that scan angle and flying height have an influence on the magnitude of these parameters, and that their variation may be systematic for some vegetation properties (Morsdorf *et al.*, 2006a). Yu *et al.* (2004) showed that ALS tree height underestimation, as shown by Gaveau and Hill (2003), was larger for higher flying heights and that fewer trees were detected as flying height increased. Ahokas *et al.* (2005) showed that tree height estimates varied with scan angle. Næsset (2004) analysed the influence of flying height on the distribution of vegetation canopy structure estimates from ALS and found little impact for flying heights lower than 1000 m above ground level. However, it is expected that estimates of vegetation density will be influenced by variations in incidence angle, since the distance the laser pulse travels through the canopy will increase with scanning angle.

MEASURING VEGETATION CANOPY STRUCTURE USING AIRBORNE AND GROUND BASED LASER SCANNER DATA

ALS systems have been widely used for standwise derivation of structural parameters (Lovell *et al.*, 2003; Means *et al.*, 2000; Lefsky *et al.*, 1999), often by means of regression methods choosing some ALS predictor variables (e.g. height percentiles) for ground based measures of vegetation structure (Næsset, 2002, 2004; Cohen *et al.*, 2003; Andersen *et al.*, 2005). Using small footprint laser data with high point density, the derivation of single tree metrics becomes possible (Hyyppä *et al.*, 2001; Andersen *et al.*, 2002; Morsdorf *et al.*, 2004). Several studies have derived LAI and vegetation cover from discrete return laser scanning data including Riano *et al.* (2004) who used a relationship from Gower *et al.* (1999) to compute LAI from the gap fraction distribution derived by means of airborne laser scanning. Small footprint, full waveform sensors are just becoming commercially available. Still, discrete returns contain valuable information about the vegetation density and structure at a high spatial resolution, usually in the order of less than one metre. A number of studies have shown that first and last returns can be used to model stand properties such as basal area, biomass and LAI. For example White *et al.* (2000) compared different methods for field LAI estimation with airborne laser altimetry along transects for arid ecosystems and Lovell *et al.* (2003) computed plant area profiles from ALS data and compared them to estimates based on crown models and those derived by a TLS.

Although there is now a large body of research on the interaction of ALS pulses with forest canopies, there are relatively few studies that have used TLS to examine vegetation structure. Most of these have been concerned with the measurement of forest stand variables, including tree height, stem taper, diameter at breast height and planting density (e.g. Hopkinson *et al.*, 2004; Thies *et al.*, 2004; Watt and Donghue, 2005; Henning and Radtke, 2006a). There are however a small number of innovative studies that have attempted to use TLS for characterizing canopy variables like leaf area index, canopy cover and the vertical distribution of foliage. An early advance in this area was the work of Tanaka *et al.* (1998) who developed a laser-based method to measure foliage profile by coupling a laser source and a digital CCD camera. This system was later extended (Tanaka *et al.*, 2004) to use two different laser wavelengths in an attempt to map the distribution of woody and green material in the canopy. Radtke & Bolstad (2001) used a laser range finder to conduct a point quadrat survey in order to determine vertical foliage profile in a broad-leaved forest. LAI estimates from the laser survey were not significantly correlated with LAI derived from litter fall surveys however, and problems with such hand-held systems were highlighted. Lovell *et al.* (2003) used a tripod-mounted laser scanner to determine directional gap fraction in woodland stands of different species and found close correlation with data from hemispherical photography. They also successfully estimated LAI of the stands by using gap fraction data. Henning and Radtke (2006b) tested the application of a TLS to measure a wide range of variables in a mixed species broad-leaved woodland and published the first comparisons of single site multi-temporal TLS data for vegetation canopies. Estimation of plant area index and LAI was based on the computation of laser hits within voxels of 0.5 m, a method also used by Hosoi & Osama (2006).

Most ALS studies to estimate LAI and canopy cover have been based on the assumption that the ration of ground returns to total returns is equivalent to the gap fraction in the zenith direction. TLS have the advantage of sampling canopy gap fraction in multiple directions and in this case the Poisson model Equation (13.1) can be used to derive LAI (Van der Zande *et al.*, 2006). With TLS it is the ratio of laser 'shots' to laser 'hits' in a given direction that yields gap fraction estimates and these estimates are likely to provide more accurate estimates of LAI than the vertical sampling of an ALS.

Airborne laser scanning vegetation LAI and canopy cover in the Swiss National Park

The authors conducted a series of experiments to measure vegetation canopy structure using ALS in the Swiss National Park, in eastern Switzerland. The field site is a forested area close to the Ofenpass valley at an altitude of approximately 1900 m. The dominant tree species is mountain pine (*Pinus mugo*) with some stone pine (*Pinus cembra*). The average tree height in the study area, which was determined from field measurements, was 12 m, and the stem density was 1200 trees ha^{-1} (Koetz *et al.*, 2004). Airborne laser scanner data were collected from a helicopter using a Toposys Falcon II Sensor in October 2002. The system is a push-broom ALS recording both first and last pulse. The flight altitude was 850 m, providing a point density of over 10 points per square metre. A subset of the area (0.6 km^2) was flown at 500 m providing a point density of over 20 points per square metre. LAI and canopy cover (fCover) were estimated from the ALS data based on a gap fraction measurements (Morsdorf *et al.*, 2006b). fCover was computed as the fraction of laser vegetation hits over the number of total laser echoes per unit area, analogous to the concept of contact frequency. An effective LAI proxy was estimated by the fraction of first and last echo types inside the canopy. Validation was carried out using 83 digital hemispherical photographs of canopies in test plots taken with a Nikon Coolpix 4500 with a calibrated hemispherical lens and georeferenced with centimetre accuracy by differential GPS. Gap fractions were computed over a range of zenith angles using the Gap Light Analyzer (GLA) software (Frazer *et al.*, 2000). LAI was computed from gap fraction estimations at zenith angles of 0 to 60 degrees. Different circular ALS data trap sizes, from 2 to 25 m were used to compute fCover and LAI. For fCover, a data

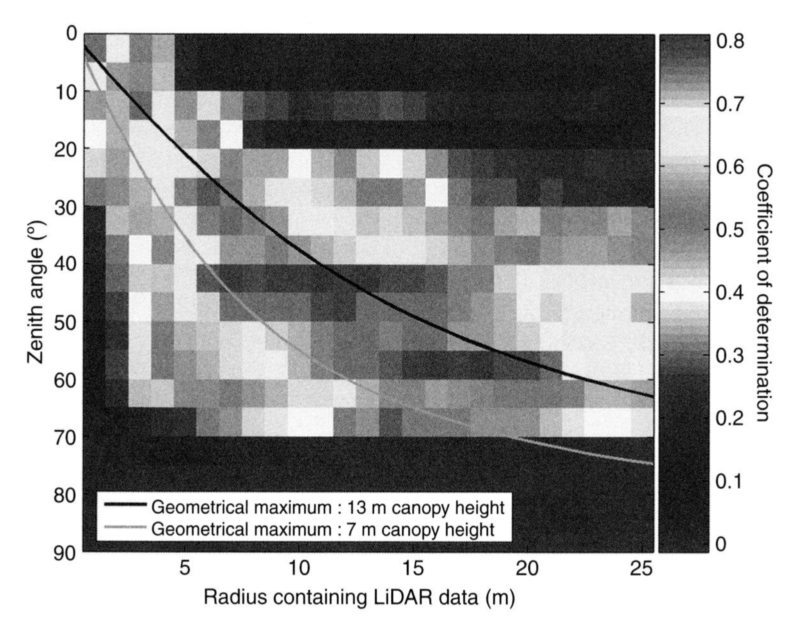

Fig. 13.4 Matrix of coefficients of determination for regression of gap fraction from hemispherical photographs and ALS-derived fCover. See Plate 13.4 for a colour version of this image.

trap size of 2 m radius was used, whereas for LAI a radius of 15 m provided the highest correlation with field measurements.

In Figure 13.4 the coefficients of determination (R^2) between ALS-derived fCover and canopy cover from the hemispherical photographs is displayed as a function of both zenith angle (hemispherical photographs) and ALS data trap size. The band of high R^2 values (orange/red) shows that the geolocation of the hemispherical photographs was clearly sufficient to link field and ALS estimates even at small radii of 2 m. fCover was estimated both from first and last echo data, with first echo data overestimating field fCover and last echo data underestimating field fCover. A multiple regression of fCover derived from both echo types with field fCover showed no increase of R^2 compared to the regression of first echo data, and thus only first echo data were used for fCover estimation.

R^2 for the fCover regression was 0.73, with an RMSE of 0.18. For the ALS LAI proxy, R^2 was lower, at 0.69, while the RMSE was 0.01. For LAI larger radii (>15 m) provided the strongest correlations due to the importance of a larger range of zenith angles (0–60 degrees) in LAI estimation from hemispherical photographs. Based on the regression results, maps of fCover and LAI were computed for the study area and compared qualitatively to equivalent maps based on imaging spectrometry, revealing similar spatial patterns and ranges of values (Morsdorf *et al.*, 2006b).

Terrestrial laser scanning of canopy directional gap fraction

The same field site in the Swiss National Park was used to test the application of a TLS for measuring vegetation canopy directional gap fraction (Danson *et al.*, 2007). The instrument used was a Riegl LMZ210i TLS which has a two-axis beam-scanning mechanism and a pulsed time-of-flight laser range finder with a range of about 350 m. A rotating polygon mirror deflects the beam in a plane orthogonal to the main axis of the instrument and the rotation of the optical head of the scanner plane perpendicular the axis provides frame scanning. Angular step width in both line and frame scan directions may be determined by the user within the limits of the instrument. With this instrument the return data are recorded as either first return or last return, or a combination of both.

A 260 m transect with eight sampling locations was employed. At each sampling location a single hemispherical photograph was taken using upward-looking digital hemispherical photography. The hemispherical photographs were again analysed using the Gap Light Analyzer software (Frazer *et al.*, 2000). Gap fractions were computed for zenith angles 0–90° in 5° bands and averaged over all azimuth angles. The TLS was mounted on tripod at an inclination angle of 90°, and a single scan was recorded with a line scan angle of approximately 80° and a frame scan angle of 180° (Figure 13.5). The TLS was then rotated by 90°, and a second orthogonal scan recorded. The resolution was set at 0.108° in line and frame scan directions, and the first return data were recorded. The two orthogonal scans comprised approximately two million points collected in around six minutes.

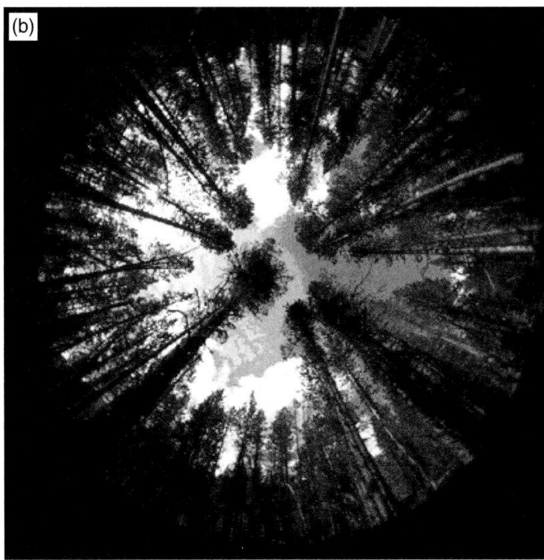

Fig. 13.5 TLS data of forest stand. (a) Intensity of returns in a cylindrical projection (scan range 180° × 80°). (b) Digital hemispherical photograph at the same location (equiangular projection).

Comparison between the gap fractions determined from the hemispherical photography and from the laser scanner showed general agreement. For the canopy shown in Figure 13.6a, there were differences at zenith angles between 15° and 25°, with the photography indicating a gap fraction of 40% and the laser scanner a gap fraction of

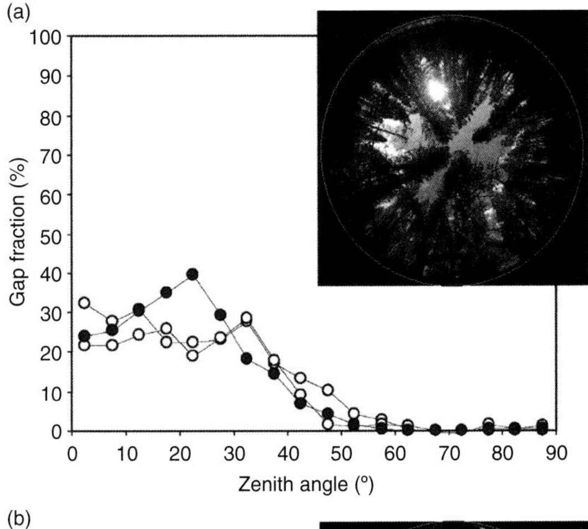

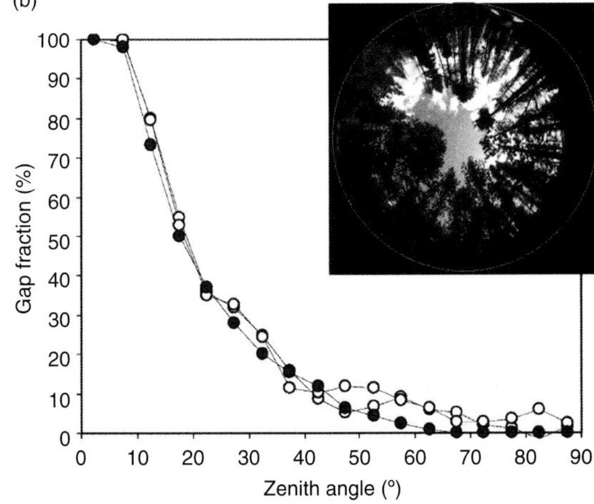

Fig. 13.6 Comparison of TLS-derived and hemispherical photograph-derived directional gap fraction.

20%. It is hypothesised that this may have been due to the glare of the Sun seen in the photograph or errors related to the thresholding of the digital hemispherical photograph.

The data from the two orthogonal scans of the TLS indicated consistency in the measurements, with no evidence of underestimation of vegetation cover with the TLS. In the example shown in Figure 13.6b, the fit between the photography and TLS data is closer, but at zenith angles larger than 40°, the laser data appear to show higher gap fraction than the photography, which shows zero gap from 60° zenith and larger. This is a feature of data from all of the eight plots sampled and suggests that the laser is hitting low reflectance targets at far range, so that the return intensities may be too low for detection. Gap fractions measured at large zenith angles from hemispherical photography were close to 0% because of the overlap of the tree stems for any given line of sight. This study showed that TLS may be used to rapidly estimate directional gap fraction in coniferous forest stands and, given the similarity of the gap fraction data from the hemispherical photography, LAI estimates from both systems would be very close. The measurement of gap size distribution and the separation of leaf and woody material in the TLS scans is currently being explored in order to realize the full potential of these measurements.

MODELLING INTERACTION OF LASERS AND VEGETATION CANOPIES

As the previous review and case studies illustrate, the response signal of a laser scanner in a vegetation canopy is mainly governed by the complex canopy structure, the physical interaction processes taking place within the canopy, the sensor specifications and the measurement configuration. Physically-based radiative transfer models (RTM) may be used to relate the laser signal to the observed vegetation canopy characteristics by considering both the nature of the incident radiation as well as the radiative transfer processes that take place within the canopy. Radiative transfer modelling may therefore be a useful method to understand and interpret the signal recorded by such a system in order to use them to measure vegetation canopy structure.

RTM simulations allow sensitivity studies to be undertaken in order to explore how a wide range of canopy characteristics, instrument types and measurement configurations affect the measurement of canopy structure. Such studies may lead to an improved understanding of the remote sensing signal as well as to optimized instrument design of future Earth Observation systems. A number of RTM have been developed specifically for the simulation of the laser signal over vegetation canopies (e.g., Goodwin *et al.*, 2007; Ni-Meister *et al.*, 2001; Sun & Ranson, 2000). These models describe the radiative transfer of the laser pulse within a canopy as a function of the complex three-dimensional canopy structure defined by the geometry, position and density of canopy elements as well as the optical properties of the scatterers.

Several studies based on RTM simulations have shown the sensitivity of small and large footprint ALS relative to vegetation canopy properties such as tree height, fractional vegetation cover and canopy structure (Ni-Meister *et al.*, 2001; Sun & Ranson, 2000; Kotchenova, 2003; Morsdorf *et al.*, 2007). For the retrieval of such canopy properties, based on ALS or TLS data, a RTM has to be inverted against the observed signal. A prerequisite of a successful inversion is the choice of a validated and appropriate RTM, which correctly represents the radiative transfer within the observed medium (Pinty & Verstraete, 1992). For modelling laser scanner signals a realistic 3D representation of the canopy is therefore essential. RTM model inversion is generally an 'ill-posed problem' since the numbers of variables in the models is normally greater than the number of independent spectral wavebands. A range of numerical approaches, each with strengths and weaknesses, have been developed to solve this problem (Kimes *et al.*, 2000; Tarantola, 2005). The inversion of a RTM provides a method for quantitative retrieval of vegetation properties and, given their physically-based nature, RTM inversions should show greater robustness and accuracy over time and space compared to empirical approaches (Verstraete & Pinty, 1996; Kimes *et al.*, 2000).

A conceptual study based on such a RTM retrieval algorithm showed the feasibility to characterize the canopy structure of a forest canopy exploiting large footprint LiDAR waveform data (Koetz *et al.*, 2007). The methodology was developed and tested on a synthetic dataset generated by the ZELIG forest growth model for a wide range of forest stands (Urban, 1998). ZELIG simulations over time and for different sites in changing environmental settings described in detail highly variable canopy attributes, such as the canopy structure of the studied forest stands (Ranson *et al.*, 1997). Forward simulations of a radiative transfer model subsequently generated the laser signature of the forest stands as observed by the airborne large footprint Laser Vegetation Imaging Sensor (LVIS) (Figure 13.7), an airborne, wide-swath mapping system developed at NASA's Goddard Space Flight Center capable of recording the full waveform over 25 m diameter footprints (Blair *et al.*, 1999).

The retrieval of the canopy properties was based on the inversion of a 3D waveform RTM which simulates large footprint waveforms as a function of forest stand structure and sensor specifications (Sun & Ranson, 2000). The model constructs a 3D-representation of the observed forest stand taking into account the number and position of trees, tree height, crown geometry and shape as well as the shape of the underlying topography. The crown itself is described as a turbid scattering medium parameterized by its leaf area density, the Ross–Nilson G-factor and the foliage reflectance. Finally, the ground reflectance needs to be defined for an accurate waveform simulation. The inversion of the waveform model was based on a lookup table approach in two stages, (i) the generation of the look-up table itself and (ii) the selection of the solution that corresponds most closely to a given measurement. A comprehensive look-up table was generated by simulating waveforms for 100,000 canopy simulations, while considering the sensor configuration. The input model parameters were sampled randomly within defined ranges and followed a uniform distribution.

The inversion of the LiDAR waveform RTM showed reliable results for vegetation properties describing the horizontal as well as the vertical canopy structure of the simulated forest stands. The two variables describing the vertical canopy

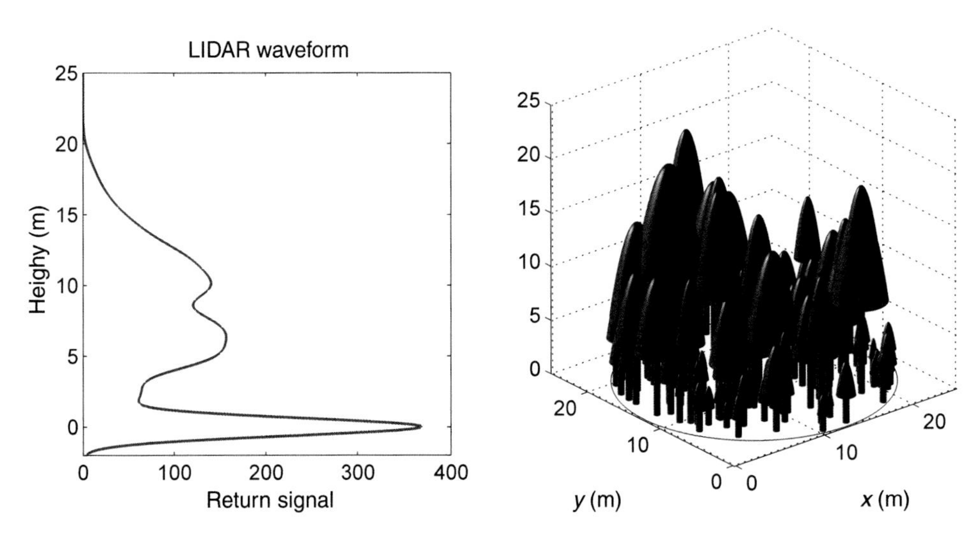

Fig. 13.7 Simulated waveform of forest canopy derived from radiative transfer model.

structure, maximum tree height and vertical crown extension, were estimated with high correlation coefficients of 0.97 for both variables, and a RMSE of 1.46 and 1.72 respectively. Part of the observed error was due to an underestimation of both variables, most likely caused by missing the signal start of the highest tree top within the retrieval. The horizontal structure represented by the canopy fractional cover was retrieved with a correlation coefficient of 0.92 and RMSE of 0.08. The estimates of the fractional cover showed some underestimation for low values as well as an overestimation for high values. Some of this behaviour could be attributed to compensation of the fixed LAI parameter. The results of this conceptual study were further supported by similar findings based on real data acquired by small footprint ALS observations aggregated to large footprint waveform LiDAR data (Koetz *et al.*, 2006). It is clear that the inversion of a waveform RTM provides a novel way to retrieve robust and quantitative forest canopy properties that exploits the way in which the LiDAR waveform represents the complex 3D structure of vegetation canopies.

RESEARCH PRIORITIES

The key research priority for laser scanning of vegetation canopy structure is to develop, deploy and exploit full waveform airborne and terrestrial laser scanning systems. For airborne laser scanning there are now several operational systems acquiring full waveform data (e.g. TopEye Mark II, LiteMapper 5600, Riegl LMS Q560 and Optech ALTM 3100) and it is likely that the full potential of these will be realized over the next few years. There are still some significant challenges with the large data volumes produced by such systems and with the interpretation of the waveform for complex targets like vegetation canopies. The inevitable year-on-year increases in computer power, and the application of RTM for information extraction, will help overcome such problems. Exploitation of the intensity information in ALS data is also a major area for future research. Laser intensity, which may be extracted directly from full waveform data, may provide useful information on the physical properties of the scatterers since it is partly related to their reflectance and density (Höfle & Pfeifer, 2007). However, the proper use of intensity information will require range and sensor-dependent corrections to be made and it is not certain that such corrections will be robust over vegetation canopies without the use of *a priori* information on the vegetation structure. The development of multi-spectral laser scanners, using wavelengths chosen to differentiate leaves and woody material, would reduce the dimensionality of this problem.

Today most LiDAR applications are based on airborne platforms collecting data at a local to regional scale. The development of large footprint LiDAR systems on space-borne platforms, such as the GLAS sensor on ICESat, will allow the up-scaling of LiDAR observations of vegetation canopy structure to a global scale (Dubayah & Drake 2000; Levsky *et al.*, 2005). There are a number of issues still to be resolved, like the interpretation of spaceborne large footprint LiDAR signals where there are complex returns caused by terrain variations within the footprint (Harding *et al.*, 2005; Rosette *et al.*, 2008). Space-borne LiDAR observations will provide spatially discontinuous samples of the Earth surface, which could be interpolated using multi-angular spectral observations to provide spatially continuous global measurements of forest canopy structure (Hese *et al.*, 2005; Kimes *et al.*, 2006).

TLS measurements provide complementary data that may be used to interpret the results of ALS or spaceborne LiDAR surveys of vegetation canopy structure. The high point density of TLS will allow detailed local models of vegetation structure to be built and these may be used to test ALS estimates of vegetation structure. TLS may also provide a means to rapidly acquire the sort of *a priori* information on canopy structure, for example clumping and gap size distributions, required for successful model inversion using ALS data. The development of full waveform TLS is likely to further enhance the role of such systems in measuring vegetation canopy structure. One such system (Echidna) is currently being developed by the CSIRO, Australia, and this initiative is likely to lead to the requirement for new methods of data analysis, probably based on RTM inversion.

REFERENCES

Ahokas E, Yu X, Oksanen J, Hyyppä J, Kaartinen H, Hyyppä H. 2005. Optimization of the scanning angle for countrywide laser scanning. In *International Archives of Photogrammetry, Remote Sensing and Spatial Information Sciences*, Vol. XXXVI, Part 3/W19, Vosselman G, Brenner C (eds), September 12–14, Enschede, Netherlands: International Society for Photogrammetry and Remote Sensing. 115–119.

Andersen H-E, Reutebuch SE, Schreuder GF. 2002. Bayesian object recognition for the analysis of complex forest scenes in airborne laser scanner data. In *International Archives of Photogrammetry, Remote Sensing and Spatial Information Sciences*, Vol. XXXIV, Part 3A, Leberl F (ed.), September 9–13, 2002, Graz, Austria: International Society for Photogrammetry and Remote Sensing. 7pp.

Andersen H-E, McGaughey RJ, Reutebuch SE. 2005. Estimating forest canopy fuel parameters using LiDAR data. *Remote Sensing of Environment* **94**: 441–449.

Axelsson P. 1999. Processing of laser scanner data – algorithms and applications. *ISPRS Journal of Photogrammetry and Remote Sensing* **54**: 138–147.

Baltsavias EP. 1999a. Airborne laser scanning: basic relations and formulas. *ISPRS Journal of Photogrammetry and Remote Sensing* **54**: 199–214.

Baltsavias EP. 1999b. Airborne laser scanning: existing systems and firms and other resources. *ISPRS Journal of Photogrammetry and Remote Sensing* **54**: 164–198.

Blair JB, Rabine DL, Hofton MA. 1999. The Laser Vegetation Imaging Sensor: a medium-altitude, digitisation-only, airborne laser altimeter for mapping vegetation and topography, *ISPRS Journal of Photogrammetry and Remote Sensing* **54**: 115–122.

Bonan GB. 1993. Importance of leaf area index and forest type when estimating photosynthesis in boreal forests. *Remote Sensing of Environment* **43**: 303–314.

Boyd DS, Danson FM. 2005. Satellite remote sensing of forest resources: three decades of research development. *Progress in Physical Geography* **29**: 1–26.

Chasmer L, Hopkinson C, Treitz P. 2006. Investigating laser pulse penetration through a conifer canopy by integrating airborne and terrestrial LiDAR. *Canadian Journal of Remote Sensing* **32**: 116–125.

Cohen WB, Maiersperger TK, Gower ST, Turner DP. 2003. An improved strategy for regression of biophysical variables and Landsat ETM+ data. *Remote Sensing of Environment* **84**: 561–571.

Colombo R, Bellingeri D, Fasolini D, Marino CM. 2003. Retrieval of leaf area index in different vegetation types using high resolution satellite data. *Remote Sensing of Environment* **86**: 120–131.

Cox PM, Betts RA, Jones CD, Spall SA, Totterdell IJ. 2000. Acceleration of global warming due to carbon-cycle feedbacks in a coupled climate. *Nature* **408**: 184–187.

Danson FM, Rowland CS, Baret F. 2003. Training a neural network to estimate crop leaf area index. *International Journal of Remote Sensing* **24**: 4891–4905.

Danson FM, Bowyer P. 2004. Estimating live fuel moisture content from remotely sensed reflectance. *Remote Sensing of Environment* **92**: 309–321.

Danson FM, Hetherington D, Morsdorf F, Koetz B, Allgower B. 2007. Forest canopy gap fraction from terrestrial laser scanning. *IEEE Geoscience and Remote Sensing Letters* **4**: 157–160.

Dubayah RO, Drake JB. 2000. LiDAR remote sensing for forestry. *Journal of Forestry* **98**: 44–46.

Frazer GW, Canham CD, Lertzman KP. 2000. Gap light analyzer (GLA), Version 2.0: Image-processing software to analyze true-color, hemispherical canopy photographs. *Bulletin of the Ecological Society of America* **81**: 191–197.

Gaveau D, Hill R. 2003. Quantifying canopy height underestimation by laser pulse penetration in small-footprint airborne laser scanning data. *Canadian Journal of Remote Sensing* **29**: 650–657.

Goodwin NR, Coops NC, Culvenor DS. 2007. Development of a simulation model to predict LiDAR interception in forested environments, *Remote Sensing of Environment* **111**: 481–492.

Gower ST, Kucharik CJ, Norman JM. 1999. Direct and indirect estimation of leaf area index, f(APAR), and net primary production of terrestrial ecosystems. *Remote Sensing of Environment* **70**: 29–51.

Harding DJ, Carabajal CC. 2005. ICESat waveform measurements of within-footprint topographic relief and vegetation vertical structure. *Geophysical Research Letters* **32**: L21S10.

Henning JG, Radtke PJ. 2006a. Detailed stem measurements of standing trees from ground-based scanning LiDAR. *Forest Science* **52**: 67–80.

Henning JG, Radtke PJ. 2006b. Ground-based laser imaging for assessing three-dimensional forest canopy structure. *Photogrammetric Engineering and Remote Sensing* **72**: 1349–1358.

Hese S, Lucht W, Schmullius C, Barnsley MJ, Dubayah R, Knorr D, Neumann K, Riedel T, Schroter K. 2005. Global biomass mapping for an improved understanding of the CO_2 balance – the Earth observation mission Carbon-3D. *Remote Sensing of Environment* **94**: 94–104.

Höfle B, Pfeifer N. 2007. Correction of laser scanning intensity data: Data and model-driven approaches. *ISPRS Journal of Photogrammetry and Remote Sensing* **62**: 415–433.

Hofton MA, Minster JB, Blair JB. 2000. Decomposition of laser altimeter waveforms. *IEEE Transactions on Geoscience and Remote Sensing* **38**: 1989–1996.

Hopkinson C, Chasmer L, Lim K, Treitz P, Creed I. 2006. Towards a universal LiDAR canopy height indicator. *Canadian Journal of Remote Sensing* **32**:139–152.

Hopkinson C, Chasmer L, Young-Pow C, Treitz P. 2004. Assessing forest metrics with a ground-based scanning LiDAR. *Canadian Journal of Forest Research* **34**: 573–583.

Hosoi F, Omasa K. 2006. Voxel-based 3-D modeling of individual trees for estimating leaf area density using high-resolution portable scanning LiDAR. *IEEE Transactions on Geoscience and Remote Sensing* **44**: 3610–3618.

Hyyppä J, Kelle O, Lehikoinen M, Inkinen M. 2001. A segmentation-based method to retrieve stem volume estimates from 3-d tree height models produced by laser scanners. *IEEE Transactions on Geoscience and Remote Sensing* **39**: 969–975.

Jonckheere I, Fleck S, Nackaerts K, Muys B, Coppin P, Weiss M, Baret F. 2004. Review of methods for in situ leaf area index determination – Part I: Theories, sensors and hemispherical photography, *Agricultural and Forest Meteorology* **121**: 19–35.

Jutzi B, Stilla U. 2003. Laser pulse analysis for reconstruction and classification of urban objects. *International Archives of Photogrammetry and Remote Sensing* **34**: 151–156.

Kimes D, Knyazikhin Y, Privette JL, Abuelgasim AA, Gao F. 2000. Inversion Methods for Physically-based Models. *Remote Sensing Reviews* **18**: 381–439.

Kimes DS, Ranson KJ, Sun G, Blair JB. 2006. Predicting LiDAR measured forest vertical structure from multi-angle spectral data. *Remote Sensing of Environment* **100**: 503–511.

Koetz B, Sun G, Morsdorf F, Ranson KJ, Kneubuhler M, Itten K, Augöwer B. 2004. Fusion of imaging spectrometer and LiDAR data over combined radiation transfer models for forest canopy characterization. *Remote Sensing of Environment* **106**: 449–459.

Koetz B, Morsdorf F, Sun G, Ranson KJ, Itten K, Allgöwer B. 2006. Inversion of a LiDAR waveform model for forest biophysical parameter estimation. *IEEE Geoscience and Remote Sensing Letters* **3**: 49–53.

Kötz B, Schaepman M, Morsdorf F, Bowyer P, Itten K, Allgöwer B. 2004. Radiative transfer modeling within a heterogeneous canopy for estimation of forest fire fuel properties. *Remote Sensing of Environment* **92**: 332–344.

Kotchenova SY, Shabanov NV, Knyazikhin Y, Davis AB, Dubayah R, Myneni RB. 2003. Modeling LiDAR waveforms with time-dependent stochastic radiative transfer theory for remote estimations of forest structure. *Journal of Geophysical Research* **108** (D15): 4484.

Lefsky MA, Harding DJ, Keller M, Cohen WB, Carabajal CC, Espirito-Santo FD, Hunter MO, de Oliveira R. 2005. Estimates of forest canopy height and aboveground biomass using ICESat. *Geophysical Research Letters* **32**: L22S02.

Lefsky MA, Cohen WB, Acker SA, Parker GG, Spies TA, Harding D. 1999. LiDAR remote sensing of the canopy structure and biophysical properties of Douglas-fir western hemlock forests. *Remote Sensing of Environment* **70**: 339–361.

Lovell JL, Jupp DLB, Culvenor DS, Coops NC. 2003. Using airborne and ground-based ranging LiDAR to measure canopy structure in Australian forests. *Canadian Journal of Remote Sensing* **29**: 607–622.

Maas H-G, Vosselman G, 1999. Two algorithms for extracting building models from raw laser altimetry data. *ISPRS Journal of Photogrammetry and Remote Sensing* **54**: 153–163.

Means JE, Acker SA, Fitt BJ, Renslow M, Emerson L, Hendrix C. 2000. Predicting forest stand characteristics with airborne scanning LiDAR. *Photogrammetric Engineering and Remote Sensing* **66**: 1367–1371.

Morsdorf F, Frey O, Koetz B, Meier E. 2007. Ray tracing for modelling of small footprint airborne laser scanning returns. International Archives of Photogrammetry and Remote Sensing, XXXVI, 294–299 Part 3/W52.

Morsdorf F, Frey O, Meier E, Itten K, Allgöwer B. 2006. Assessment on the influence of flying height and scan angle on biophysical vegetation products derived from airborne laser scanning. In: *3d Remote Sensing in Forestry*, 14–15. February 2006, Vienna, Austria.

Morsdorf F, Kötz B, Meier F, Itten KI, Allgöwer B. 2006a. Estimation of LAI and fractional cover from small footprint airborne laser scanning data based on gap fraction, *Remote Sensing of Environment* **104**: 50–61.

Morsdorf F, Meier E, Kötz B, Itten KI, Dobbertin M, Allgöwer B. 2004. LiDAR- based geometric reconstruction of boreal type forest stands at single tree level for forest and wildland fire management. *Remote Sensing of Environment* **3**: 353– 362.

Næsset E. 2002. Predicting forest stand characteristics with airborne scanning laser using a practical two-stage procedure and field data. *Remote Sensing of Environment* **80**: 88–99.

Næsset E. 2004. Effects of different flying altitudes on biophysical stand properties estimated from canopy height and density measured with a small-footprint airborne scanning laser. *Remote Sensing of Environment* **91**: 243–255.

Nilsson M. 1996. Estimation of tree heights and stand volume using an airborne LiDAR system. *Remote Sensing of Environment* **56**: 1–7.

Ni-Meister W, Jupp DLB, Dubayah R. 2001. Modeling LiDAR waveforms in heterogeneous and discrete canopies. *IEEE Transactions on Geoscience and Remote Sensing* **39**: 1943–1958.

Pfeifer N, Briese C. 2007. Geometrical aspects of airborne laser scanning and terrestrial laser scanning. In *International Archives of Photogrammetry, Remote Sensing and Spatial Information Sciences*, Vol. XXXVI, Part 3/W52, Rännholm P, Hyyppä H, Hyppä J (eds), September 12–14 2007, Espoo, Finland: International Society for Photogrammetry and Remote Sensing.

Pinty B, Verstraete MM. 1992. On the design and validation of surface bidirectional reflectance and albedo models. *Remote Sensing of Environment* **41**: 155–167.

Radtke PJ, Bolstad PV. 2001. Laser point-quadrat sampling for estimating foliage-height profiles in broad-leaved forests. *Canadian Journal of Forest Research* **31**: 410–418.

Ranson KJ, Sun G, Weishampel JF, Knox RG. 1997. Forest biomass from combined ecosystem and radar backscatter modelling. *Remote Sensing of Environment* **59**: 118–133.

Riano D, Valladares F, Condes S, Chuvieco E. 2004. Estimation of leaf area index and covered ground from airborne laser scanner (LiDAR) in two contrasting forests. *Agricultural and Forest Meteorology* **124**: 269–275.

Rosette JAB, North PRJ, Suarez JC. 2008. Vegetation height estimates for a mixed temperate forest using satellite laser altimetry. *International Journal of Remote Sensing* **29**: 1475–1493.

Schlerf M, Atzberger C. 2006. Inversion of a forest reflectance model to estimate structural canopy variables from hyperspectral remote sensing data. *Remote Sensing of Environment* **100**: 281–294.

Schnadt K, Katzenbeisser R. 2004. Unique airborne fiber scanner technique for application-oriented LiDAR products. In *International Archives of Photogrammetry, Remote Sensing and Spatial Information Sciences*, Vol. XXXVI, Part 8/W2, Thies M, Koch B, Spiecker H, Weinacker H (eds), October 3–6, Freiburg, Germany: International Society for Photogrammetry and Remote Sensing. 19–23.

Sun GQ, Ranson KJ. 2000. Modeling LiDAR returns from forest canopies, *IEEE Transactions on Geoscience and Remote Sensing* **38**: 2617–2626.

Tanaka T, Yamaguchi J, Takeda, Y. 1998. Measurement of forest canopy structure with a laser plane range-finding method – development of a measurement system and applications to real forests. *Agricultural and Forest Meteorology* **91**: 149–160.

Tanaka T, Park H, Hattori S. 2004. Measurement of forest canopy structure by a laser plane range-finding method improvement of radiative resolution and examples of its application, *Agricultural and Forest Meteorology* **125**: 129–142.

Tarantola A. 2005. *Inverse Problem Theory and Methods for Model Parameter Estimation*. Paris: Society for Industrial and Applied Mathematics.

Thies M, Pfeifer N, Winterhalder D, Gorte BGH. 2004. Three dimensional reconstruction of stems for assessment of taper, sweep and lean based on laser scanning of standing trees. *Scandinavian Journal of Forest Research* **19**: 571–581.

Urban DL. 1990. *A Versatile Model To Simulate Forest Pattern: A User Guide to ZELIG*, Version 1.0, Charlottesville, VA: Environmental Science Department, University of Virginia.

Van der Zande D, Hoet W, Jonckheere L, van Aardt J, Coppin P. 2006. Influence of measurement set-up of ground-based LiDAR for derivation of tree structure. *Agricultural and Forest Meteorology* **141**: 147–160.

Verstraete MM, Pinty B. 1996. Designing optimal spectral indexes for remote sensing applications, *IEEE Transactions on Geoscience and Remote Sensing* **34**: 1254–1265.

Wagner W, Ullrich A, Ducic V, Melzer T, Studnicka N. 2006. Gaussian decomposition and calibration of a novel small-footprint full-waveform digitising airborne laser scanner. *ISPRS Journal of Photogrammetry and Remote Sensing* **60**: 100–112.

Warren-Wilson J. 1963. Estimation of foliage denseness and foliage angle by inclined point quadrats. *Australian Journal of Botany* **11**: 95–105.

Watt PJ, Donoghue DNM. 2005. Measuring forest structure with terrestrial laser scanning, *International Journal of Remote Sensing* **26**: 1437–1446.

Weiss M, Baret F, Smith GJ, Jonckheere I, Coppin P. 2004. Review of methods for in situ leaf area index (LAI) determination – Part II: Estimation of LAI, errors and sampling, *Agricultural and Forest Meteorology* **121**: 37–53.

White MA, Asner GP, Nemani RR, Privette JL, Running SW. 2000. Measuring fractional cover and leaf area index in arid ecosystems: Digital camera, radiation transmittance, and laser altimetry methods. *Remote Sensing of Environment* **74**: 45–57.

Yu X, Hyyppä J, Kaartinen H, Maltamo M. 2004. Automatic detection of harvested trees and determination of forest growth using airborne laser scanning. *Remote Sensing of Environment* **90**: 451–462.

14 Flood Modelling and Vegetation Mapping in Large River Systems

IAN C. OVERTON[1], ANDERS SIGGINS[2], JOHN C. GALLANT[1], DAVID PENTON[1] AND GUY BYRNE[1]

[1]CSIRO Division of Land and Water, Urrbrae, Australia
[2]CSIRO Division of Sustainable Ecosystems, Clayton South, Victoria, Australia

INTRODUCTION

In large low-land floodplain rivers, like the River Murray in Australia, there is a requirement for detailed understanding of channel flows and their connection with floodplain inundation to be able to describe the ecological outcomes that might result from different environmental water allocations (Oliver *et al.*, 2007). To achieve this requires detailed floodplain elevation modelling, flood extent, vegetation mapping and condition assessment over the river floodplain. In large rivers, scale factors make such detailed data difficult to obtain and process and a costly exercise using traditional field based mapping and survey techniques.

Large river floodplains are also highly complex environments with small elevation differences across the floodplain creating a multitude of different flooding regime habitats. The need for better flood modelling and vegetation mapping dictate the use of a precise and detailed Digital Elevation Model (DEM). Flows across floodplains are driven by water level differences of a few centimetres, so the DEM must be able to capture height differences of that order. Important connecting channels can be as small as a few metres across. This leads to a specification of a DEM with vertical precision of about 5 cm and horizontal resolution of about 1 m. This is a very demanding specification that is far beyond the capabilities of readily available DEMs derived from traditional topographic mapping in large areas. The only technologies that provide DEMs approaching this accuracy are airborne laser scanning (ALS, also known as LiDAR (Light detecting and ranging and laser altimetry) and digital photogrammetry from high resolution photography. LiDAR has the added advantage of providing information on vegetation canopy height and density.

This chapter explores the development of a detailed elevation model from LiDAR data captured over the River Murray in South-Eastern Australia. The LiDAR was processed to produce sufficient quality for elevation and flood modelling. A number of issues such as non-ground hits caused by dense vegetation and no returns caused by penetration of water bodies are discussed.

Modelling the movement of water across large floodplains is a complex task and data for hydrodynamic model development and calibration

Laser Scanning for the Environmental Sciences,
1st edition. Edited by G.L. Heritage and A.R.G. Large.
 ISBN 978-1-4051-5717-9

is limited. An existing satellite image-derived flood extent model, the River Murray Floodplain Inundation Model (RiM-FIM) (Overton *et al.*, 2006), will be improved using the new LiDAR DEM described here. The elevation data improves the predictive capability of the model and provides flood depths and volumes.

Existing vegetation maps of the River Murray are either at a scale that is too small to provide relationships between flood extent and vegetation condition, or provide sufficient detail for only a part of the system. State based (the River Murray covers three states) and local maps all use different approaches to vegetation description. A consistent map across the whole of the River Murray valley is needed to make comparable assessments and to inform environmental flow strategies which include trade-offs between different parts of the system. Remote sensing provides an economical and repeatable method for vegetation assessment. Traditional remote sensing using satellite imagery gives poor results due to changes in canopy density, changes in understorey exposure, similar signals from different tree species and varying health of trees. The LiDAR data is again used to aid vegetation mapping by combining LiDAR derived vegetation canopy height and density with satellite imagery for classification of vegetation communities. Other LiDAR derived indices such as vegetation texture can assist in identifying vegetation health condition with satellite image analysis.

River Murray, Australia

The Murray-Darling Basin occupies one seventh of the Australian continent and contributes to 70% of Australia's irrigated agriculture. This basin is the catchment for Australia's largest river, the River Murray. The River Murray is over 2000 km long and has a floodplain that covers an area in excess of 600,000 h (Figure 14.1). The River Murray is one of Australia's key environmental assets, as well as being a working river providing a vital water supply to Adelaide and other towns in the basin and providing irrigation for agriculture.

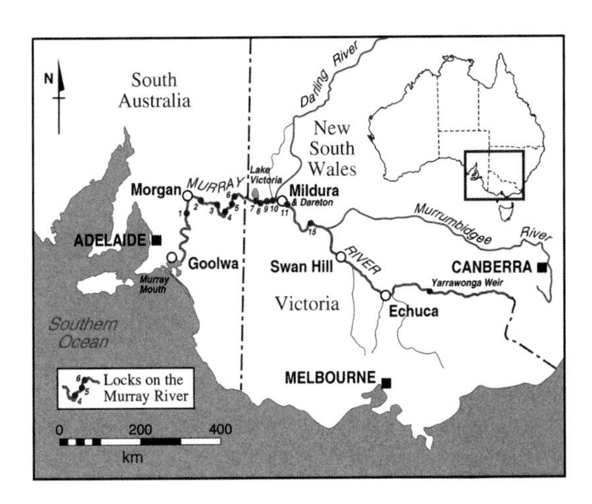

Fig. 14.1 The River Murray in Australia.

Management of the River Murray has been implemented to mitigate large floods and to protect infrastructure, while maintaining storages for regular water supply to irrigators. The river falls from 192 m above sea level at Hume Dam to 3.3 m above sea level at Lock 1 and is highly regulated by 14 weirs and Locks and several storage facilities. Current extraction rates of approximately two-thirds of natural flows have created a severely degraded environment and a river mouth that has closed.

River regulation has caused environmental degradation over the last century and the rate of loss of ecological function is increasing (MDBC, 2003). In the lower half of the River Murray, naturally saline groundwater has risen close to the floodplain surface in many areas as a result of Locks, broad-scale clearing of native forests and irrigation water applied to surrounding lands. Rising groundwater has led to large scale floodplain salinisation and further decline of ecological condition. Concerns have been raised over the river's health and recent attention has been focused on environmental flow strategies. One of these strategies includes periodic floodplain inundation to provide environmental benefits to the floodplain and wetland environments and in-stream water quality.

Many studies have been conducted on the longitudinal (upstream and downstream) dimension of the river, examining the ecological impacts of river regulation on native flora, fauna and the physical changes occurring in the littoral zone. It has also been established that long term modification of flow rates, changes to the frequency of flooding events and alteration to the timing of flows have caused degradation beyond the littoral zone into both the lateral (floodplain extent) and vertical (soil profile) dimensions. These anthropogenically-induced changes have contributed to two processes; land degradation throughout the region and increased accession of saline groundwater to the river (Jolly 1996). The semi-arid nature of the River Murray floodplain exacerbates these man-made ecological imbalances. More extreme and less predictable sized floods and variability in timing of flows in semi-arid regions compared to other riverine environments, possibly intensify the salinity problems (Walker & Thoms, 1993).

Fundamental to this is a detailed flood extent prediction tool and vegetation map to determine the extent, depth and frequency of vegetation inundation from given flow management options. A predictive tool for modelling the extent and depth of floods is required to provide decision support for environmental flows in the River Murray.

DATA CAPTURE AND ELEVATION MODELLING

LiDAR data capture

LiDAR is an active form of remotely sensed data, typically collected from an airborne platform. High frequency (25–100 kHz) pulsed laser light is projected from a scanning sensor mounted beneath an aircraft, and return signals from the terrain beneath the aircraft flight path are recorded. The positions of these returns on the landscape (measured in northing, easting, and height above sea level) are calculated from the current Global Positioning System (GPS) location of the aircraft, the orientation of the aircraft, the scan angle of the sensor, and the path time of the pulse to return to the sensor. Taking a random two-dimensional slice through a LiDAR dataset would produce results similar to Figure 14.2 in forests and woodlands.

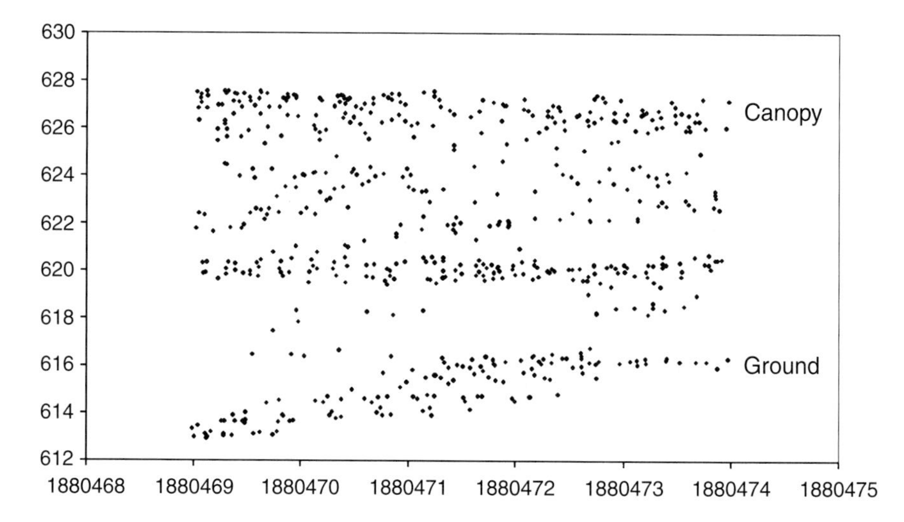

Fig. 14.2 LiDAR data – a random slice through a LiDAR dataset collected over vegetation terrain showing canopy, understorey and ground responses. The X axis shows distance in metres and the Y axis shows the height above sea level in metres. Tree heights of approximately 12–14 m (628 minus 616) can be identified from the return signals.

A LiDAR dataset consists of points that can be thought of as falling into three broad categories:

- Canopy hits: Laser pulses that reflect from the canopy of the vegetation;
- Non-ground hits: Laser pulses that make it through the canopy but are reflected by vegetation above the ground; and
- Ground hits: Laser pulses that penetrate through the canopy, and any other vegetation to be reflected by the ground beneath.

By grouping the points into these three categories a number of products can be derived that describe both the terrain and vegetation properties of the data. For example, by considering only the ground hits, the points can be used to produce an interpolated ground surface, or DEM. DEMs created by this process are usually of very high resolution (one pixel in the DEM may represent one square metre on the ground, or even less), but may contain erroneous regions due to the misclassification of low lying vegetation as ground points. A similar product can be derived for the canopy hits, producing a canopy elevation model (CEM). By subtracting the DEM from the CEM, the height of the vegetation above the ground can be calculated.

The LiDAR datasets

A portion of the River Murray floodplain from Hume Weir to Robinvale was mapped using LiDAR provided by the Murray-Darling Basin Commission (Figure 14.3). The Chowilla floodplain has also been mapped using LiDAR. This data approaches the quality required for flood modelling, although there are some issues that need to be resolved as described below. Most of the remainder of the River Murray floodplain is covered by coarser resolution data of varying quality.

The main River Murray LiDAR dataset consists of approximately 5300 individual tiles, acquired in 2001. Each tile represents a two kilometre square planar area on the ground, and has associated ground, non-ground and canopy classified files as determined at the time of data acquisition. In total, the dataset consists of approximately 1.2 billion LiDAR data points. Managing this amount of data requires specialist software. This data

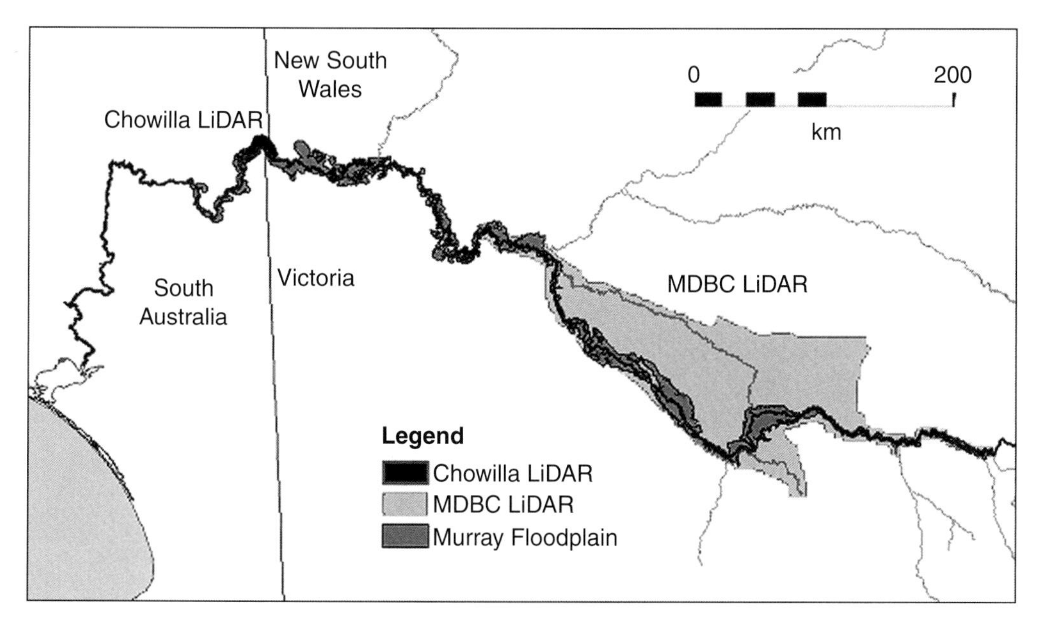

Fig. 14.3 The Murray River from Hume Weir to the terminal lakes near the river mouth showing the extent of the MDBC LiDAR data acquired for the project.

covers much more than the immediate floodplain around the River Murray and includes the Edwards River. In some small areas the LiDAR data does not reach to the edge of the maximum floodplain extent of the River Murray. Across the extents of the dataset the point density varies, but points are typically spaced between two and three metres apart. Figure 14.4 shows a section of data interpolated into a DEM of one metre resolution around the junction of the Murrumbidgee and Murray Rivers, and illustrates the degree of detail captured by LiDAR data.

Variations in point density across the LiDAR dataset can have a direct effect on the quality of derived surfaces. By generating density surfaces (Figure 14.5), regions of the dataset can be identified that may produce issues on further processing. The light coloured irregular stripes are overlaps of adjacent flight lines while the sinuous dark stripe is the River Murray where few returns are

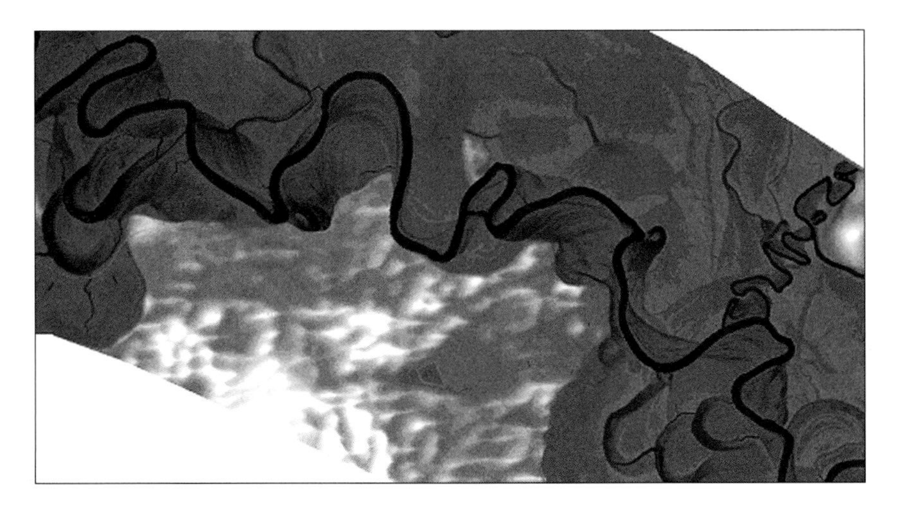

Fig. 14.4 LiDAR elevations (dark = low, bright = high) around the junction of the Murray and Murrumbidgee Rivers.

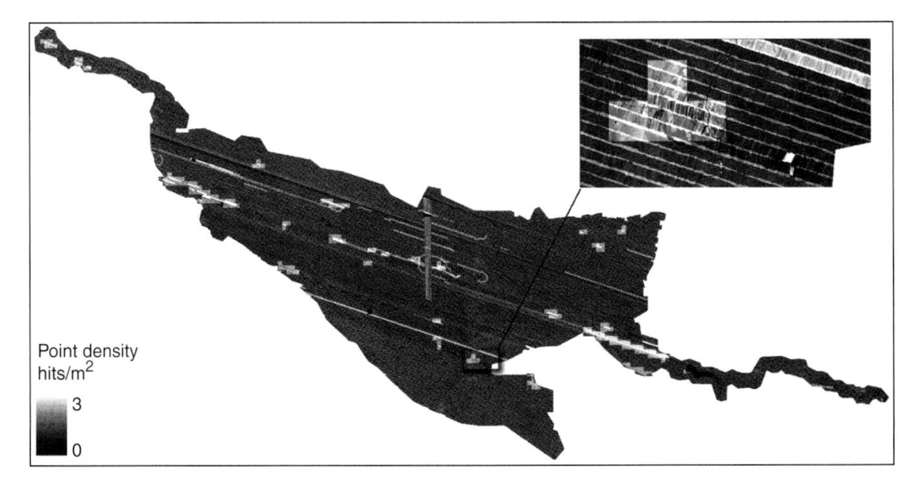

Fig. 14.5 Density (25 m) of above ground classified points for subset of MDB LiDAR data extent.

acquired as the laser signal is absorbed or scattered rather than reflected. Water directly beneath the aircraft in the centre of the swath may produce a return by reflection, providing intermittent measures of water height in the river. Over most of the land surface the point density is between 2 and 5 points per square metre. In some areas (not shown in Figure 14.5) differences in point density between adjacent flight lines are apparent, suggesting different LiDAR sensor parameters used in different flights or variations in flying height or post-processing. Another feature of interest is the banding running roughly perpendicular to the flight lines (see inset to Figure 14.5). The most likely explanation for these bands is turbulence or crosswinds experienced by the data acquisition aircraft as it collected data. Directly impacting on the quality of the dataset is an area of low point density in the west of the subset region. Once again this feature appears directly linked to the attitude of the data acquisition aircraft, and is possibly due to severe turbulence causing the LiDAR instrument to increase scan angle to a point where no data was returned.

Figure 14.6 shows a small area as a shaded relief image of the one metre resolution DEM derived from ground points. While much of the land areas contain considerable detail, the elevations in the river channels are clearly dominated by artefacts of the interpolation. The elevations used to interpolate river heights are mostly from the banks adjacent to the river and the height is affected by the location of the last laser strike before the river on the bank. These erratic height variations in the river channels and near other open water areas need to be removed before the data can be used for the purposes flood modelling, where estimating overbank flow is critical.

The areas where adjacent swaths overlap are just visible in Figure 14.6 as stripes with more detailed texture resulting from the higher point density. In some areas residual height differences between the two swaths exaggerates the texture in the overlap areas.

A further problem occurs in wetlands as dense reed beds produce laser returns that are interpreted as ground points, resulting in elevations about one metre higher than ground level. This misidentification will contribute to errors in estimates of volumes of water stored in wetlands and, where there are contiguous rings of reeds around deeper water, will prevent the detection of wetland filling at the appropriate water height. Figure 14.7 highlights this problem.

These features are all present across the entire dataset in varying degrees, and it is important to

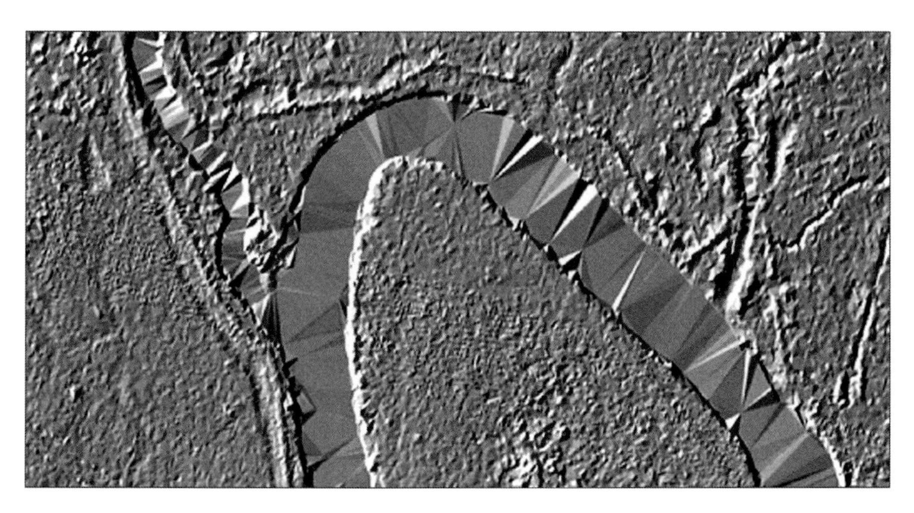

Fig. 14.6 Shaded relief image of the Murray and Edwards Rivers. Note triangulation artefacts on the rivers and the more subtle cross-hatch artefacts diagonally across the centre of the image.

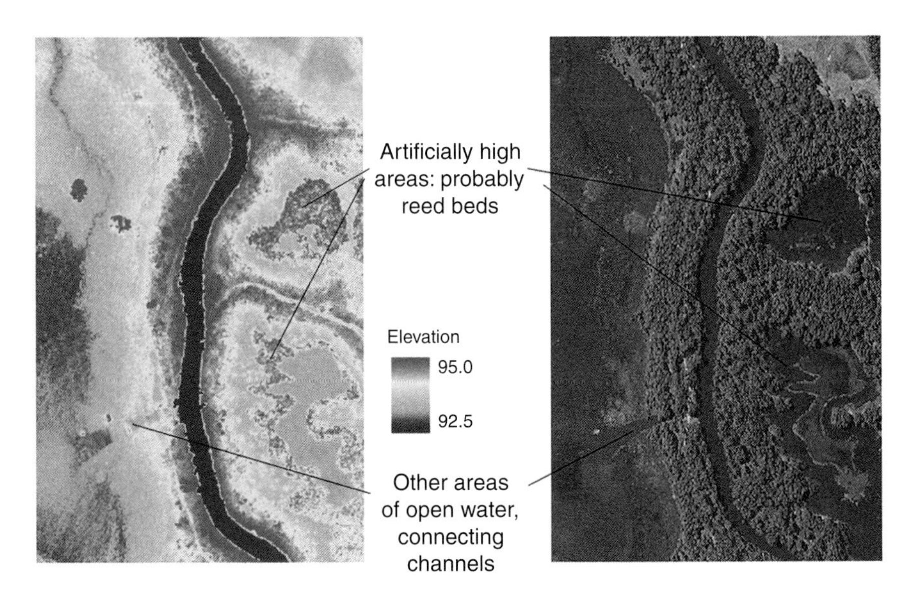

Fig. 14.7 Reed beds in and around floodplain wetlands have been interpreted as ground elevations. See Plate 14.7 for a colour version of these images.

know the effects they might have on derived terrain and vegetation layers. The DEM originally derived from this LiDAR data has been used for hydrodynamic modelling in the Barmah forest. This exercise proved to be very difficult and the LiDAR data had to be re-flown because of errors in overlapping flight paths.

Analysis suggests that the banding between flight paths, and the overlap of flight paths, will not overly influence the terrain with respect to deriving vegetation layers. However, these minor height differences may influence modelled floodplain behaviour. Regions missing data will, of course, will be greatly influenced. Small holes in the data can be overcome by producing surfaces at lower resolution (surfaces calculated at 25 m resolution exhibit fewer holes than the 2 m surfaces). Regions that are missing large amounts of data (possibly due to instrument malfunctions or turbulent acquisition conditions) are considered 'no data' for the purposes of further analysis.

Overall LiDAR point densities (the total of ground, non-ground and canopy) are restricted by the equipment used to collect the LiDAR data, however, the ratio of ground to above ground point densities are dependant on the vegetation present at any given location. At one extreme, bare earth will have all collected points classified as ground. At the other extreme, very dense vegetation may result in a high number of points per square metre of above ground points, but low ground point densities. Low ground point densities will have a lower impact on the fidelity of terrain and vegetation surfaces in regions of gradual slope.

The accuracy of the LiDAR data was tested at five study sites and the root-mean-square error was found to be about 17 cm, with 95% of measured points being within 33 cm of the actual height. Errors are typically higher under dense canopy and in areas of dense low vegetation where differentiation of laser returns from ground and non-ground points is difficult.

LiDAR coverage of the Chowilla area has produced a DEM of 2 m resolution and the vertical precision of this data is similar to that of the MDBC LiDAR data. Unlike the MDBC data, areas of water have been set to 'no data' in this DEM, preventing the problems of improper elevations along the channels. The Chowilla LiDAR data has already been used for hydraulic modelling and, while there were some concerns over the height accuracy in the beds of small channels, other quality constraints were not apparent.

Remedial quality improvement

The LiDAR data required substantial work to resolve the quality issues noted above which include: recovery of realistic elevation surfaces for the channels, removal of artificially high elevations induced by misclassification of reed beds as ground, and adjusting adjacent swaths to remove height mismatches. Fixing these problems using traditional GIS software required hours of computing time for each tile, to generate a new DEM from the point data. A dedicated software package was used that provided facilities for incorporating break-lines to prevent the triangular artefacts on rivers and for re-classifying points to deal with the reed beds. As well as supporting the improvements in quality, this software is able to generate a 2 × 2 km tile of DEM in seconds instead of hours. Re-generation of the DEM within the River Murray floodplain was therefore feasible.

Realistic elevations for the river channel required the construction of breaklines along the edge of the water surface. Three dimensional breaklines (lines with elevation attributes) and the breaklines from boundaries were used for the triangulation of the surface. Elevations must be chosen carefully for these breaklines to produce a suitable result.

The first step was to selectively construct lines along the centre of the channel using elevations from the mid-swath patches of data where they provided a reliable elevation value. These centrelines were constructed in segments in a downstream direction and forced to have a smoothly decreasing elevation along each segment. Lines of constant elevation were then constructed across the channel at regular intervals along the channel, taking their elevations from the centreline. The bank breaklines then linked these cross-section lines using a smooth change in elevation from one cross-section to the next. This procedure resulted in bank breaklines that had consistent and reasonable elevations declining gradually in the downstream direction. Elevation points between the bank breaklines were then discarded so that elevations in the river would be derived solely from the well-constrained breakline elevations.

Following construction of the river bank breaklines, the DEM was recreated at 10 m resolution. This resolution was chosen as a compromise between fidelity to the original points and the quantity of data. Ground heights were recalculated for the entire dataset, the height of the ground at the centre of each 10 × 10 m cell calculated using an inverse distance formula as specified in Equation (14.1).

$$\text{elevation} = \frac{\sum_{i=0}^{n} z^{i}/d_i^{w}}{\sum_{i=0}^{n} 1/d_i^{w}} \qquad (14.1)$$

where, n, number of points in bucket; z^i, height of ith ground point (measure from sea level); d^i, distance of ith point from centre of 10 × 10 m bucket; w, weighting coefficient (set to 0.5).

Figure 14.8 shows the resulting elevation model for the entire MDB dataset generated at 10 m resolution.

Figure 14.9 illustrates the elevations before and after correction. The image rendering has been adjusted to highlight variations in the elevations of the river.

FLOOD MODELLING

To model flood behaviour a floodplain inundation model was developed using a spatial information system framework (Penton & Overton, 2007). Spatial decision support is one of the main roles of a GIS; it provides an excellent framework for the integration of multi-criterion, spatial datasets (Taylor *et al.*, 1999; Jankowski *et al.*, 2001). Therefore a GIS was recognised as the best method for integrating non-spatial

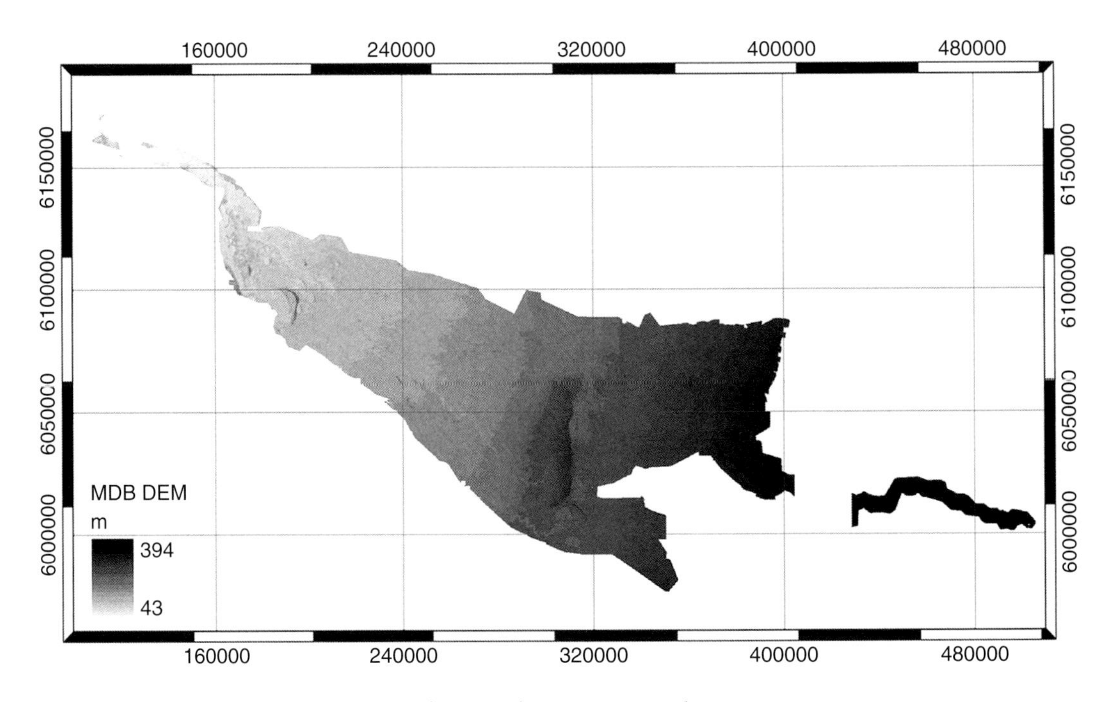

Fig. 14.8 Digital Elevation Model calculated at 10 m for MDB LiDAR dataset.

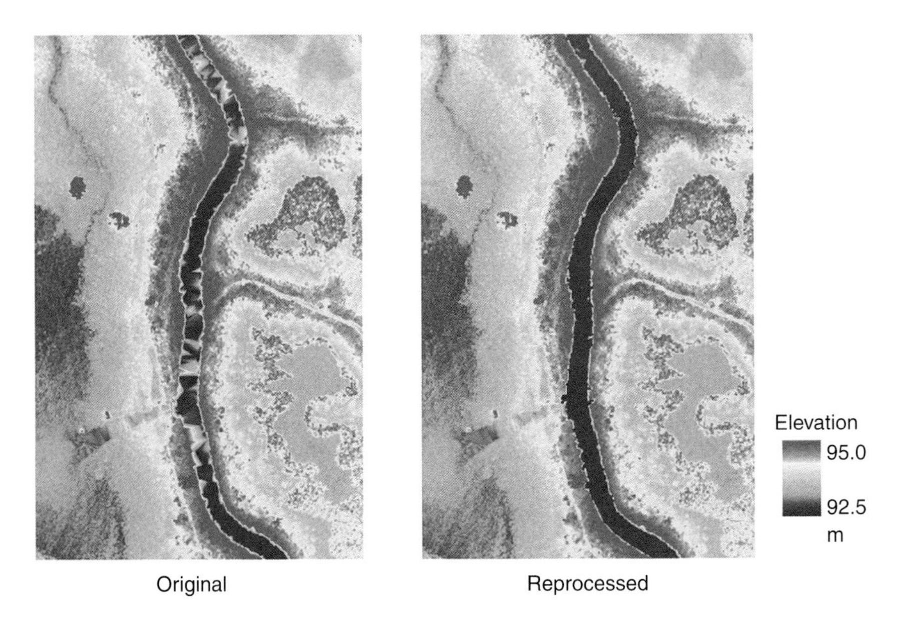

Fig. 14.9 Elevations before (left) and after (right) the addition of river breaklines. See Plate 14.9 for a colour version of these images.

river flow models with mapped flood inundation extents and other related spatial datasets used in this project.

Mapping flood extent

Floodplain inundation mapping has previously been conducted on sections of the River Murray in South Australia (Overton *et al.*, 1999; Overton, 2005). Flow in the river is dependent on releases from the Hume Dam, inflows into the river from major tributaries and outflows from irrigation and water storage extractions. The flow in each reach of the river is dependent on the management of associated locks and weirs.

Remote sensing is particularly useful for monitoring flood extents because it provides basic data more cheaply and efficiently than ground based methods (Whitehouse, 1989). Previous studies to determine the aerial extent of flooding have commonly involved optical satellite image analysis (Walker *et al.*, 1986; Townsend & Walsh, 1998, Overton *et al.*, 1999; Shaikh *et al.*, 2001; Sheng *et al.*, 2001; Frazier *et al.*, 2003), radar remote sensing (Townsend & Walsh, 1998) or an integration of remote sensing and GIS (Brivio *et al.*, 2002). These methods have proved to be very useful and economical for large area flood analysis. More detailed studies have used digital elevation models to create a floodplain surface that can be inundated at certain river heights (Townsend & Walsh, 1998). Elevation methods are particularly useful for predictive studies of changing flow paths through the floodplain by manipulating flow barriers. However, surface modelling alone may not give the best representation of flood inundation, as there are numerous impediments and small channels across a predominantly flat floodplain. The modelling of flood inundation from surface elevation also requires detailed information on stage heights, backwater curves, flow impedances and roughness coefficients as simple height levels are insufficient in this dynamic environment.

The mapping of flood extents using satellite imagery is described by Overton (2005). Landsat TM imagery was used with a resolution of 25 m. Water was detected by isolating the pixels within the image that have a very low reflectance value for Band 5, indicating that the light was absorbed and not reflected. However, other features, especially shadow, also have very low reflectance in this region of the light spectrum and this problem is compounded by the slightly higher reflectance of turbid or shallow water leading to these features being detected as non-water. Separating water from dark shadow was not possible using a single band image and the assessment on the cut-off value of percentage reflectance for determining surface water was a judgement made by the analyst based on the histogram and image characteristics. However, Frazier *et al.* (2000) found that a simple density slice on Band 5 achieved an overall accuracy of 96.9% when compared to aerial photographic interpretation and was as successful in delineating water as a six-band maximum likelihood classification (a multi-band statistical analysis commonly used in remote sensing for image classification). Despite the problems involved in individual pixel misclassification, the method of density slicing a single mid-infrared band to detect surface water has been successfully used in previous studies and represents an economical method for determining flooding over large areas (Sheng *et al.*, 2001). Whilst satellite images were obtained to provide a range of flood events (flow intervals), it was necessary to consider flows between these events to provide a more continuous predictive model. Interpolation between the discrete flow intervals was initially performed to produce finer intervals of flood extent using an image morphological process called a 'marker-based watershed segmentation algorithm'. A map of flood growth was successfully completed by interpolating between known flood event boundaries identified from satellite imagery. Figure 14.10 shows three flood masks from satellite images south of Barmah on the River Murray taken between 1990 and 2000. Figure 14.10d is a combined flood map with just the observed values. The original predicted flood growth map using the 'marker-based watershed segmentation algorithm' is shown in Figure 14.10e.

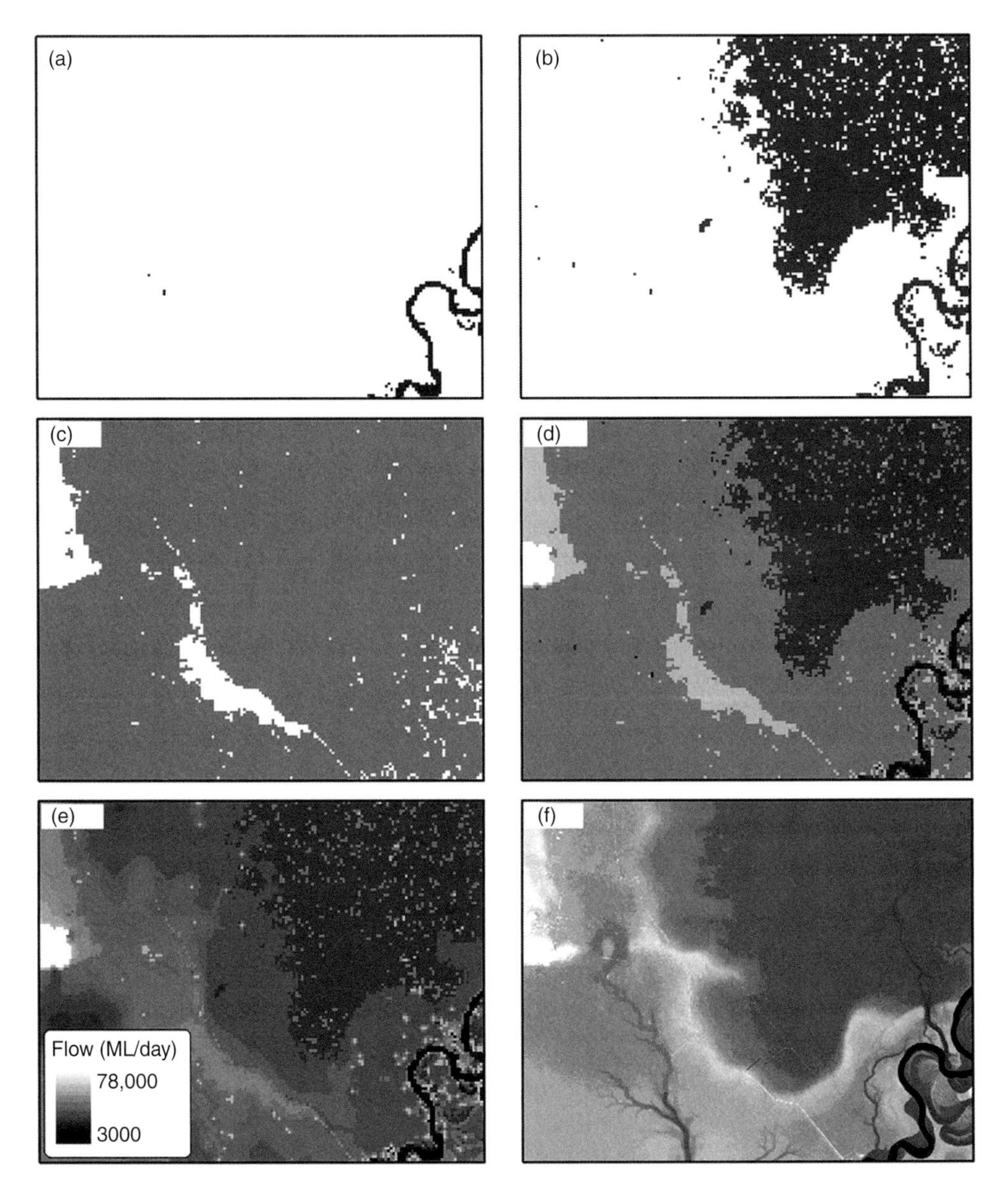

Fig. 14.10 (a), (b) and (c) show flood masks for 23,000, 30,000 and 56,000 ML/day flows. (d) combines the flood masks into an inundation map. (e) predicts inundations by contouring between the observed masks. (f) predicts inundations using information from the high resolution LiDAR DEM.

The result of the interpolation stage is a raster grid of cell values that represent the commence-to-flow based on the flows on the day of the satellite images (e.g. Figure 14.10e). This flood growth map was linked to a hydrological model of the River Murray to determine the commence-to-flood river flow levels (discharge) and river heights.

The resulting model was version 1 of the River Murray Floodplain Inundation Model (RiM-FIM)

(Overton, 2005; Overton *et al.*, 2006). It provided a research and decision support tool for environmental flow management. The floodplain inundation model predicts the extent of flooding on infrastructure, wetlands and floodplain vegetation and allows for spatial and quantitative analysis of the flood extents. The RiM-FIM model provides a series of GIS layers showing areas inundated at different flow rates but does not provide information on water depths, water volumes or water flow paths. Another limitation of the model is the extent of the largest satellite image available. A further issue is that the interpolation between images was based on a model of flood growth and not based on detailed elevation data that would control the growth. This additional information requires the use of a sufficiently accurate digital elevation model (DEM) to obtain the difference between water height and land height, giving depths of inundation, estimates of water volumes, and better describing the nature of connections between the main channel and wetland areas. Using information on water depths and flooded areas obtained by the combination of the DEM and RiM-FIM, water balance approaches can be used to estimate the rate at which water is lost from wetlands by evaporation and infiltration after floods have receded, in order to describe wetting and drying regimes for wetlands.

To address these limitations the LiDAR DEM was then used in conjunction with RiM-FIM to produce a new version with improved inundation maps, estimates of depths and volumes of water on the floodplain, and descriptions of water movement across the floodplain. Figure 14.10f shows the interpolation between observed flood masks using a digital elevation model. Combining the satellite imagery-based inundations with the DEM permits interpolation between mapped levels from elevations while retaining the actual flood extents from a series of known events. This provides more reliable inundation extents than those mapped from satellite imagery alone or derived purely from elevation data, and also provides flood depth information for vegetation response modelling.

Hydrodynamic techniques work well for modelling the River Murray channel. However, for high resolution modelling of the outer extents of the River Murray corridor, application of hydrodynamic techniques become expensive monetarily and computationally. Obstacles for modelling the outer extents using hydrological models include poor soils and geological information, difficulty building river cross-sections and features poorly defined in DEMs, such as levee banks and below ground channels. As such, different methods for calculating water depth in the floodplain are necessary. One such method, described here, calculates the depths of flood waters using information from a series of observed flood masks.

Integrating LiDAR DEM for flood elevation

The initial objective of integrating a DEM with flood masks was to calculate the depth of water in flooded areas. Depth is not directly observable from a DEM or a flood mask. DEMs of the River Murray describe the elevation of the landscape relative to the Australian Height Datum (AHD), which is an estimate of height above sea level. Flood masks indicate whether a particular location is inundated. However, depth can be calculated from these datasets given two assumptions. The first assumption is that at the edge of an inundated area the depth of the water will be zero and the water will be the same height as the landscape. The second assumption is that in a local region, water maintains approximately the same height, which allows the water height at a point in the landscape to be predicted. The depth of an inundated location is calculated as the water height less the landscape height.

The technique for producing inundation maps involves calculating spot heights for inundations observed in satellite imagery, predicting the water height across the entire satellite image and generating inundation maps for observed and unobserved inundations. The result is a high resolution set of images that describe the water height for different flow scenarios.

The magnitude of the River Murray corridor necessitates the automated identification of accurate water heights. The water height of an inundation is greater towards the top of a zone than towards the bottom. However, for most inundated areas on the River Murray, the water height differs little over areas approximately 300 m in length. Since the size of the inundation influences its properties, water patches were divided into three categories based on the area inundated: small pools and thin streams; medium sized areas and lakes; and large areas of inundation.

Small disconnected patches are common in low flow flood masks and often represent thin channels, the edge of inundated areas or errors in the flood mask classification. The water height of a small discontinuous patch is calculated by comparing the landscape height of the inundated area with the landscape height of adjacent dry cells. The inundated area should be on lower ground than the dry cells. The resulting water heights are averaged with other small patches to produce a best prediction.

Medium sized areas and lakes are bodies of water where the surface should have a constant height. At the edge of the area, the surface should have the same height as the landscape. The inner edge of an inundated area should be under water and the outer edge, which is just outside the inundated area, should be dry. The mean inner and outer heights represent upper and lower bounds on the actual heights. The standard deviation of the values in each range indicates the amount of noise in the calculation. The water height for the medium sized area is the mean of the inner and outer edges.

Large areas are usually either main river sections or large flooded areas. Often large flows inundate the entire corridor in the satellite image. The areas are large enough that the water height varies across the surface. The large inundated areas often contain small patches of water that are misclassified as dry land due to forest cover or classification mistakes. These islands are easy to identify based on their area and can be removed. Since the slope of the surface is not always linear, there are different water heights for different locations in the landscape. As with medium sized inundations, water heights can be calculated from the landscape height at the edge of the inundation. However, first the area must be split into smaller areas, for example 250 m, where the water height is near constant.

Generating an accurate water height surface requires finding enough accurate water heights to produce an interpolated surface. The river flows downstream at a slope of approximately 1 m in every 4,000 m. The slope of the water height varies laterally from the channel. Assuming the lateral slope were the same as the downstream slope, if one accurate point were used to represent every 250 m × 250 m square of inundation, the vertical accuracy would be approximately 0.05 m. Accurate water heights can be found in many flood masks. However, there are areas without a water height measurement including areas where the inundation covers kilometres without break. Furthermore, water height points are inaccurate where the waters edge meets vertical structures such as river banks, levee banks and cliffs. In addition, imperfections in the flood mask can produce erroneous flood edges. For example, the islands in the water masks from very dense tree cover must be removed to avoid false low points. Note also that the water height of permanently inundated wetlands is only relevant when the wetlands are connected to the river. These were removed by comparing the satellite images taken at different times and referencing existing wetlands databases. Since the water height points cannot be determined for every part of every inundation observed in a flood mask, the water height layer must be generated using an interpolation technique.

An interpolated surface was fitted to the water heights using a kriging technique with linear trend removal. This surface needed to match the decline in heights along the river and out into the floodplain. In addition, the interpolation needed to smooth over erroneous points and handle twists in the river. Trend surfaces alone were unable to adequately describe the local variability in some parts of the floodplain. Some spatially adjacent areas have different water sources and, therefore, unrelated heights. These areas were split then interpolated separately to avoid smoothing over the discontinuities. The surface was predicted beyond the extent of the observed inundation to produce an estimate of how much higher the water would need to rise to inundate other areas. Once the interpolated surface had been fitted to the observed water heights the flood height and a revised inundation map were produced as shown in Figure 14.11a for a 23,000 ML/day flow and Figure 14.11b for a 30,000 ML/day flow.

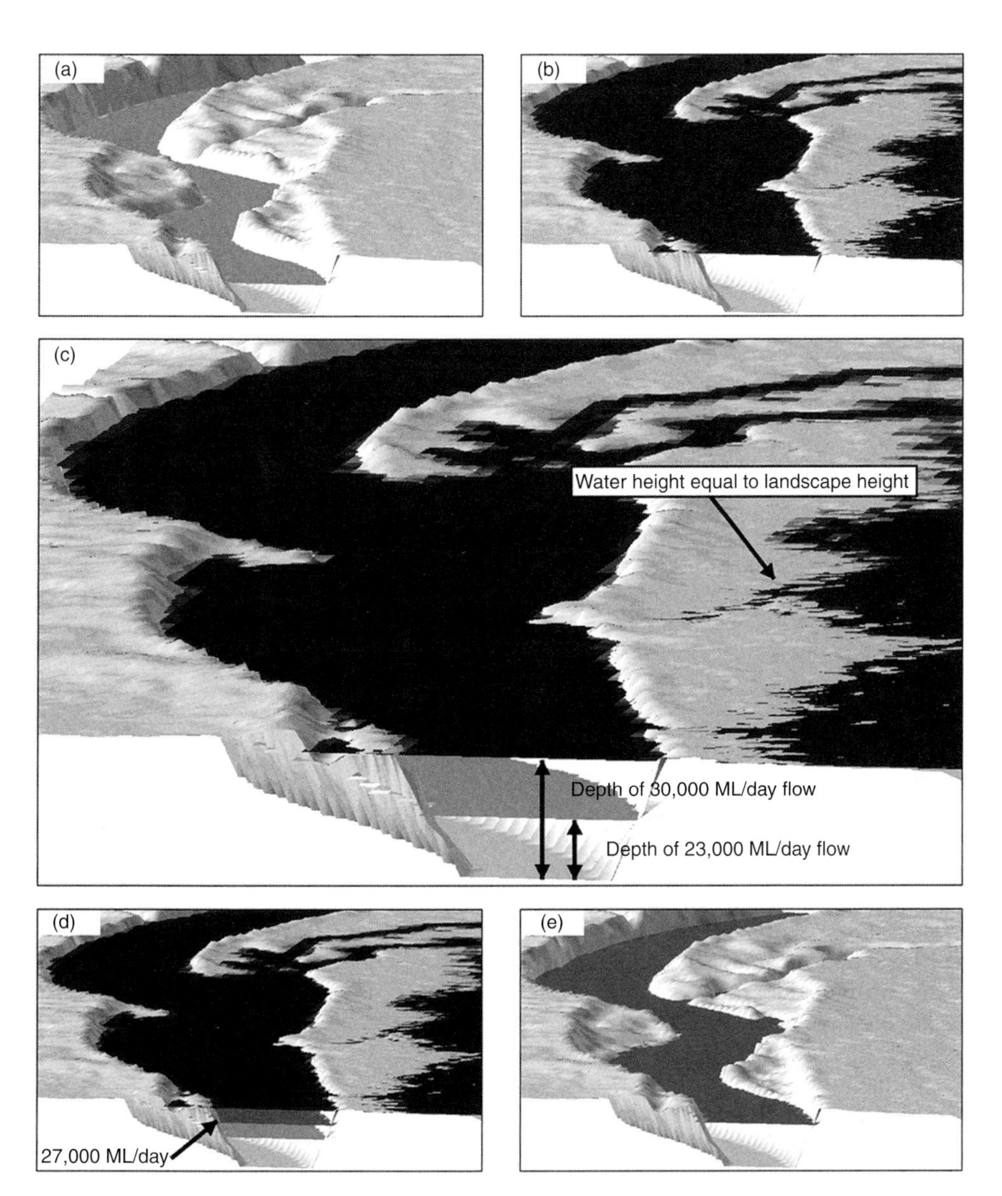

Fig. 14.11 (a) and (b) show flood heights for 23,000 and 30,000 ML/day flows. (c) compares the heights for each flow. (d) adds a predicted height surface for a 27,000 ML/day flow using a linear interpolation between depths. (e) shows the predicted height layer for the 27,000 ML/day flow.

Producing a continuous connect-to-fill surface for comparison to hydrographs required the prediction of unseen flows. The depth of the observed floods is equal to the water height less the landscape height as shown in Figure 14.11c, with the edge of flooding defined as the point where water height equals the landscape height. New depths were created by linearly interpolating between observed water heights. For example, consider a point in the landscape depicted in Figure 14.11 where the water height of the 23,000 ML/day flow was 93 m AHD and the water height of the 30,000 ML/day flow was 94.4 m AHD. The water height of a 27,000 ML/day flow for that location was predicted as 93.8 m AHD. The surfaces were generated at 1,000 ML/day intervals from the smallest regular flow to the highest recorded flow. A flood mask defines the outer boundary of the River Murray floodplain derived from aerial photography, and was recorded as the size of the 1956 flood of 250,000 ML/day which was the largest flood in recent history. The predicted height surface for a 27,000 ML/day flow is shown in Figure 14.11e.

Areas where the height surfaces predict inundation at a given flow can be different to those areas predicted by the satellite derived flood masks. Often the height surface incorrectly indicates that water would inundate an area, but the water cannot inundate the area because there is no connection to the river. For example, physical barriers such as a levee banks and river banks restrict the distribution of water across the landscape. The inundated areas were removed if there was no water observed in that contiguous area by a flood mask for the next highest observed flow. This removed water which had a barrier between the river and the inundation. Inundations not within 60 m of an observed inundation were removed to guard against barriers such as levee banks that were not evident in the high resolution DEM. This process removed disconnected areas to predict water heights for the actual area inundated.

The commence-to-fill at each point was calculated as the lowest predicted flow to inundate each point. Hydrographs from gauging stations along the River Murray show the frequency at which each flow has been observed. These frequencies were mapped directly onto the commence-to-fill layer to produce an inundation frequency layer for the river.

Discussion of flood modelling

Figure 14.12 shows the results of integrating a high resolution DEM derived from the LiDAR data into the RiM-FIM. The resolution is significantly improved from a contour-based interpolation and localised errors in the image classification are removed. Difficulties include handling residual water in the landscape, calculating the height of within bank flows and deriving heights in areas with cliffs, banks and levees. The accuracy above the highest observed flood is poor as the height of levee banks is unknown. In addition, kriged surfaces do not always accurately represent the flood patterns. The assumed linear interpolation between observed heights is simplistic but provides meaningful results that could be recalibrated with additional information such as further satellite scenes. The result is a high resolution set of images that describe the water height for different flow scenarios. The surfaces provide river managers, conservationists, scientists and others with a predictive model of floodplain inundation for the River Murray floodplain.

VEGETATION MAPPING

Assessing floodplain vegetation structure from LiDAR

In addition to providing high resolution terrain information from the last returns of the LiDAR signal, LiDAR data also contains structural vegetation information that can be derived from the first and intermediate returns.

Maximum vegetation height

LiDAR data in forested areas has been shown to be very reliable for measuring tree height

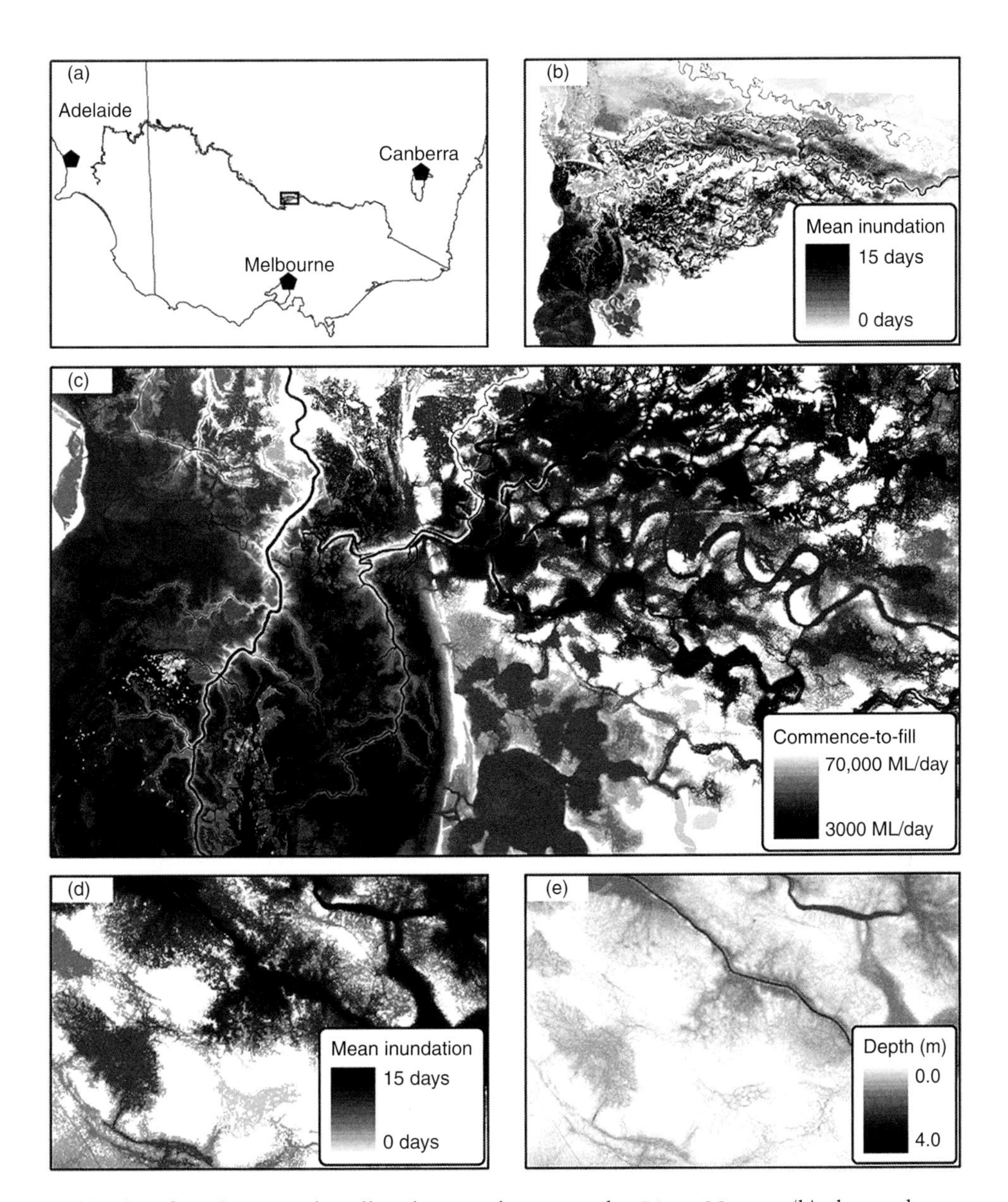

Fig. 14.12 (a) identifies the Barmah-Millewah state forest on the River Murray. (b) shows the mean days connected in August for the region. (c) shows the commence-to-fill using the DEM-based interpolation. (d) shows the mean days connected in August for a smaller region. (e) shows the depth of a 113,000 ML/day flood for that region.

(Anderson *et. al.*, 2006). Maximum vegetation height was calculated by subtracting the inverse distance weighted ground height from the maximum above ground height for the centre of each output cell Equation (14.2). Outputs were calculated at 10 m pixel resolution.

$$\text{Max. Veg. Height} = \text{Max. Above Ground Elevation} - \text{Ground Elevation} \qquad (14.2)$$

Figure 14.13a shows the results of Equation (14.2) applied to the River Murray LiDAR dataset for a subset of the area in the Barmah region.

Vegetation cover

Vegetation cover is the ratio of above ground points to ground points. To limit the effect of understorey vegetation, only points greater than 3 m above the ground (calculated using Equation (14.1)) are considered as 'above ground' points Equation (14.3). Outputs were calculated at 10 m pixel resolution.

$$\text{Vegetation Cover}_{\text{above 3m}} = \frac{\text{Number of Points Above 3m}}{\text{Total Number of Points}} \qquad (14.3)$$

Figure 14.13b shows the results of applying Equation (14.3) to the River Murray LiDAR dataset for a subset of the area in the Barmah region.

Vegetation texture

The Moran's I index (Moran, 1950) compares the differences between neighbouring pixel values and the local mean to provide a measure of local homogeneity. High positive values indicate a strong positive autocorrelation, 0 indicates spatially uncorrelated data, and low negative values a strong negative spatial autocorrelation. As the Moran's I index is a texture filter, it is not generated directly from the LiDAR data, but rather an image already generated from the LiDAR data. Running the Moran's I algorithm over the maximum height surface Equation (14.2) produces a layer that indicates the level of height homogeneity in the canopy surface. Figure 14.13c shows the results of applying the Moran's I texture filter to the River Murray LiDAR dataset for the Barmah-Millewa Forest area.

Information on vegetation characteristics derived from LiDAR was tested in the Barmah-Millewa region for its capacity to improve the identification of vegetation types from remote sensing (see later section). It appears that the LiDAR data can help improve on traditional remote sensing techniques based on spectral analyses to differentiate vegetation patterns.

LiDAR data has the potential to assist floodplain vegetation community identification using:

- Biomass: determining the volume of vegetation present using the density of LiDAR data through the vertical canopy profile;
- Vegetation heights: restricting possible community types through the height of the canopy;
- Structure: through analysis using the vertical vegetation cover profile to differentiate between single or multi-storey communities;
- Cover by strata: determining the percentage of vegetation present at different height intervals below the canopy (understorey, mid-storey, canopy);
- Health: linking the health of an identified vegetation community to vegetation cover and multi-spectral data. Vegetation vigour indices such as the Normalised Difference Vegetation Index (NDVI) require an understanding of vegetation density; and
- Temporal datasets: linking canopy condition to vegetation cover changes over time.

Satellite image classification

The initial LiDAR products described above were used in conjunction with SPOT satellite imagery to produce a vegetation community map for the extents of the MDBC LiDAR dataset.

The dominant tree vegetation on the River Murray floodplain are Eucalyptus species. As a genus, the structural form of Eucalypts are

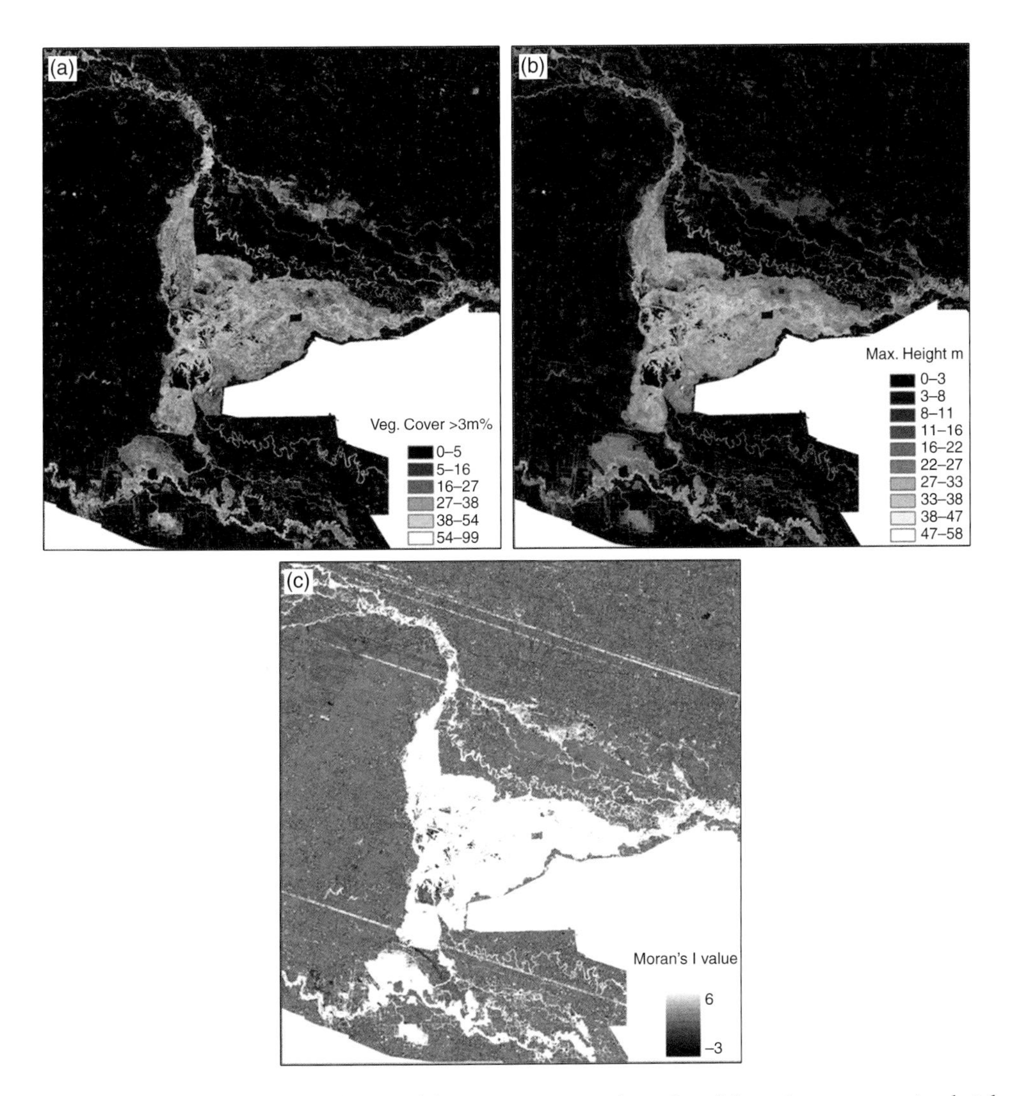

Fig. 14.13 The Barmah-Millewa forest showing (a) vegetation cover above 3 m, (b) maximum vegetation height, and (c) Moran's I vegetation texture.

unique. They exhibit an indeterminate growth habit – meaning the shape of the canopy is unpredictable and the tallest canopy elements are not necessarily centred relative to the crowns diameter, unlike conifers and many broad leafed trees. Secondly, they have a very open architecture when viewed from above and high levels of the ground and sub-canopy layers are often visible. Further to this, the complex canopy structures of mixed species forests make the delineation of distinct forest crowns difficult. This difficulty compounds with decreasing pixel size.

10 m SPOT multi-spectral image data and LiDAR canopy height data have been merged and classified to produce a series of floristically coherent and structurally partitioned vegetation classes. These results are consistent with the claims of Hyde *et al.* (2006) who report that

'combining information from multiple sensors, or data fusion has yielded promising results for the estimation of forest structural parameters' and Wulder and Seemann (2002), who state that, while optical remote sensing is good at mapping the vegetative content and horizontal structure of forest, it does nothing to describe the vertical elements.

The forests being mapped are dominated by the large eucalyptus hardwoods, River Red Gum (*Eucalyptus camaldulensis*) and Black Box (*Eucalyptus largiflorens*). The crown diameter of a large River Red Gum may be greater than 20 m. The advantage of using data from of a medium resolution sensor data such as SPOTs 10 m pixel is the canopies of these large trees are essentially under sampled by the imagery, needing only two, three or four pixels to cover a crown. In comparison very high resolution data (less than 5 m pixels), will mean the crown is over sampled – with many pixels imaging the crown. These examples represent the Low and High Resolution remote sensing models described by Strahler *et al.* (1986).

At the beginning of the classification, the LiDAR canopy height data is essentially an unrelated series of points with no explicit canopy context. In the case of this study the initial LiDAR data was imaged at around 0.4 m sampling interval, and represents a very high resolution dataset. Given the crown diameter of a large River Red Gum may be greater than 20 m, it is not unusual for the heights recorded 'within' a crowns footprint to come from the top of canopy, the canopy branches, the underlying shrub layer or the ground. The LIDAR data creates a linkage between a neighbourhood of LiDAR canopy heights and individual tree crowns as imaged by a remote sensing device. This requires algorithms that allow the boundaries of a crown to be delineated, and that delineation needs to be transferable when interpreting the LiDAR height data.

One approach is to develop crown delineation techniques that allow the boundaries of individual canopies to be defined and preserved in the processing model as discrete objects. Bunting and Lucas (2006) have successfully developed crown delineation techniques as part of a tool set for the mapping of forest structural types in mixed forests of central Queensland using very high resolution data (1 m) data from the Compact Airborne Spectrographic Imager (CASI).

When using medium resolution data such as SPOT, where individual crowns cannot be clearly resolved, an alternative method is to use an object-oriented approach. This approach aggregates groups of pixels within the image, based on the spectral and spatial relationships of those and neighbouring pixels.

Object-oriented image classification

An object-oriented image classification software was used to create a vegetation classification for the dominant canopy elements along the River Murray. Object-oriented image classification offers advantages over traditional pixel based classification techniques in that the spatial and spectral relationships of neighbouring pixels within the image are preserved and enhanced. At its simplest level this results in a smaller number of image objects being fed into the classification process, but at a more functional level the spatial and spectral relationships that exist between the objects that make up the image are used to build a meaningful understanding of the patterns of variance within the image data (Definiens, 2006).

Fundamental to the paradigm is the segmentation or clustering of the image pixel data into discrete larger objects using a combination of scale and weighting factors. These scale and weighting factors allow either the spectral or spatial relationships of groups of neighbouring pixels to have greater importance in the formation of a new object. Defining the most suitable scale and weighting factors is an iterative process and depends on (a) the resolution of the image, (b) the thematic data layers being used and (c) the spatial and spectral characteristics of the objects to be mapped. Repeated testing has shown these forests and woodlands are best described by scaling and segmentation weights that favour spectral variability over final object shape. In general, this means the resulting image objects are more fractal in nature, than smooth and rectilinear. This is consistent with the observable patterns of variations within these landscapes.

The classification methodology first involved fusing or merging the raw 10m SPOT image bands, and a number of transformed SPOT bands, with the LiDAR canopy height data which has been re-sampled to a 10m resolution. This fusion image was then imported into the object-oriented software where a number of image segmentations and classifications were performed. The first involved a very fine segmentation of the LiDAR data. In the processing, model data layers were ordered by virtue of the object sizes they contain, that is finer objects exist below courser objects whilst the boundaries of these objects are always conformable.

LiDAR canopy height classification

The primary segmentation of the LiDAR canopy height layer was then run through an ordinal classification whereby heights were binned into a number of florsitical significant height classes. The specification of these was based on McDonald *et al.* (1990) and given in Table 14.1. Examples of the segmentation of the LiDAR canopy heights are given in Figures 14.14a,b.

Table 14.1 Vegetation height classes.

Class	Description	Height (m)
Not trees		
	Nulls	<0
	Dwarf shrubs or low grasses	0–0.25
	Low shrubs or medium high grasses	0.25–0.5
	Very tall shrubs or extremely tall grasses	0.5–1
	Very tall shrubs	1–2
Trees		
	Low tree	2–6
	Medium tree	6–12
	Tall tree	12–20
	Very tall tree	20–35
	Extremely tall tree	>35

The image data to be classified was also segmented into spatial and spectrally coherent objects, but using different scaling and weighting factors. Those used resulted in larger objects than those created by segmenting the LiDAR canopy heights (Figure 14.14c). This is appropriate because the patch variation in the imagery is lower than in the canopy height data. Figure 14.16 shows the height distribution for a section of the Barmah forest in NSW.

Prior to classifying the segmented SPOT imagery, a second classification layer was created by partitioning the LiDAR height classes into two height groups – those pixels above two metres and those below two metres. Whilst some authors have argued that a 'tree' is defined as woody vegetation with a minimum top of canopy height equal to three metres we have applied the cut off at two metres, which is consistent with that used by Tickle *et al.* (2006) who describe that within the Australia context, a tree is any woody vegetation with a top height equal to or greater than two metres.

The SPOT image pixels that conform to the above two metres classification coverage were then classified using a modified Nearest Neighbour classification algorithm. This supervised classification starts by seeding the most significant 5–10 forest and woodland classes and running the classifier with a very narrow 'gate' or acceptance value (set 0.9). The resulting image was examined for unclassified pixels and further classes were then activated and seeded. This iterative process was repeated until all of the significant forest, woodland and tall shrubland regions are described by the class set. In the final iteration the gate value was widened to 0.7 to allow binning of outlier pixels. Figure 14.15 represents an example of the resultant product.

A crown density layer also generated from the LiDAR data was used at this point to help identify the varying patterns of tree density that occur in the floodplain. Interim class labelling then occurred by way of reference to a number of historic land cover classifications and field observation. These interim classes were grouped into 'super' classes in a second cover classification

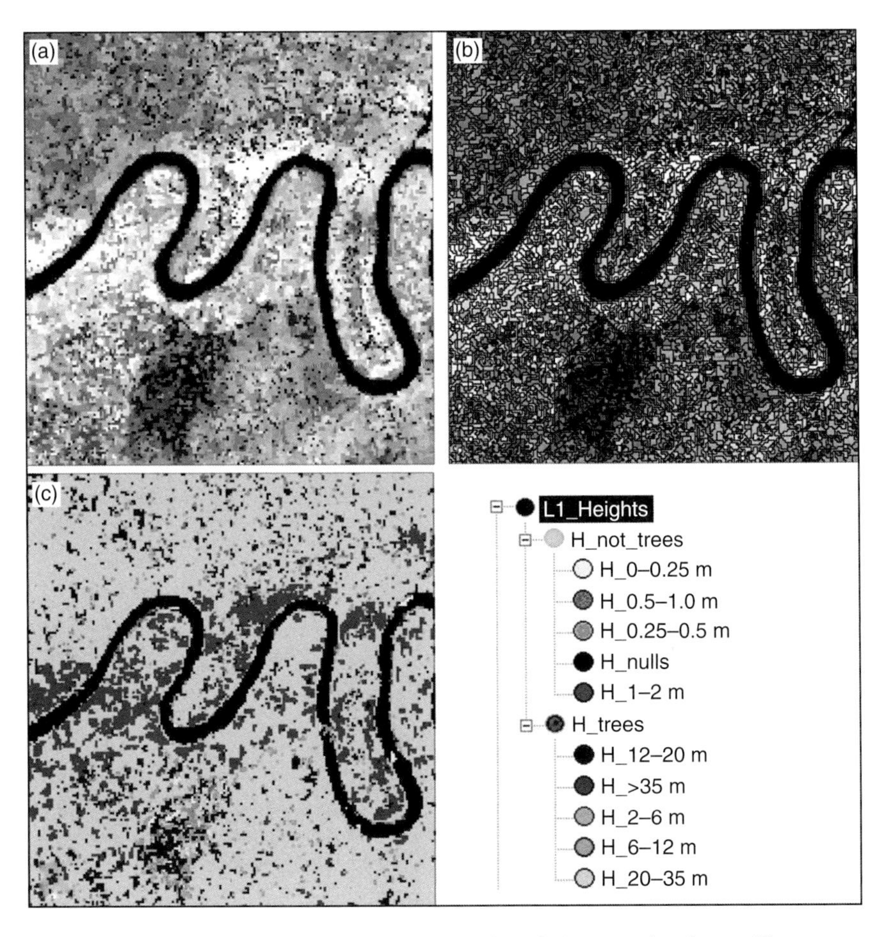

Fig. 14.14 (a) canopy height data, (b) segmentation image, (c) classified canopy heights and key.

that occurs on a lower level of the class hierarchy. This level has the same segmentation parameters to those used to segment the LiDAR data, and each new super class is further partitioned by reference to the LiDAR height classification. This results in forest and woodland classes that can be examined with respect to the height variations of the pixels that have been mapped into it (Figures 14.15c,d).

Some landcover classes show an expected range of heights within its membership, due to the varying age and vigour of members within the class. Other classes show either a predominance of one height class or an unexpected mix of heights. Careful analysis of these distributions allows for a third partitioning of the classes into more realistic cover and structural groupings (Figure 14.15d). Spectral classes with high proportions of each height class are likely to be mixed classes and will benefit from further analysis.

Discussion of vegetation mapping

The SPOT data used in this study was collected in the summer of 2005. This region, as for much of the south-east of Australia, was drought declared in the months leading up to the

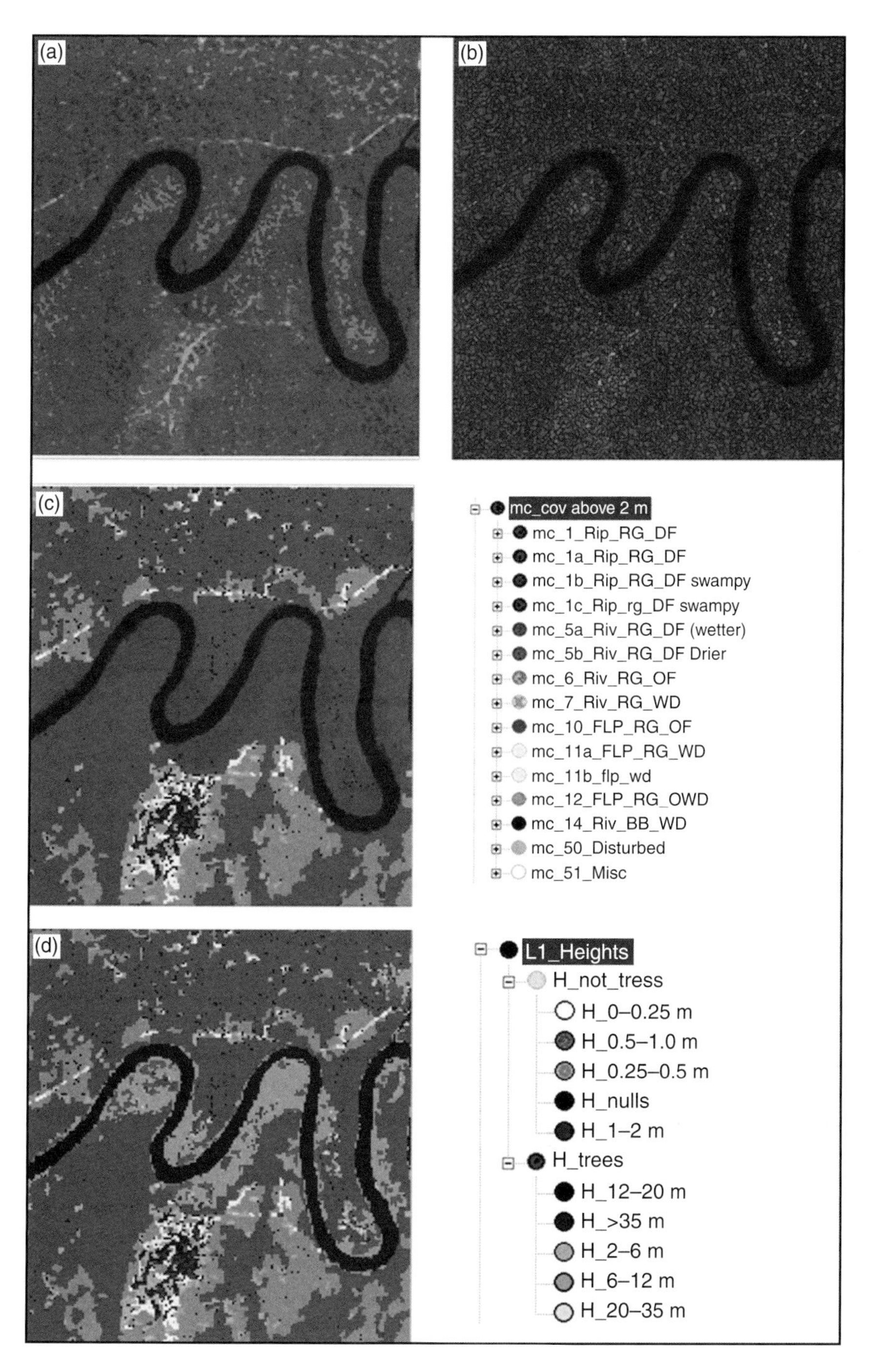

Fig. 14.15 Spot processing products: (a) raw data rgb, (b) segmentation image, (c) NN classification canopy, (d) merged canopy and LiDAR classification showing the height sub classes of the Forest type 1.

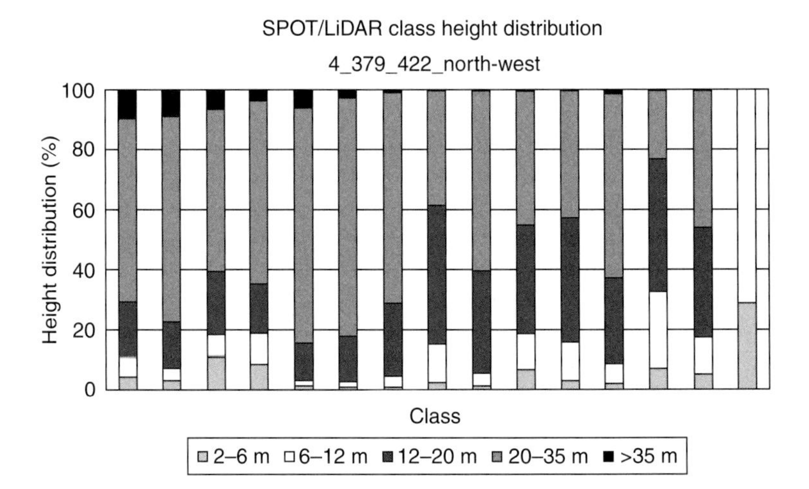

Fig. 14.16 Height distribution within the SPOT canopy classes.

December image capture dates, so much of the ground cover is senescent or non-photosynthetic vegetation (NPV). The canopy layer in many areas is also very stressed due to the drought and longer-term water management policies that have affected flow and flooding rates along the river valley. Eucalypts ultimately respond to such stresses by dropping their leaves and thereby 'opening' the canopy significantly. As a result, more of the ground layer is visible to the imaging sensor.

The canopy height data on its own cannot distinguish these stressed forest regions, whilst the spectrally driven nearest neighbour classification of the SPOT data does map these stands. By comparing the height classes, the forest density classes and the SPOT Nearest Neighbour (NN) classes, the appropriate differentiation can be made between healthy and stressed regions of dense forest (Figure14.16).

CONCLUSIONS

The LiDAR data provides detailed elevation data that can be used to produce DEMs at sufficient resolution for flood and vegetation mapping. The complex and low relief environment of a large river floodplain requires detailed and accurate elevation data. A number of issues exist with the raw LiDAR data, such as irregularities in areas of water bodies and dense vegetation. These issues can be resolved and DEMs built to required specifications using a combination of manual and software processing.

The LiDAR produced elevation model has improved the floodplain inundation model by providing a better base for interpolation between observed flood events. The LiDAR data was used to calculate the height of the water observed in different satellite image flood extents and the interpolation between flood extents was based on generating new water surface elevation levels. The elevation data also provides water depth for each flood event and allows calculation of water volumes required for environmental flow strategies.

The LiDAR also provided information for a system-wide consistent vegetation map of the River Murray. Using the vegetation canopy height and canopy density provided important information for an integrated classification based on satellite imagery. The density and height are critical factors that distinguish species and their health on the floodplain and are not available from multi-spectral satellite imagery alone. LiDAR-derived canopy height information strengthens the reliability and information content of land-cover mapping products created using image classification techniques. The perennial errors of omission and commission that result in cover classes being mixed are greatly reduced when a LiDAR-derived height filter is applied in pre-classification data processing. The post-classification analysis is also enhanced as the height distribution within classes can be used to further refine class homogeneity and give meaningful labels to a range of floristic and structural vegetative cover types.

Previous studies have shown that vegetation density of river floodplain forests can be predicted using terrestrial (as opposed to airborne) laser scanning instruments (Warmink *et al.*, 2006). However, commercially available terrestrial laser scanners are optimised for engineering applications and are not ideally suited to accurately characterising the structure of complex vegetation canopies. This limitation can be overcome using technology known as full-waveform LiDAR, which is designed specifically for characterising the structure of dispersed media. Using a full waveform terrestrial laser scanner, a number of representative vegetation communities on the floodplain could be sampled to characterise their impedance to flow across the floodplain. These values can then be related empirically to airborne LiDAR data in order to scale up point-based estimates of flow impedance across the landscape. The derived flow spatial layers would then provide valuable information to inform existing hydraulic models and improve the accuracy of flood event simulations.

These datasets and tools provide the required information for analysing the benefits of environmental flow strategies. Detailed modelling of flood movement and its extent of vegetation inundation is required for flow strategies to be developed on the River Murray, in order to allocate the finite resource of water to the most appropriate locations on the floodplain, and by the most appropriate method.

REFERENCES

Andersen HE, Reutebuch SE, McGaughey RJ. 2006. A rigorous assessment of tree height measurements obtained using airborne LiDAR and conventional field methods. *Canadian Journal of Remote Sensing* **32**: 355–366.

Brivio PA, Colombo R, Maggi M, Tomasoni R. 2002. Integration of remote sensing data and GIS for accurate mapping of flooded areas. *International Journal of Remote Sensing* **23**: 429–441.

Bunting P, Lucas R. 2006. The delineating of tree crowns in Australian mixed species forests using hyperspectral Compact Airborne Spectrographic Imager (CASI) data. *Remote sensing of Environment* **101**: 230–248.

Definiens Professional 5 Users Guide. 2006 Definiens AG, Munchen, Germany.

Frazier PS, Page KJ, Louis J, Briggs S, Robertson A. 2003. Relating wetland inundation to river flow using Landsat TM data. *International Journal of Remote Sensing* **24**: 3755–3770.

Hyde P, Dubayah R, Walker W, Blair JB, Hofton M, Hunsaker C. 2006. Mapping forest structure for wildlife habitat analysis using multi sensor (LiDAR, SAR/InSAR, ETM+, Quickbird) synergy. *Remote Sensing of Environment* **102**: 63–73.

Jankowski P, Andrienko N, Andrienko G. 2001. Map-centred exploratory approach to multiple criteria spatial decision making. *International Journal of Geographical Information Science* **15**: 101–127.

Jolly ID. 1996. The effects of river management on the hydrology and hydroecology of arid and semi-arid floodplains. In *Floodplain Processes*, Anderson MG, Walling DE, Bates PD (eds), New York: John Wiley & Sons Ltd.

McDonald RC, Isbell RF, Speight JG, Walker J, Hopkins MS. 1990. *Australian Soil and Land Survey Field Handbook*. Inkata Press.

MDBC. 2003. *Preliminary Investigation into Observed River Red Gum Decline Along the River Murray Below Euston.* Technical Report 03/03, Murray-Darling Basin Commission, Canberra.

Moran PA. 1950. Notes on continuous stochastic phenomena. *Biometrika* **37**: 17–23.

Penton DJ, Overton IC. 2007. Spatial Modelling of Floodplain Inundation Comnining Satellite Imagery and Elevation Models. In Oxley, L, Kulasiri, D. (eds) MODSIM 2007 International Congress on Modelling and Simulation. Modelling and Simulation Society of Australia and New Zealand, December 2007, pp. 1464–1470.

Oliver R, Bickford S, Byrne G, Gallant JC, Matveev V, Overton IC, Penton D, Ryan S, Watson, G. 2007. A biophysical framework for assessing environmental water allocation and delivery in the Murray River. *Proceedings of the International Conference on Environmental Flows*, Brisbane, September 2007.

Overton I, Newman B, Erdmann B, Sykora N, Slegers S. 1999. Modelling floodplain inundation under natural and regulated flows in the Lower River Murray. In *Proceedings of the Second Australian Stream Management Conference: the Challenge of Rehabilitating Australia's Streams*, Adelaidc. Rutherfurd I, Bartley R (eds), Melbourne: CRC for Catchment Hydrology.

Overton IC. 2005. Modelling floodplain inundation on a regulated river: Integrating GIS, remote sensing and hydrological models. *River Research and Applications* **21**: 991–1001.

Overton IC, McEwan, K, Gabrovsek C, Sherrah J. 2006. *The River Murray Floodplain Inundation Model (RiM-FIM)*. CSIRO Water for a Healthy Country National Research Flagship Technical Report. 127p.

Shaikh M, Green D, Cross H. 2001. A remote sensing approach to determine the environmental flows for wetlands of the Lower Darling River, New South Wales, Australia. *International Journal of Remote Sensing* **22**: 1737–1751.

Sheng Y, Gong P, Xiao Q. 2001. Quantitative dynamic flood monitoring with NOAA AVHRR. *International Journal of Remote Sensing* **22**: 1709–1724.

Strahler AH, Woodcock CE, Smith JA. 1986. On the nature of models in remote sensing. *Remote Sensing of Environment* **20**: 121–139.

Taylor K, Walker G, Abel D. 1999. A framework for model integration in spatial decision support. *International Journal of Geographical Information Science* **13**: 35–55.

Tickle PK, Lee A, Lucas RM, Austin J, Witte C. 2006. Quantifying Australian forest floristics and structure using small footprint LiDAR and large scale photography. *Forest Ecology and Management* **223**: 379–394.

Townsend PA, Walsh SJ. 1998. Modelling floodplain inundation using an integrated GIS with radar and optical remote sensing. *Geomorphology* **21**: 295–312.

Walker KF, Thoms MC. 1993. Environmental effects of flow regulation on the Lower River Murray, Australia. *Regulated Rivers: Research and Management* **8**: 103–119.

Walker RD, Aubrey MC, Fraser, SJ. 1986. *Remote Sensing in Australia*. Melbourne: Canbec Press Pty Ltd.

Warmink J, Middelkoop H, Straatsma M. 2006. *A paper presented for the Vienna Workshop on 3D Remote Sensing in Forestry*, Vienna, Austria, February 2006.

Whitehouse G. 1989. Flood monitoring and floodplain studies. In *Remote Sensing in Hydrological and Agrometeorological Applications*, de Vries D (ed.), Canberra: COSSA Publications.

Wulder MA, Seemann D. 2002. Forest inventory height update through integration of LiDAR data with segmented Landsat imagery. *Canadian Journal of Remote Sensing* **29**: 536–543.

15 Laser Scanning Surveying of Linear Features: Considerations and Applications

MICHAEL LIM[1], JON MILLS[2] AND NICHOLAS ROSSER[1]

[1]Department of Geography, Institute of Hazard and Risk Research, University of Durham, Durham, UK
[2]School of Civil Engineering and Geosciences, Newcastle University, Newcastle Upon Tyne, UK

INTRODUCTION

Terrestrial laser scanning (TLS) is increasingly viewed as a revolutionary new surveying technique (Slob & Hack, 2004). The development of sensors able to rapidly collect 3D surface information has enabled high-density measurements to be made across landscapes that are unsuited to more conventional approaches due to their inaccessibility, hazardous nature or spatial extent. In addition, the increasing application of TLS systems and the ease of data capture they offer have enabled non-specialist operators from outside traditional surveying disciplines to efficiently generate detailed information in evermore challenging and complex environments. However, there is now a mismatch between the ability of TLS to provide data at an ever higher resolution, and the proficiency with which such data is utilised. Therefore, although point clouds provide attractive visualisations, the full value of TLS cannot be realised without employing a rigorous survey workflow. This chapter explores the synergies between established surveying practice and the effective application of TLS to linear features in the landscape.

Linear features, such as transportation links and coastlines, present particular challenges to surveying because they typically require consistent collection and integration of data from multiple viewpoints. Ultimately, there is a need to establish a surveying protocol for TLS for complex applications. It is argued that a change in emphasis is required from TLS as a method of data capture to an ethos of laser scanning survey (LSS) in which the collected information is constrained in a defined workflow from the planning and capture of scans, to processing and interpretation.

THE APPLICATION OF LSS TO LINEAR FEATURES

Linear or corridor landscapes range from rivers and coastlines to pipelines and transport infrastructure, and commonly provide the focus for surveys, often coinciding with natural or

Laser Scanning for the Environmental Sciences,
1st edition. Edited by G.L. Heritage and A.R.G. Large.
 ISBN 978-1-4051-5717-9

socioeconomic boundaries which need to be documented, assessed, monitored and maintained. The approach used to document linear features will reflect a balance between the objectives of the survey and the financial and practical constraints of the project. For example, a time consuming but precise control traverse has conventionally been used to accurately plan road and railway routes. Where the transport links require a more detailed assessment and/or monitoring, the point-based transect information can now be supplemented with kinematic dGPS, photogrammetry, airborne LiDAR and TLS, each with a balance between the time, cost, accuracy, resolution and coverage of the data generated. Of all the survey techniques available, TLS has the least standardised control practices and error assessments (Lichti *et al.*, 2005). This is due to both the relative infancy of TLS as a survey tool and the apparently complete and satisfactory outputs it provides. A direct, positive relationship between data quantity and data quality is often assumed in the recording of surfaces, and there is a tendency to accept high-resolution data as being the most accurate representation. Although terrestrial laser scanners hold the potential both to reduce human error through automation and to increase the global accuracy of a survey through data redundancy, they remain subject to many issues that will generate inaccurate, misleading or inappropriate information if not considered. Error in TLS measurement is spatially variable, given the variation in survey range, spot size and incidence angle onto the target surface. The combination of separate scans, either spatially or through time, has the potential to introduce inconsistencies in the orientation, resolution and positioning of individual surveys. Therefore, users of TLS can benefit greatly from the considerations commonly associated with more traditional surveying practices. The quality of TLS data is fundamentally connected to the survey planning, field procedures and data processing within which it is constrained. The term LSS is used here to emphasise the value of methodological survey practice over the use of TLS as a data acquisition technique.

LSS: planning

The idea of planning a TLS survey may appear to be redundant, given the limited number of components involved. A major advantage of TLS is the ability to survey in an indiscriminate manner in the field and identify common features for registration at a later stage during data processing. However, there may be issues concerning the suitability of the selected scanning system to the required task. Simply put, the planning stage of LSS relates to the *relative accuracy* produced by the scanner. This term refers to the ability of the scanner to differentiate patterns and features upon a surface; its sensitivity to the topographic changes on the surface and variations within the data. A significant limitation to many terrestrial laser scanners is the lack of a specific calibration certificate, or a comparable measure of its precision on different surfaces. Often, scanners are issued with a generic specification sheet relating to the precision and accuracy of the system over a range of distances and angles onto an arbitrary target surface. The instrument specification provides an important basis for assessing the capabilities of the scanner and is particularly useful when control points (prisms or targets) are relied upon for error measurement. However, when the interest of the survey is in point coordinates across an uncalibrated surface, specific validation and performance testing should be conducted. For example, if the sag in overhead power cables is the focus of the survey, the accuracy and repeatability of scanned point measurements on cables at a set of angles and distances to the system representative of those found in the field is required. The reflective characteristics of the target surface, in addition to its distance from and orientation to the incident beam, all influence the strength of the return (Lichti & Harvey, 2002). Simple, practical tests can guide the selection of an appropriate scanning system with respect to these influences in order to most effectively meet the survey objectives.

Calculating the time taken for a complete survey may be necessary if there are constraints on accessibility to a site, for example in between tidal cycles or where roads are closed for a limited period of maintenance. The time taken for different scanners to collect a set number of points varies significantly and the divergence between the fastest and slowest systems increases dramatically with the complexity and resolution of the required survey. In addition to variations in the set-up and collection speeds of different systems, the size, weight and durability of scanners requires consideration. Where the subject area prevents vehicular access the portability of the scanner and its power supply become critical factors in its ability to be used for survey. Many scanners are transported in bulky protective cases, mounted on survey grade tripods and require external batteries and computers. More ergonomic designs generate greater flexibility in the range of environments in which they can be applied, but are often limited by battery life and their data manipulation and storage capacity. Scanner manufacturers have recognised the demand for more practical instruments with the introduction of scanners that meet Ingress Protection 66 specifications, which require the scanner to be water and dust resistant. While this adds greatly to the potential of TLS as a survey tool in testing outdoor environments such as open cast mines, separate error assessments are required if the data is collected in sub-optimal weather conditions. This is particularly critical if the data is to be collected in wet conditions because TLS systems function at wavelengths (500–1500 nm) that can be absorbed into atmospheric particles or moisture, degrading the strength of the returning radiation, although not necessarily the accuracy of the point measurement. Therefore, careful consideration is needed in selecting the most appropriate scanner for the survey application.

LSS: field procedures

Field procedures using LSS are concerned with *absolute accuracy* and the ability to locate the scanned data correctly in either local or global space. Whilst there are several studies into the calibration of individual TLS systems and the influences that can determine relative point measurements (see e.g. Clark & Robson, 2004; Kersten *et al.*, 2005), their application as a practical survey tool has been limited. The ability to accurately (re-)locate scanners over known survey station control points is often limited by the lack of a survey grade plummet, placing renewed emphasis on the procedures used to constrain and match point cloud information directly using external control. Separate point clouds are aligned with the use of control targets, which derive a specific reflectance signature and geometry that can be recognised in several scanning systems. The positioning of targets should be chosen according to the specific requirements of the survey to allow multiple tie points between scans, with redundancy. Linear surveys require that targets need to be visible from view points parallel to the survey area, exerting an upper limit on how oblique the angle of incidence can be from the targets to the scanner position. Where the optimal scanning positions are limited, solid geometry targets – such as white spheres – may be more beneficial, relying on adjustment theory to position the control rather than reflective contrast on flat targets. It is also essential that the target positions are as secure and stable as possible during the data collection; it is assumed that targets do not move between scans, although isolated erroneous points can be identified and removed during processing.

The positioning of the scanning stations is constrained by the field of view achievable by the instrument. The operator must be aware of the concentration of data coverage that diffuses towards the edge of the scan section, particularly on sloped surfaces; this can be minimised if equal survey legs (the baseline distance between the scan positions) are maintained. Scans of continuous linear features produced by systems with a fixed field of view, commonly exhibit triangular patches of data loss if the survey leg is too large. In addition to their use in combining point cloud datasets, control targets can also be used to assess the errors incurred during the registration process. A representative distribution of check points, known to a greater accuracy than the scanning data (most specifications require accuracy improvements of an order of magnitude), is needed for a rigorous investigation of error.

Variability in ambient conditions during the data collection should be minimised in order to generate a consistent survey. As examples, sub-zero temperatures can affect the ability of the hardware to function and changes in lighting can degrade the signal to noise ratio in the information received by the scanner and limit its ability to use referenced images to obtain colour information for each point. Scanners which can be combined with high-resolution external cameras tend to cope better with problems of over exposure and strong contrast within the scanned scene than those with bore mounted CCDs.

LSS: data processing

The data processing elements of a LSS workflow should reflect the considerations made throughout the planning and collection of TLS data. The first task is to identify gross errors or blunders within the point cloud data. These can be produced by both instrumentation error and moving features within the survey scene, such as cars or people passing by. The removal of erroneous points involves either manual point cloud editing or automated threshold procedures designed to reflect the desired product of the survey. For example, if the aim of the survey is to record the extent of vegetation encroachment onto a railway line then the majority of isolated points may be assumed to relate to overhanging tree branches and thus be retained in the scan data. By contrast, if the aim is to analyse the effective height of river levee defences, the critical data may only be the returns from the ground surface, necessitating the removal of a large number of points associated with vegetation and fences.

Point cloud data produced from TLS are typically collected in arbitrary coordinate systems that calculate the distance between the target surface and the optical centre of the instrument. Separate point clouds need to be related or registered to a common data space to form a complete, georeferenced survey. A wide variety of techniques exist in order to transform and associate scan data to consistent frames of reference. The approaches basically fall into point-to-point, surface matching and external control methods (see Eggert *et al.* (1998) for a comprehensive review of iterative closest point algorithms). The three-dimensional conformal transformation of calibrated control targets is the ideal registration method of associating scan data to a geodetic datum. TLS software often incorporates this procedure into the data capture and processing workflow. Where this is not possible, perhaps due to on-site inaccessibility or hazards, more computational iterative data matching and convergence procedures must be relied upon.

The point cloud data can be registered to either real or local coordinates depending on the requirements of the survey. When TLS data is to be used for change detection over repeated collections or in combination with other datasets such as total station, GPS or LiDAR surveys, it is often essential to work in national or global coordinates. Working in real-world coordinates can become cumbersome and inefficient due to the large numbers produced, but as with conventional survey approaches, data may be processed in local Cartesian coordinates and only the final completed survey transformed to real world coordinates. Additional consideration must also be given to the increasingly high demands on data structure, storage and archiving, particularly where surveys contain numerous dense point clouds, as is the case in linear features. The most efficient use of TLS data is governed by the application of the survey, for example, all surface information may need to be retained for change detection in river bank erosion studies, but only minimal information would be required for the extraction of solid pipe geometries for asset modelling. The higher the degree of data processing the greater the potential to incorporate error, and careful consideration should be given to the effects of data cleaning and thinning algorithms (Boehler *et al.*, 2002). Ultimately, the current emphasis on data collection over data processing needs to be redressed if the full potential of TLS as a survey tool is to be realised. Following are two case studies into the application of an LSS approach to linear features. First, the ability of TLS data to provide valuable terrain measurements from below vegetation on transport corridor earthworks is investigated. Second, TLS is used in the detection and interpretation of change to coastal cliffs.

CASE STUDY 1: LSS FOR EARTHWORK STABILITY ASSESSMENT

The effective operation of both rail and road networks is closely related to the stability of the earthworks upon which they are founded (Ridley *et al.*, 2004); even relatively minor movements can result in severe disruption, unacceptable potential risk, financial losses and occasionally injury or loss of life. The UK's Network Rail, who manage the rail infrastructure, estimate that 400,000 delay minutes resulted from geotechnical causes between 2000 and 2003 in the UK, costing £26m (Scott *et al.*, 2007). Therefore, the assessment of slope failure in transport corridors is an essential and continuous task for network operators. This is confounded by the deterioration of a now aging infrastructure, which in places is approaching and exceeding its design life. To date, the management of transport corridor environments has tended to focus on the reactive remedy of past failures rather than the proactive identification and mitigation of future problems, although prevention costs are approximately one fifth of preventative remedial works (Lloyd *et al.*, 2001). This reactive approach is inhibited by the difficulty of interpreting the potential for slope problems to develop. Currently, engineers must conduct site visits to inspect embankment condition, analysing the risk of failure through observations or occasionally employing geotechnical models and monitoring instrumentation. There remains a pressing need for consistent, quantitative tools with which to assess slope movements over a wide scale. TLS is capable of providing high-resolution data coverage of topography. The aim of this case study is to analyse whether it can be implemented within an LSS workflow to generate surface models of sufficient consistency to identify the development of embankment instability.

Earthwork failure

The potential for an earth slope to move and ultimately to fail is determined by the interaction between *in situ* characteristics such as location, geometry, vegetation cover, underlying geology, drainage and the nature of the constituent material and external factors – including the quantity and delivery of rainfall, groundwater, wind and insolation. Each element may enhance or reduce slope instability to varying degrees and this relationship may alter or evolve with time. A critical control on earthwork instability is pore water pressure (Scott *et al.*, 2007), which is linked to both seasonal slope responses (Andrei, 2000) and longer term failures during wet periods (Barker, 1995). Seasonal deformations result in a cumulative downwards and outwards movement of the slope material. Over subsequent years, progressive displacement can lead to deep seated failure when propagated along preferred shear surfaces. Vegetation can exacerbate shrink-swell movements, accelerating desiccation and suction stresses during hot, dry conditions (Blight, 1997). Seasonal variations can contribute over 50 mm of vertical ground movement during winter heave and settlement in the summer months (Scott *et al.*, 2007). The higher the water demand of the vegetation the greater the potential depth of slope material affected (Crilly & Driscoll, 2000). Horizontal displacements are generally less significant, but are thought to reach up to 12 mm where the soil is not strengthened by extended root structures (Scott *et al.*, 2007). The delayed failure of railway embankments is usually associated with grass covered slopes, while slopes that contain large trees that bind the soil against deep seated movement more commonly exhibit track deformation problems resulting from shrink-swell processes (O'Brien, 2007). These processes can operate over varying time periods, from long-term influences such as tree life cycles, through seasonal fluctuations, to daily or diurnal wetting and drying generating a complex superimposition of responses (Copin & Richards, 1990).

Earthwork assessment

Due to the complexity of influences driving slope behaviour, which vary spatially and temporally, network operators are often limited to the identification of indicators of potential movement in order to evaluate asset stability. In such an approach the most effective indicators of slope instability are the features upon the slope surface. Details such terracing, bulging and cracking of the slope surface are all signs of movement. In the natural environment, differences in slope gradient have been strongly correlated with the potential occurrence of mass movements such as landslides (Montgomery & Dietrich, 2004) and debris flows (He *et al.*, 2003). In transport corridors, earthwork gradients are engineered to be generally consistent and even. Consequently, the subtle contrasts both within and between different earthworks, often exacerbated over time due to settlement, degradation and localised remediation of the slope. can become the critical descriptors of instability (see e.g. Al-Homoud and Al-Masri, 1999). Therefore, the ability to gain an accurate and quantifiable measure of slope gradient, and the changes in slope position or angle, is fundamental to determining the risk of significant earthwork failure.

A critical concern for the application of TLS to earthwork assessment is the ability to generate models of the slope surface of sufficient consistency and accuracy to detect movements greater than the maximum total displacement expected due to seasonal variations. In addition to the challenges of optimising the quality of the TLS survey over an area of concern that may stretch for several hundred metres, is the problem of vegetation cover, which may obscure the view from the laser to the slope surface itself; indeed a dense mat of vegetation often blurs the physical division between the slope surface and vegetation cover. Such effects have long imposed a severe restriction on terrestrial laser scanning for slope monitoring, limiting applications to surfaces free from vegetative influence. Filtering processes are commonly applied to data from airborne laser scanning in order to extract ground surface returns from below objects on the surface, such as vegetation and buildings (Sithole & Vosselman, 2004). The extraction of ground returns from airborne laser scanning data relies on the assumption that some of the radiation sent out is able to pass through the first object it hits and continue on to the next (Kraus & Pfeifer, 1998). The ability of the sensor to record more than one return from each pulse generates the possibility that the last signal received is produced from the ground surface. By contrast, the majority of TLS systems do not capture more than one return from each emitted pulse. This reduction of the full waveform of the returning signal into a condensed single point echo significantly simplifies the data recorded from terrestrial scanners. In many instances, the point measurement may relate to objects on the surface that produce a stronger return than any of the signal that does reach the ground below. However, although there is a lower chance of deriving ground information from a pulse that first hits vegetation, the resolution of TLS is an order of magnitude finer than the data produced in airborne surveys, and consequently portions of the ground surface will be captured wherever there is a direct line of sight to an area of ground larger than the laser footprint. The potential for ground surface reconstructions from TLS survey data, particularly in winter conditions where there are greater gaps in deciduous vegetation, presents the potential to utilise TLS as a tool for assessing earthwork condition. A LSS has been conducted in order to explore the effect of vegetation on the use of TLS as a tool for the assessment of earthwork stability.

LSS planning and data collection for earthwork stability assessment

A test embankment was selected with distinct areas of different vegetation cover types considered representative of those found in typical UK transport corridors. The survey area of the embankment is approximately 21 m high and stretches for over 90 m. The overall slope is inclined at an average angle of 20° but alters from 18° to 30° in places. The site was surveyed with the use of a Leica TCRA 1103 total station during winter conditions to minimise the vegetation cover. An initial traverse was performed, sighting on to tripod-mounted prisms leveled over the adjacent stations from each survey position; each station comprised a Feno survey marker georeferenced with differential GPS. The separation between the survey stations was calculated from the 40° field of view of the scanner (defined below) to be used and the minimum station distance to the slope face in order to ensure that the control targets would be clearly visible in overlapping scans. Based on a minimum station to slope distance of 32 m, the stations were located 18 m apart, allowing an overlap within which to locate the control targets. The data from the station traverse were processed in Star Net (Star Plus, 2000) and the best fit solution used to fix the position of the survey stations. Ground points were collected from the base stations over the area corresponding to the laser scanner field of view from each scanning position. A ground sampling distance of approximately 2 m was measured using the total station in automatic target recognition mode for the efficient and consistent collection of data. Problems were encountered when mature vegetation, over the 1.7 m staff height, interrupted the line of sight to the instrument resulting in a loss of lock that was not always remedied by an automatic search for the reflector. Despite limited coverage in the areas of densest vegetation the total station, calibrated to 2 mm ±2 ppm/1.0 sec, provided the most accurate measure of 342 control points across the slope surface.

Following the establishment of the control dataset, TLS was used to survey the test embankment. A Leica 2500 terrestrial laser scanner was selected in order to investigate the performance of one of the industry's most widely distributed systems commonly used to survey transport infrastructure. Its ability to survey at fine resolutions provided a greater chance of penetration through vegetated layers to the embankment surface. Targets were positioned in a zig–zag formation down the slope, within the overlapping area between scans in order to ensure that adequate three-dimensional control was obtained. The targets used for registering the point clouds were calibrated to give a specific reflectance, allowing them to be identified and accurately positioned by the scanner. The capture of target positions was semi-automated; areas matching the target reflectance characteristics were validated by a manual operator check before the target was located in a high-resolution scan. A minimum of six control points was used in order to provide sufficient redundancy to ensure separate scans could be registered and assessed. Each target was mounted on to rigid white boarding in order to aid the identification of the targets within the relatively low resolution (480 × 480 pixels) image produced by the Leica 2500 for framing the scan area. The white boards were secured into the slope with stakes enabling an efficient and practical field data collection. Following the high-resolution scanning of the targets, each embankment section was scanned at 0.01 m resolution producing an average of 2.7 million points in each scan. The scans took an average of 45 minutes to capture, but more time was needed when targets were occasionally not identified automatically and required rescanning before they could be located. Problems were also encountered with the low winter temperatures that meant the system, as a thermally stabilized laser, was close to its minimum operating temperature, and consequently it took significantly longer to warm sufficiently for data collection. Fewer scans could have been used to survey the earthwork by taking several scans from each station, but it was found that the optimum data for penetrating vegetation was achieved from an orthogonal view point and the matching of oblique scans produced lower measurement accuracies. The seven scans collected were registered separately in Leica's Cyclone 5.8, which was found to be more effective than referencing all scans into the same registration. The maximum RMSE of the registrations was found to be 0.021 m. The final survey generated over 15 million points across the site.

LSS data processing for earthwork stability assessment

The station coordinates derived from adjusted differential GPS measurements were used to georeference both the registered embankment scan and the total station survey. Although the total station measurements related directly to the earthwork surface, a significant percentage of points within the laser scanned point cloud relate to vegetation on the slope and thus are of little use in assessing the slope itself. The georeferenced TLS survey was imported into TerraScan™, a module from the Terra Solid suite of software hosted within Bentley Mircostation. TerraScan is more commonly associated with the filtering of information from airborne laser scanning, but the facility to tailor algorithms to specific datasets allows ground-based collections to be processed. The filtering procedure was based on the premise that the desired terrain surface can bounded by quantitative criteria and excluded from non-terrain surfaces that fall outside these bounds (Sithole, 2003). Discriminant functions were used to determine whether a particular point represented the terrain surface, assessing its elevation and angle to neighbouring points in order to identify and exclude gross outliers, discontinuities and breaks of slope, similar to the robust interpolation methods suggested by Kraus and Pfeifer (1998) and Pfeifer *et al.* (2001). The processing allowed the terrain surface to be extracted from beneath even highly vegetated surfaces. The points classified as terrain were then used to produce a slope model from below the vegetation, utilising a critical element of the information produced by TLS for the specific application of assessing earthwork condition (Figure 15.1).

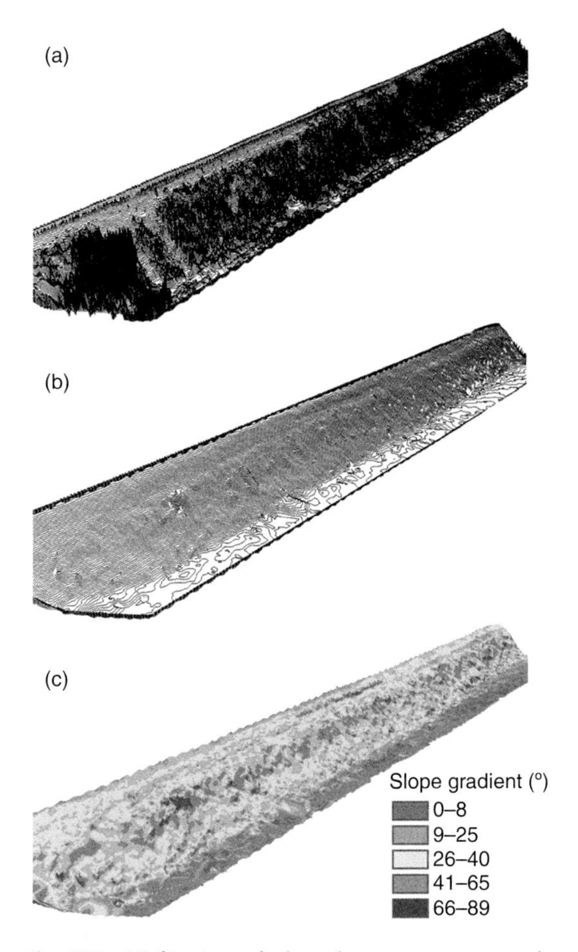

Fig. 15.1 Utilisation of algorithms more commonly associated with aerial LiDAR allow vegetation on surfaces (a) to be removed (b) from TLS data and subsequent slope analysis to be performed (c), providing an effective tool for analysing slopes under vegetation. See Plate 15.1 for a colour version of these images.

Results and discussion: survey quality assessment

The application of TLS data to earthwork slope assessment demonstrates the potential for providing a detailed representation of slope conditions throughout transport corridors. Even after the removal of vegetation returns, the TLS survey provided a significantly denser and more continuous coverage of measurements than could be achieved by conventional survey practices. Although this data source may provide an important new method of recording continuous earthwork surfaces, a key concern is how accurately the slope has been generated and whether this accuracy varies with different types of vegetation cover. The discrete ground surface measurements surveyed with the total station have been used to investigate the degree to which slope monitoring can be reliably conducted under different vegetation types. The elevation differences between the surveyed ground surface and the slope model derived from the filtered TLS information were analysed for each of the three distinctive vegetation types found on the test embankment (Figure 15.2). The results indicate that the ability of TLS to provide accurate information on slope condition is strongly correlated with each vegetation type. The most accurate surface information was obtained from the grass covered areas of the earthwork, producing an RMSE of 0.075 m and relatively low levels of scatter. A similar error pattern was recorded from shrub covered slopes, here defined as having vegetation between 0.2 m and 3 m above the ground surface. The relatively consistent RMSE of 0.096 m produced by shrub covered slopes would still be able to detect changes beyond the seasonal shrink-swell processes that affect earthworks. A significantly less accurate and more variable surface was produced from tree covered slopes, reflected in a high degree of scatter and a RMSE of 0.575 m.

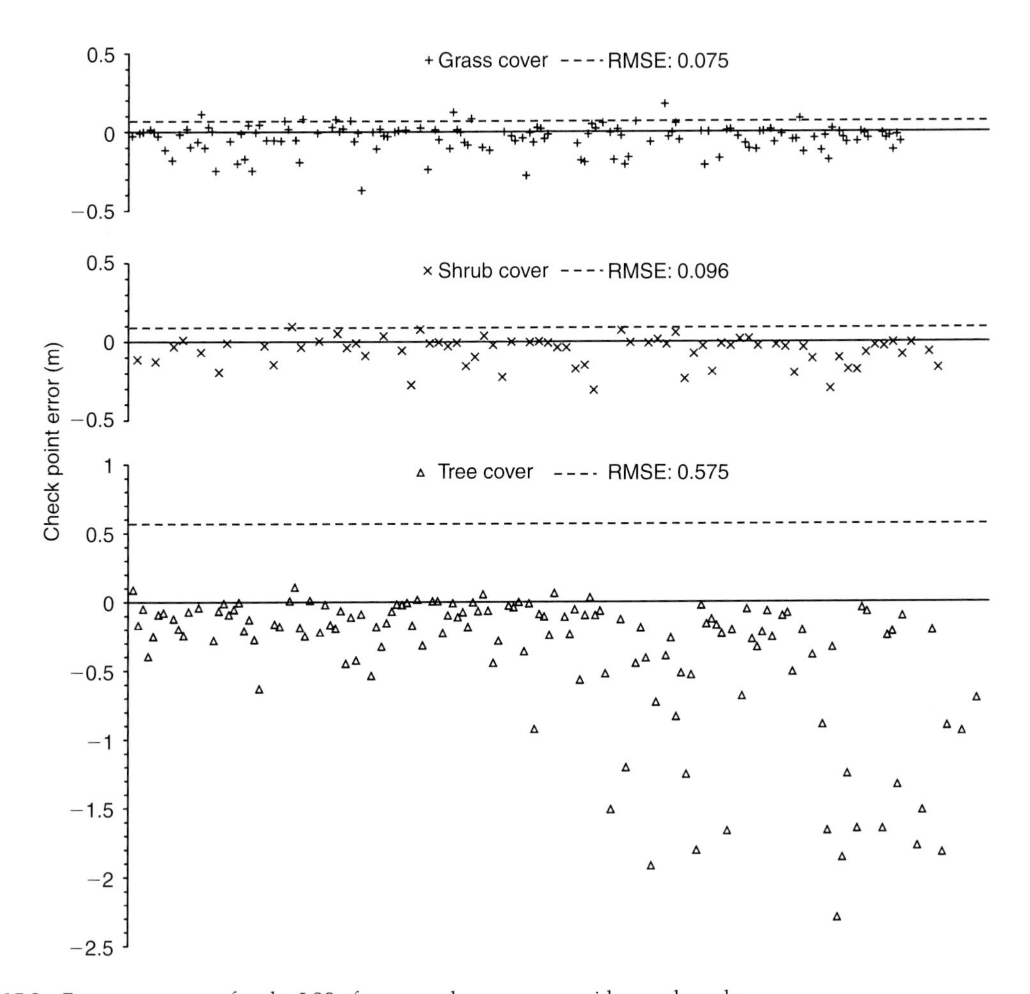

Fig. 15.2 Error assessment for the LSS of vegetated transport corridor earthworks.

The results from this case study demonstrate the extent to which TLS can improve the assessment of linear earthwork structures. It has been shown to provide consistent high-density surface information that is valuable to asset managers, when considered within the limitations of the technique. Optimal data can only be achieved within a disciplined workflow from the preparation and planning of the survey to its implementation and processing. Furthermore, it should be noted that TLS does not provide *'the definitive answer'* in this instance. In a significant proportion of earthwork environments it has generated surface models of higher resolution and with greater efficiency than can currently be obtained from other techniques such as total station surveys or aerial LiDAR. However, its application remains limited where the density of vegetation is sufficient to prevent a representative coverage of ground hits being obtained, such as when slopes are obscured by dense mature trees; total station surveys may need to be used in such areas. Therefore, the most effective use of TLS for earthwork assessment may be in combination with other survey methods. For example, LiDAR surveys could be used to classify areas deemed to be at risk of instability by vegetation type and hence allocate TLS or total station survey resources accordingly. Careful consideration must be given to the effect of vegetation in earthwork surveys (Marsland *et al.*, 1998). Vegetation has the potential to influence both the quality of the survey measurements and the propensity of the slope to move. Therefore, further investigation is required into the types and scales of movement that can reliably be detected by TLS and other survey methods in different vegetative environments.

CASE STUDY 2: LSS FOR MONITORING COASTAL CLIFF EROSION

Coastal environments are increasingly under pressure from the interplay between the need for land and the action of the natural processes of erosion and cliff retreat. For example, in Europe it is estimated that the population living in coastal areas has risen to over 70 million inhabitants, despite the loss of an estimated 15 km^2 of land and approximately 100 houses each year due to erosion, with a further 3000 suffering a reduction in their market value (European Commission, 2004). Recent studies have estimated that coastal erosion in Europe will cost an average of 5,400 million euros each year by 2020 (Salman *et al.*, 2002). Therefore, recording, interpreting and predicting change in coastal landforms has become a critical concern in the effective management of coastal areas. Coastal cliffs are one of the most poorly documented of all coastal environments, despite accounting for approximately 75% of coastlines worldwide. A key element in the management of coastal cliffs is the variability of mechanisms through which they erode, ranging from gradual adjustments in profile form to catastrophic whole-cliff collapses. The hazardous nature of cliff erosion is confounded by the current unpredictability of coastal behaviour. This case study explores the use of TLS for understanding the contemporary evolution of coastal cliffs with an LSS monitoring program.

Recording cliff failure

The inaccessibility and hazardous nature of coastal rock slopes, and their complex interaction with environmental processes, make them a particularly challenging environment from which to extract quantitative surface topography. Cliff-line surveys from aerial laser scanning have achieved some success in recording the position of the cliff top and occasionally the cliff toe, overlaying cliff lines derived from successive surveys to form aggregate rates of retreat (Young & Ashford, 2006). Despite recording submetre resolutions over long coastal stretches, the perspective associated with aerial data capture is limited in the assessment of short-term (subdecadal) volume changes. The assumption that recorded changes within the cliff top and/or toe can be used as an accurate representation for the behaviour of the rock slope remains questionable.

Whilst it is evident that long-term patterns of evolution, particularly in the case of near-vertical cliff sections, may ultimately be identified, the smaller scale iterative adjustments to the cliff face may not be reflected in the behaviour and response of the cliff top. Furthermore, the errors incurred in the identification of the cliff line position often mean that the typically low rates of retreat detected are within the error tolerances of the measurement technique, casting doubt on data validity and requiring longer time intervals between surveys to achieve measurements with significance.

TLS now provides an opportunity to supplement understanding of failure in coastal cliffs with detailed surface monitoring of rock faces from a face-normal perspective, although there remain many challenges associated with the measurement of rock surfaces from a terrestrial perspective that need to be addressed before this can be achieved. The potential of TLS to provide high spatial and temporal resolution monitoring of coastal rock cliffs is explored within an LSS workflow.

LSS planning for monitoring coastal cliffs

The Red Nabbs headland (NZ 776188) north of Staithes, North Yorkshire (UK), has been selected to investigate the potential of LSS to monitor the behaviour of near-vertical coastal cliffs in high spatial and temporal resolution. The headland is formed from interbedded Liassic mudstones, shales, ironstones and cretaceous sandstones of the Staithes Sandstone Formation (Rawson & Wright, 2000). The variable strength characteristics of the rocks combined with local-scale variations in exposure to marine activity generates various failure mechanisms across the cliff section. The cliff face varies between 33 m and 38 m in height, with a 3.7 m thick mudstone base at the toe. The cliff mass is sparsely jointed and capped by up to 20 m of glacial till that reaches angles of over 50° in places. The till commonly stains the rock cliff below red-brown following wetter periods, concealing the scars from previous failures and making recent activity on the cliff face difficult to identify from visual inspection alone. The rock platform in front of the headland is exposed by a 6 m tidal cycle, producing between 70 m and 300 m of foreshore, which is littered with weathered boulders that appear to be mostly rockfall debris.

A Trimble GS200 scanner was selected for its rugged design and the clean, accurate data produced at high resolutions. Resistance to rain and saltwater intrusion from sea spray was also essential for the use of the scanner in the coastal environment. The equipment was transported in frame backpacks to enable the operators to walk on the foreshore uninhibited. Although the scanner had a 350 m range, the high-resolution required to record the potentially small scale changes from the rock face restricted the maximum range to 60 m; and due to the variations in the topography of the headland, three scans were required to complete each survey without data loss due to occlusions from protruding areas. Although monthly monitoring intervals were required, the need to collect a complete survey within a single low tide restricted both the date of the surveys and how wide an area could be covered at the desired resolution. In order to record the full range of changes thought to influence the evolution of the cliff face, the scan resolution from each station was set at a point every 0.01 m^2, calculated from the distance from the scanner to the centre of the monitored area. After comparing the time taken for transporting the scanner to the field site, setting up and collectingthe data at the required resolution, from the three scanning positions, with the survey window available from a mean low tide, a 100 m stretch of coastal cliffs was established for the monitoring surveys.

LSS field collection for monitoring coastal cliffs

The scan stations were positioned using the criteria established in the LSS planning stage to identify the appropriate areas and then selecting suitable sites on the rock platform on which to locate the scanner. The scan positions were marked with a semi-permanent survey nail that was tied to the Ordnance Survey Great Britain reference system with differential GPS. The inaccessibility of the cliff face prevented control targets from being located directly and evenly within the monitoring area and the steep angle of the glacial till above the rock meant that they could not be positioned on the cliff top. Therefore, control target stations were installed into the foreshore in the overlap between adjacent scans. Each target station comprised of a calibrated Trimble control target backed with Perspex and mounted on a Manfrotto camera tripod, which was levelled over a survey nail at a measured height. A total of six stations were used to tie each pair of adjacent scans together to provide the data redundancy necessary when survey nails were removed as a result of the dynamic coastal environment. The main restriction on locating the target stations was the angle at which they could still be recorded from adjacent scanning positions; although this was maximised by the 360° horizontal field of view in the Trimble GS200, the flat face of the targets still meant they had to be located close to the cliff to reduce the angle of incidence to the scanner. The position of the control targets introduced two limitations to the survey. First, the proximity of the targets to the cliff base meant that they were within an area of poor GPS satellite visibility produced by the cliff, preventing the precise determination of their locations with differential GPS. This was overcome with a total station survey from each of the scanning stations in order to tie the control to National Grid coordinates. Second, the omission of control points in a representative range of distances within the scanned area meant that another method of assessing the quality of the data was required. Therefore, a quality test was conducted by surveying the headland completely and then immediately resurveying it during a period for which no rockfalls were recorded. There should have been no difference between the two laser scans, but small differences of up to 0.021 m were detected around ledges and outcrops within the rock face, which is attributed to the location of point hits on the rock face surface varying between scans.

The LSS of the Red Nabbs headland was conducted every month for a 20 month period. The laser scanning system was tested in a wide range of operating conditions ranging from bright sunshine through to rain and subzero temperatures, but all adjacent laser scans were combined with a maximum RMSE of 0.02 m. The scan rate of up to 1000 points per second proved to be a critical factor in enabling the practical survey of the cliff within a single tidal cycle. The loss in resolution toward the edges of each individual scan was found to be less of an issue than in the survey of earthwork slopes, for example, because the near-vertical cliff face maintained a relatively normal angle to the scanner location. Higher or more gradually inclined coastal slopes may require information from the extremities of each scan to be ignored depending on the specifications of the survey. Strong areas of contrast at the division between the cliff top and the sky beyond limited the ability of the integrated video within the scanner to resolve detailed colour information on areas of the cliff face. Therefore, a calibrated Kodak Pro 14n digital camera was used to generate orthoimages of the cliff face for which contrast effects could be adjusted before combining with the TLS data. Optimal survey conditions were found to be uniformly overcast skies and relatively warm operating temperatures. The 532 nm wavelength of the green laser beam was found to be absorbed by green algae on the rock face, which led to data loss due to low reflectivity in certain areas. Although this was not a significant factor in the monitored cliff area, it could provide a critical limitation in more sheltered coastlines where algae is more prolific; the effect of the scanned surface properties on the information produced should always be tested at the site in question before a scanner is chosen for a particular survey.

LSS data processing for monitoring coastal cliffs

The scans were individually filtered for outlying points or clusters of points that exceeded a threshold distance of 0.08 m from surrounding data. In order to achieve this, the laser scans were processed with error detection algorithms generated in TerraScan (Bentley Microstation). Error detection became particularly important for the data collected during the summer months, when breeding seabirds were frequently recorded as they passed through the monitored area of rock face. The orthoimage colour information associated with each point during the collection enabled the white pixels around nesting seabirds to be differentiated from the dark browns and greys of points following genuine losses of cliff material. The laser scans were then registered to a complete survey file in real coordinates using Trimble's field software PointScape. Although individual monthly surveys were accurately positioned planimetrically, the omission of control points within the scanned area, particularly towards the top of the cliff face, generated occasional mismatches between surveys. Therefore, a final iterative convergence procedure was performed on the scans in Demon (Archaeoptics) to refine the final outputs. Each survey was converged to the previous and the initial scan with a maximum difference of 0.03 m, much of which was accounted for by genuine change in the cliff between collections, or point location differences between scan.

The ability to use laser scanning data for more than just visualisation of a surface is directly dependent on how well it has been constrained within an LSS work flow. The rigorous and consistent collection of surveys within real coordinates has allowed the volumetric changes to the cliff occurring each month to be quantified. Each successive cliff surface model has been differenced from the previous month and regions exceeding 0.0002 m^3 selected for analysis, even though smaller changes were detected. The minimum threshold for volumetric detection reflects the potential error contained both within and between individual surveys. Over the 20-month monitoring period, in excess of 30,000 rockfalls were recorded from an area of over 3,000 m^2 of rock face. Blocks of rock ranging in size from 0.0002 m^3 to over 136 m^3 have been quantified and each individual failure related to the time and geomorphological setting from which it occurred.

Results and discussion: understanding coastal cliff behaviour

The LSS of the cliff face has enabled changes to be detected and quantified at a level of detail that is not possible with other techniques. The short-term development of the rock face can now be analysed in high spatial and temporal resolution (Figure 15.3). Much of the change detected was focused on the left hand side (easterly) of the monitored area, which corresponds to the most seaward extent of the headland. A few patches of localised change were found at the base of the cliff, although much of the cliff toe remained unchanged throughout the monitoring period. This demonstrates the importance of material properties; the competence of the mudstone cliff base was sufficient to resist the pounding effect of waves that is often thought to control the behaviour in rock cliffs. A band of small scale change was located on a vertical step in the rock face, suggesting exposed edges may be worn down iteratively over time wherever they are on the cliff face (Figure 15.3, Inset a). A particularly large failure was detected towards the front of the headland in February 2004 (Figure 15.3, Inset b). The failure consisted of a protruding lobe of material, partially separated from the rest of the rock mass by a tension crack running parallel to main cliff face. Difference analysis also revealed a small loss in the cliff toe during the same month (Figure 15.4). The smooth, sheared surface and location of the truncated toe indicated the mechanism of the larger block failure above was likely to have been sliding, rather than a toppling or falling movement for which the trajectory of the failed mass would not have passed through the toe. Therefore, the ability of LSS to detect the complete range of failure sizes and styles occurring from a cliff section may significantly improve understanding into the nature of the changes.

Fig. 15.3 Orthoimage of the Red Nabbs headland overlaid with a shapefile locating the spatial extent and position of changes recorded over the monitoring period. The headland was relatively stable throughout the monitored period with concentrations of change on vertical steps in the cliff face (Inset a) and towards the seaward edge of the site where the largest failure occurred (Inset b). See Plate 15.3 for a colour version of this image.

CONCLUDING REMARKS

This chapter has highlighted a divergence between the use of TLS in an increasingly wide variety of applications and the manner in which this relatively new survey approach is applied. Inherent within all systems that automate the collection of data is the requirement for the user to understand the operation being carried out to a degree that is sufficient to maximise the system performance and data quality. As TLS systems mature from state-of-the-art instruments to become standard practice for recording complex topography, a change of emphasis is required from TLS as a method of data capture to its use within an LSS framework in order to achieve its full potential (Table 15.1). The importance of locating scanned data within a complete workflow from survey planning and field collection, to data processing and interpretation, has been discussed with reference to linear features within the landscape. Linear environments commonly provide the focus for surveys but pose particular challenges to the accurate collection and interpretation of information.

From the examples presented above, several conclusions can be drawn on the development and application of LSS. It is evident that TLS does not necessarily provide a complete answer to the requirements of surveying projects; instead it is most effectively used in combination with other datasets such as GPS referencing and total station traverses. Careful consideration should be given to the appropriate scales used to record and analyse surface information with respect to the purpose of the study or survey. For example, the term 'high-resolution' is frequently used in remote sensing in reference to a wide range of scales, but ultimately reflects the achievement of data coverage over and above that required for the specific application. Consequently, the resolution of the survey should be defined by the requirements of the survey, ensuring adequate coverage and concentration is achieved to reconstruct the minimum sized features of interest (Mills & Barber, 2004), often determined by approaches such as the Nyquist frequency, in which the survey interval in space is a direct function of the size dimension of the features under investigation.

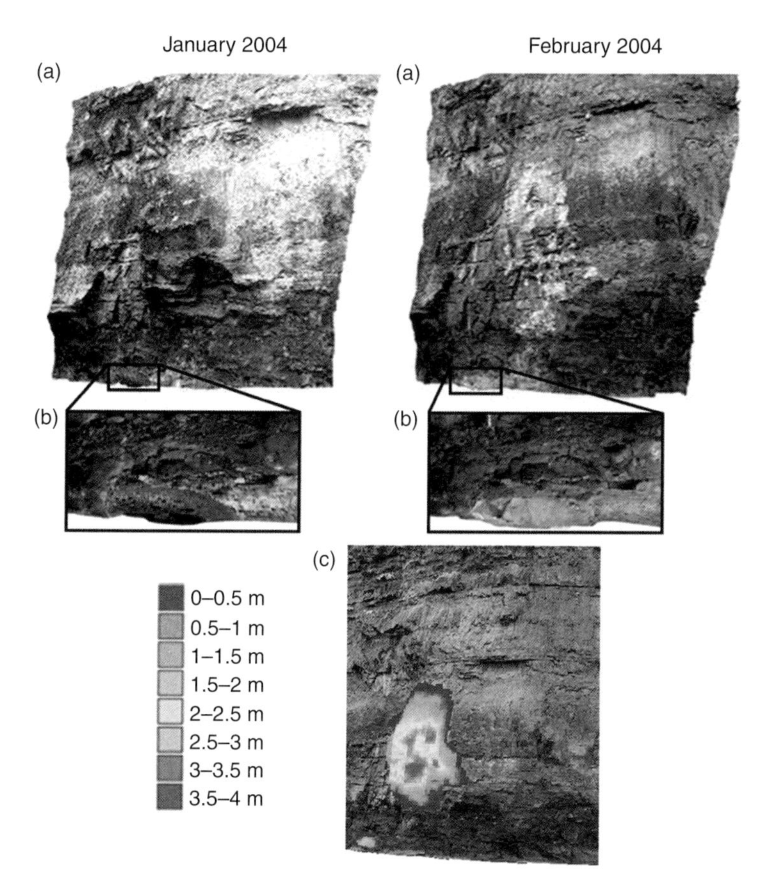

Fig. 15.4 Sections of the surveys collected during January and February 2004 (a) illustrate a significant failure from undercut cliff material. An additional, smaller failure was revealed (b) through high-resolution differencing analysis (c). The LSS approach therefore indicates that the larger loss failed through a tangential sliding mechanism, truncating the adjacent cliff toe below in the process. See Plate 15.4 for a colour version of these images.

Preparation and planning are essential to maximising the success and efficiency of an LSS, considerations that are becoming ever more important as the scale and complexity of the desired outputs increases. The recognition that laser scanning is a survey tool, and consequently subject to many of the same constraints on data quality faced by other survey methods, highlights the value of employing established survey principles in its application. A critical concern to the operators of TLS should be the minimisation, identification and removal of errors, which ultimately requires a global standard for characterising the performance of these instruments. The methods with which to analyse and interpret the information provided by LSS are not as developed as those for capturing the data. Significant advances are required in the processing of LSS information in order to benefit more fully from the capabilities now offered by TLS systems. Ultimately, the principles of LSS can provide an effective framework for circumventing many of the problems that have long limited alternative methods for recording surfaces, increasing our understanding and ability to predict change in linear features and other such challenging environments.

Table 15.1 Suggested stages in a laser scanning survey for optimising data quality.

	Laser scanning survey workflow
Planning	Test a range of scanners (or those available) to the suitability of the specific application
	Tests should reflect the practicality of the scanner in the survey environment in addition to its effectiveness at producing the required data
Field procedures	Calculate ideal scanning locations based on the subject surface parameters and the optimal field of view produced by the scanner
	Position the control targets according to the overlapping field of views between the regularly spaced scan stations
	Tie scanning stations and control points into a common coordinate system with a dGPS or a total station survey
	Level and orientate the instrument
	Scan the surface consistently at the most effective resolution
Data processing	Georeference the survey using the known positions of the scan stations and control targets
	Clean clouds either with manual search for blunders or automated assessments which perform well on georeferenced datasets
	Register scans based on control target positions
	Assess the final survey against known check points excluded from the registration and use to eliminate consistent errors

ACKNOWLEDGEMENTS

The authors would like to acknowledge the contribution and support provided by EPSRC for funding the research into remote asset inspection for transport corridor environments (EP/D023726/1) and Cleveland Potash Limited for supporting the research into coastal cliff behaviour.

REFERENCES

Adams JC, Chandler JH. 2002. Evaluation of LIDAR and medium scale photogrammetry for detecting soft-cliff coastal change. *Photogrammetric Record* **17**: 405–418.

Al-Homoud AS, Al-Masri GA. 1999. CSEES: An expert system for analysis and design of cut slopes and embankments. *Environmental Geology* **39**: 75–89.

Andrei A. 2000. Embankment stabilisation works between Rayners Lane and South Harrow Underground stations. *Ground Engineering* **33**: 24–26.

Barker DH. 1995. Vegetation and slopes: stabilization, protection and ecology. *Proceedings of the International Conference, University Museum, Oxford*. Thomas Telford.

Blight GE. 1997. Interactions between the atmosphere and the earth. *Geotechnique* **42**: 715–766.

Boehler W, Heinz G, Marbs A, Siebold M. 2002. 3-D Scanning Software: An Introduction. *Proceedings of the International Workshop on Scanning for Cultural Heritage Recording*. Corfu, Greece, September 1–2, 42–47, International Society for Photogrammetry and Remote Sensing, 2002.

Clark J, Robson S. 2004. Accuracy of measurements made with a Cyrax 2500 laser scanner against surfaces of known color. *Survey Review* **37**: 626–638.

Copin NJ, Richards IG. 1990. *Use of Vegetation in Civil Engineering*. CIRIA, Westminster.

Crilly MS, Driscoll RMC. 2000. The behaviour of lightly loaded piles in swelling ground and implications for their design. *Proceedings of the Institution of Civil Engineers – Geotechnical Engineering* **143**: 3–16.

Eggert DW, Fitzgibbon AW, Fisher RB. 1998. Simultaneous registration of multiple range views for use in reverse engineering of CAD models. *Computer Vision and Image Understanding* **69**(3): 253–272.

European Commission. 2004. Living with Coastal Erosion in Europe – Sediment and Space for Sustainability. *Luxembourg: Office for Official Publications of the European Communities*, 1–40.

He YP, Xie H, Cui P, Wei FQ, Zhong DL, Gardner JS. 2003. GIS-based hazard mapping and zonation of debris flows in Xiaojiang Basin, southwestern China. *Environmental Geology* **45**: 286–293.

Kersten TP, Sternberg H, Mechelke K, 2005. Investigation into the accuracy behaviour of the terrestrial laser scanning system Mensi GS100. In: Gruen A, Kahmen H. (eds), *Optical 3-D Measurement Techniques VII*, Vienna 2005, Vol. I, 122–131.

Kraus K, Pfeifer N. 1998. Determination of terrain models in wooded areas with airborne laser scanner data. *ISPRS Journal of Photogrammetry and Remote Sensing* **53**: 193–203.

Lichti DD, Harvey BR. 2002. The effects of reflecting surface material properties on time-of-flight laser scanner measurements. *Symposium on Geospatial Theory, Processing and Applications*, Ottawa, 1–9.

Lichti DD, Gordon SJ, Tipdecho T. 2005. Error models and propagation in directly georeferenced terrestrial laser scanner networks. *Journal of Survey Engineering* **131**: 135–142.

Lloyd DM, Anderson MG, Hussein AN, Jamaludin A, Wilkinson P. 2001. Preventing landslides on roads and railways: a new risk-based approach. *Proceedings of ICE Civil Engineering* **144**: 129–134.

Marsland F, Ridley AM, Vaughan PR. 1998. Vegetation and its influence on the soil suction in clay slopes. *Proceedings of the 2nd International Conference on Unsaturated Soils, Beijing, 1998* **1**: 249–254.

Mills J, Barber D. 2004. Geomatics techniques for structural surveying. *ASCE Journal of Surveying Engineering* **130**: 56–64.

Montgomery DR, Dietrich WE. 1994. A physically-based model for topographic control on shallow landsliding. *Water Resources Research* **30**: 1153–1171.

O'Brien AS. 2007. Rehabilitation of urban railway embankments – Investigation, analysis and stabilisation. *Proceedings of the 14th European Conference on Soil Mechanics and Geotechnical Engineering*, Madrid, Spain, 24–27 September 2007, 125–145.

Pfeifer N, Stadler P, Briese C. 2001. Derivation of digital terrain models in the SCOP++ environment. In: *OEEPE Workshop on Laserscanning and Interferometric SAR for Digital Elevation Models*, Stockholm.

Rawson PF, Wright JK. 2000. *The Yorkshire Coast*. The Yorkshire Coast Geologists' Association Guide No. 34, Geologists' Association, London.

Ridley A, McGinnity B, Vaughan P. 2004. Role of pore water pressures in embankment stability. *Geotechnical Engineering* **157**: 193–198.

Salman A, Doody P, Ferreira M, Trevisani F, Yperlaan J, Marquetant T, Revesz T, Taveira Pinto F, Randazzo G, Paskoff R. 2002. *Coastal Erosion Policies: Defining the issues. EUROSION Scoping Study, 2002*. European Commission contract.

Scott JM, Loveridge F, O'Brien AS. 2007. Influence of climate and vegetation on railway embankments. *Proceedings of the 14th European Conference on Soil Mechanics and Geotechnical Engineering*, Madrid, Spain, 24–27 September 2007, 659–664.

Sithole G. 2003. Filtering strategy: Working towards reliability. In: *International Archives of Photogrammetry and Remote Sensing*, Volume XXXIV 3A III, Graz, Austria, pp. 330–335.

Sithole G, Vosselman G. 2004. Experimental comparison of filter algorithms for bare earth extraction from airborne laser scanning point clouds. *ISPRS Journal of Photogrammetry and Remote Sensing* **59**: 85–101.

Slob S, Hack R. 2004. 3D Terrestrial Laser Scanning as a New Field Measurement and Monitoring Technique. In: *Engineering Geology for Infrastructure Planning in Europe*. Springer Berlin/Heidelberg, Lecture notes in Earth Sciences: Volume 104, 179–189.

Starplus Software Inc. 2000. *STAR*NET-PRO V6, implementations of the Least Squares Survey Network Adjustment Package*. Oakland, California, USA.

Young AP, Ashford SA. 2006. Application of airborne LIDAR for seacliff volumetric change and beach-sediment budget contributions. *Journal of Coastal Research* **22**: 307–318.

16 Laser Scanning: The Future

ANDREW R.G. LARGE[1], GEORGE L. HERITAGE[2] AND MARTIN E. CHARLTON[3]

[1]School of Geography, Politics and Sociology, Newcastle University, Newcastle Upon Tyne, UK
[2]JBA Consulting, Greenall's Avenue, Warrington, UK
[3]National Centre for Geocomputation, National University of Ireland, Maynooth, Ireland

LIDAR: A 'SECOND REVOLUTION' IN SURVEYING TECHNOLOGY?

As we summarise contributions to this book, and reflect on the science, it is pertinent to recall the conclusions of McCaffrey *et al.* (2005) who comment on the fact that one of the most significant changes in field-based study since the invention of the geological map has been the development of affordable digital technologies which allow the collection and analysis of georeferenced field data. Digital methods make it easier to re-use existing data during repeat survey, and the increased spatial accuracy from satellite and laser positioning systems provides access to geostatistical and geospatial analyses that can inform hypothesis testing during fieldwork. It would appear from the preceding chapters that this change in the way we can now collect data is being strongly facilitated by developments in laser scanning technology. As we approach the end of the first decade of the 21st century, it is clear that, by using laser technology, we are embarking on the second revolution in the surveying and monitoring of our physical environment (the first concerned the development of the ruby laser in the early 1960s).

The decades following the 1960s saw the introduction of portable, yet robust and accurate field equipment (EDM instruments), which enabled precise rendering of often large swaths of terrain. The 1970s saw widespread introduction of laser-based surveying primarily in the engineering and construction industries. In the 1980s and 1990s environmental scientists realised the potential for a wide range of applications in a broad range of environmental systems. The second advance in surveying innovation began in the 1990s with the development of sensitive laser based hardware, capable of generating high-resolution digital terrain surfaces accurately representing complex natural and semi-natural environments, incorporating morphological features at a range of scales. As such, LiDAR has huge attractions for the environmental scientist and the intervening period has seen a rapid take-up of the technology. The technology is still rapidly developing and we are now witnessing the introduction of smaller, more robust, faster instruments capable

Laser Scanning for the Environmental Sciences,
1st edition. Edited by G.L. Heritage and A.R.G. Large.
 ISBN 978-1-4051-5717-9

of detecting very low level radiation return to generate High Definition Survey (or HDS) data recording information on spatial position and physical properties of remotely surveyed surfaces. As a result, LiDAR instruments now offer opportunities to research landscape character and dynamics over spatial and temporal ranges not previously practicable (see Heritage & Hetherington, 2007), facilitating a potential revolution of our understanding of Earth systems (Figure 16.1) and changing paradigms in disciplines previously working with sparse information and relying on interpolation for their models and conclusions.

While it is tempting to simply say how fantastic is the technology described by the contributors to this book, it is, at the same time, pertinent to consider the very philosophy of our research approaches. More and more in an increasingly environmentally degraded world, we need techniques that allow us to unpick ecosystem function and processes – do systems that allow us to map surfaces in minute detail allow us to make allied step-changes in our understanding of system function? In river restoration, for example, a key question since the early 1990s has been: 'if we reinstate structure, does ecological function automatically follow?' That question remains essentially unanswered. Other questions are starting to appear relating to scale mismatches between ecological systems and the major legislation designed to protect these very systems. As an example, the European Water Framework Directive is designed to protect water-based ecosystem health over the next two decades, yet the spatial scale for administration of the WFD (the river basin) is extremely difficult to reconcile with the spatial and temporal scales within which the very ecology of those systems functions.

We need to interrogate the philosophy behind current patterns of data collection using the technology. If we consider advances in data acquisition over the last decade or so, we see an analogy with digital photography. Nowadays,

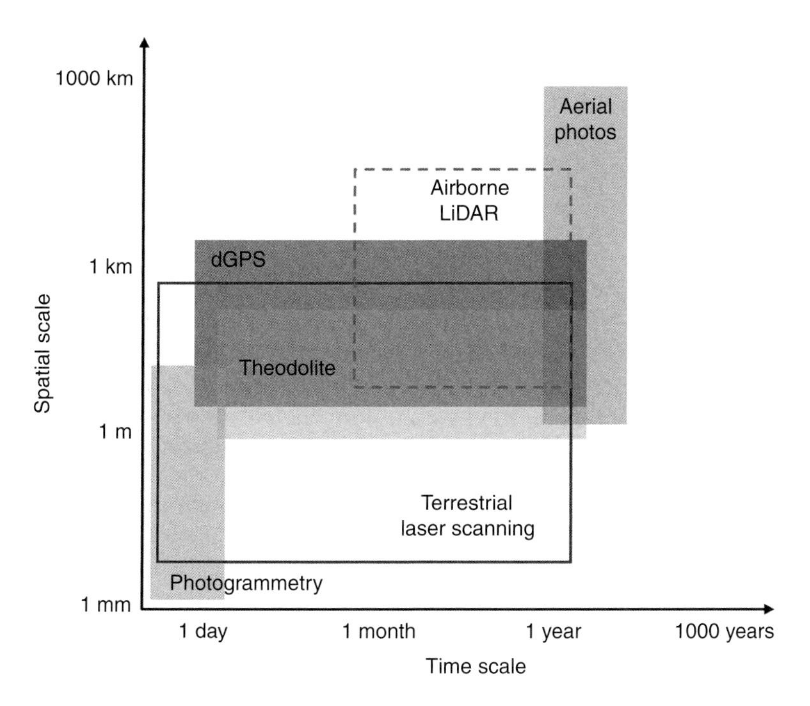

Fig. 16.1 Filling the data collection gap using LiDAR technologies (after Heritage and Hetherington, 2007).

even the most amateur of photographers can take and re-take photographs of a scene or subject to ensure that one 'perfect shot'. No longer do we have the anticipation of collecting our photos developed from film by the shop, yet on opening the packet merely find the head of our favourite grandchild chopped off on every print. Modern laser scanning machines display the data in the field in the form of a data cloud on a screen and using these as guides we happily gather vast amounts of data to render our terrain as accurately and precisely as possible. Paralleling the persistent advances in hardware and software, the debate now needs to move to what we need this level of data for, at what level of detail, and what should be the dominant protocols for data management and mining. A number of the chapters in this book have focused on these issues.

There is no doubt that a major advantage of LiDAR technology lies in its ability to deal with large swaths of landscape (Figure 16.1). In the past, the survey of large systems (or across significant swaths of systems) encountered scale factors which made detailed data difficult to obtain (see Overton *et al.*, Chapter 14). Dealing with large areas also adds to costs, meaning that attempting to survey large systems using traditional field-based mapping and survey techniques is a non-starter. Traditional methods of dealing with large-swath survey has always had the issue of pixel size (SPOT for example has a range of pixel sizes from 20 m to 2.5 m at best, while NASA's TERRA satellite imagery has spatial resolution (level of detail) of only 250 m per pixel). Traditional remote sensing using satellite imagery also gives only mixed results due to differences in vegetation: the imagery deals poorly with changes in canopy density and understorey exposure, and differentiation between species and their varying health.

Two excellent examples of LiDAR application across large areas are reported in this book. Hodgetts (Chapter 11) details the successful use of combined terrestrial LiDAR and digital imagery to map the Hamman Faraum fault block in the Gulf of Suez, developing sedimentological and stratigraphical models of the area. Overton *et al.* (Chapter 14) describe the development of high resolution DEMs from aerial LiDAR mapping of highly complex environments with relatively small elevation differences across floodplain areas creating a multitude of different flood regime habitats. This reflects the fact that elevation differences of centimetres can significantly influence flow direction and flood severity. DEMs designed to model these flows must be able to capture height differences of that order. The only technology that provides DEMs (over the large areas required by managers) approaching this accuracy is airborne laser scanning. In conducting their study, Overton *et al.* (Chapter 14) needed to resolve quality issues including recovery of realistic elevation surfaces for channels, removal of artificially high elevations induced by misclassification (in this instance, of reed beds as ground) and, less importantly adjusting adjacent swaths to remove height mismatches. It is possible to adjust for these problems using traditional GIS software, but this requires significant input of computing time for each tile to generate new DEMs from the point data.

Despite the excellent work of the chapter authors and many others in various disciplines, there remains a major dissemination issue with research being conducted using LiDAR technology. Currently too much of the science is published in the grey literature – reports and conference proceedings with a limited readership. There is a clear need for publications in international, peer-reviewed journals to encourage greater awareness and to facilitate wider critical review (see Large & Heritage, Chapter 1). Indiscriminate data collection and poor data analysis is rife in the grey literature and standards must be developed and adhered to, maximising the robustness and applicability of published research and securing a high degree of confidence within the user community of the value of the technology.

FUTURE DEVELOPMENTS IN LIDAR

Airborne LiDAR systems provide elevation data at up to decimetre resolution and across swaths up to 1–2 km in width. Often, cost factors drive a move towards widening swaths further, but the problem of lower incidence angles and dense vegetation canopies effectively places limits on the swath width achievable. One way to achieve higher swath widths is to increase the elevation from which data is captured. Space-borne instrumentation has been developed by bodies such as NASA's Goddard Space Flight Center, with examples being the Shuttle Space Altimeter and Vegetation Canopy LiDAR (SLA and VCL respectively), with recent developments including the full-waveform airborne Laser Vegetation Imaging Sensor (LVIS) which uses a 3 km wide swath with plans to increase this further. Blair *et al.* (2001) detail the development of medium-large footprint, full-waveform mapping laser altimeters to enable space-borne, wide-swath operational full-Earth mapping and dense vegetation penetration. Readers are referred to the chapter by Danson *et al.* (Chapter 13) for their views on the future of space-borne instrumentation and the potential for integration with multi-angular spectral observations. In relation to this arena, Abshire *et al.* (2004) state that LiDAR is firmly entrenched in the family of remote sensing technologies that NASA is developing and using. They conclude that the technology should continue to experience significant advances and progress, being used in each of NASA's major research themes. Abshire *et al.* state that, from research at Goddard Space Flight Center, NASA will continue to generate new LiDAR applications from ground, air and space for both Earth science and planetary exploration.

Perhaps the greatest value of airborne lasers as remote sensing tools for the future is the unprecedented accuracy with which they can measure clouds. Knowledge of the concentration and distribution of atmospheric aerosols using both airborne LiDAR and satellite instruments is a field of active research. LiDAR can penetrate thin or broken clouds in the lower atmosphere, permitting observation of its vertical structure. A space-based LiDAR can provide global measurements of the vertical structure of clouds and atmospheric gases. The latest advancements in laser remote sensing can fill the gaps we have in our understanding of how clouds reflect and absorb solar energy, and how heat and moisture are exchanged between the air, ocean and earth. As space-borne LiDAR has the potential for collecting data on a global scale, including remote areas like the open ocean. In a very short period of time this technology can help in detecting man-made trends in the concentration of greenhouse gases and aerosols related to climate change above natural variability, can aid better assessment of air quality and long-range transport pathways, allow better quantification of pollutant sources and atmospheric pathways and better predict UV intensities at the Earth's surface (Hoff *et al.*, 2008).

HARDWARE AND SOFTWARE IMPROVEMENT

Straub *et al.* (Chapter 8) note the recent rapid progress that has been made in feature delineation and automated processing due to the combination of improved computing power and processing algorithms. In general, the faster the laser, the better will be the coverage, as more areas can be covered in a given period with higher spot densities. At the same time it is desirable that we see a move towards smaller, more flexible systems. We speak from experience: moving a heavy tripod-based terrestrial laser scanner by hand across some tens of kilometres of terminal moraine in south-east Iceland, tests one's enthusiasm as well as the equipment itself! Systems of the future should be small, portable and modular in design. This in turn would allow advances to be made in the use of unmanned light aircraft. Tunable lasers would greatly aid hydrographic researchers, as would improved performance of receivers and their associated optics in overcoming the problem of 'sun glint', which becomes a noise source to the receivers in water environments.

While waveform information on vegetated landscapes can provide a detailed measure of backscatter energy from canopy surfaces and the underlying ground (e.g. Harding *et al.*, 2001), full waveform LiDAR is currently only available on some airborne systems. Several authors have commented on its potential for use from a terrestrial platform, particularly when dealing with vegetation (e.g. Lefsky *et al.*, 2002). Data volume issues must be recognised here though. In addition, where possible, LiDAR sensors should be combined with other airborne sensors. It makes economical sense to gather as much complimentary data as possible for a single airborne platform, for example: LiDAR and digital photographs, LiDAR and video footage, LiDAR and thermal imagery, synthetic aperture radar and LiDAR.

REMAINING ISSUES WITH AIRBORNE LIDAR

Digital data acquisition technologies such as LiDAR have resulted in a dramatic increase in the volume and complexity of data pertaining to natural systems. These data have to be efficiently managed throughout the collection, distribution and processing stages in order for it to be of maximum use to the scientific community. Charlton *et al.* (Chapter 3), Devereux and Amable (Chapter 4), and Lim *et al.* (Chapter 15) all comment at greater length on such issues. Indiscriminate use of the technology (particularly given the relative ease of data collection) must be discouraged. Heritage and Large (Chapter 2) present a review of the many potential errors with LiDAR survey data, providing the reader with a set of issues to consider prior to any laser scan campaign. As an example, following LiDAR survey of the northern margin of Iceland's largest ice-cap, Vatnajökull, in 2007, it became apparent on receipt of the data from the Airborne Remote Sensing Facility (ARSF) of the Natural Environment Research Council that significant areas – unfortunately those at the retreating glacier margin and thus of most interest to the scientists researching the system – were not surveyed to the same quality or accuracy as other parts of the system. The problem was exacerbated by intrinsic delays between airborne collection of the data and its release to scientists in useable form – over six months in this instance. Therefore, problems in data acquisition could not in this instance be easily be addressed by extra ground survey in areas with sparse data, as the issue only became apparent long after the 2007 field season had ended. Starek *et al.* (Chapter 10) publish a protocol from project definition to model production that is both efficient and rapid. It is clear from this research that such a controlled workflow can feed quickly and directly into management plans.

The need for surveying protocols

As Lim *et al.* (Chapter 15) conclude, the development of sensors able to collect 3D surface information rapidly has enabled high-density measurements to be made across landscapes that are unsuited to more conventional approaches due to their inaccessibility, hazardous nature or spatial extent. In addition, the increasing availability of TLS systems and the ease of data capture they offer have enabled non-specialist operators from outside traditional surveying disciplines to efficiently generate detailed information in evermore challenging and complex environments. However, Lim *et al.* (Chapter 15) highlight the mismatch between the ability of TLS to provide data at what would appear to be ever higher resolution, and the proficiency with which such data is utilised. Therefore, although point clouds provide attractive visualisations, the full value of TLS cannot be realised without employing a rigorous survey workflow. Of all the survey techniques available, TLS has the least standardised control practices and error assessments (Lichti *et al.*, 2005). Lim *et al.* (Chapter 15) also state that this is due to two separate reasons: the relative infancy of TLS as a survey tool and the apparently complete and satisfactory outputs it provides. A danger exists: a direct, positive relationship between data quantity and data quality is often assumed in the recording of surfaces, and there is a tendency to accept high-resolution data as being the most accurate representation. There is a need for the enhancement of software for modelling, feature extraction, texturing and presentation to the end-user. The last is an important issue that appears largely solved in industrial applications, but which needs considerable attention in heritage work and the natural sciences. As Hetherington (Chapter 6) correctly points out, the purpose of measurement – not instrument capabilities – should dictate survey protocol. Only a limited number of studies have dealt with the combination of LiDAR with datasets such as CASI (Compact Airborne Spectrographic Image), a high-resolution hyperspectral imaging platform which measures reflected radiation in a series of consecutive narrow wavelength bands (Goetz *et al.*, 1985), and more work clearly remains to be done here.

Many of the current developments in terrestrial laser scanning focus on automatic point cloud referencing techniques, the extraction of geometric information from point clouds and the fusion of point cloud and image data. The apparent benefit of the technology, detailed topographic representation, has inherent drawbacks (Crosby *et al.*, 2007): access to these datasets for the average user is rendered difficult due to the huge volumes of data generated. The distribution and processing of large LiDAR datasets challenge Internet-based data distribution systems and readily-available desktop software. Datasets frequently exceed billions of data points and the fact that the technology can capture several elevation points per metre over large distances can add up to multi-terrabyte datasets that present daunting computational challenges for handling, processing and systematic analysis. The result has been a slower take-up of the technology or limited analysis of the data, as valuable LiDAR topographic data is effectively kept out of reach of many earth scientists. These problems are compounded in composite datasets, that is grouped data from a number of users.

Software development

LiDAR also suffers from the same kind of software development gap that plagues other high-tech industries. The market for LiDAR is steadily growing – but is not matched by advances in LiDAR handling software. Complications are introduced by the use of multiple sensors during data acquisition and the need to merge RGB colour data or intensity data with *x-y-z* data points. These problems are most likely to occur at the data registration stage, during feature extraction, or when attributes are applied to the data so it can be useful in a GIS database. This problem is exacerbated by the rapid introduction of new and more complex hardware capable of ever more efficient and varied data collection; bespoke software packages rapidly become redundant before they can be refined and improved based on user community feedback.

Large datasets

A further area that needs and already holds the attention of software developers is optimisation of manipulation of extremely large datasets. In summarising the requirements of a data sampling protocol for terrestrial laser survey (Table 16.1; also see Table 15.1), Lim *et al.* (Chapter 15) make three significant conclusions:

- The application of terrestrial laser scanning remains limited where, for example, the density of vegetation is sufficient to prevent a representative coverage of ground hits being obtained. In such cases i.e. when slopes are obscured by dense mature trees, total station surveys may be required.
- The most effective use of TLS will thus be in combination with other survey methods including total station and differential GPS. This will have implications for survey speed.
- In essence, therefore, careful consideration should be given to the scales used to record and analyse surface information and whether these are appropriate to the purpose of the survey.

The term 'high-resolution' is frequently used in remote sensing in reference to a wide range of scales, but ultimately reflects the achievement of data coverage over and above that required for the specific application. Consequently, the resolution of the survey should be defined by the requirements of the survey, ensuring adequate coverage and concentration is achieved to determine the minimum sized features of interest.

Table 16.1 Requirements of a data sampling protocol for terrestrial laser survey (adapted from Lim *et al.*, Chapter 15).

Survey component	Issue	Requirements
Planning	Relates to the *relative accuracy* produced by the scanner.	Instrument specification provides an important basis for assessing capabilities of the scanner. Particularly useful when control points are relied upon for error measurement. When surveying, the interest of the survey is in point coordinates across un-calibrated surface, specific validation and performance testing should be conducted.
	Lack of specific calibration certificates, or comparable measures of its precision on different surfaces.	
	Time taken for different scanners to collect a set number of points varies significantly. Divergence between fastest and slowest systems increases with complexity and resolution of the survey.	Size, weight and durability of scanners requires consideration before survey.
		Scanners that meet *Ingress Protection 66* specifications, which require the scanner to be water and dust resistant may be more suitable.
	TLS systems function at wavelengths (500–1500 nm) that can be absorbed into atmospheric particles or moisture, degrading the strength of the returning radiation although not necessarily the accuracy of the point measurement.	
Field procedures	Separate point clouds are aligned with the use of control targets, which derive a specific reflectance signature and geometry.	Location of targets should be chosen according to specific requirements of the survey to allow multiple tie points between scans, with redundancy.
	The positioning of the scanning stations is constrained by the field of view achievable by the instrument.	The operator must be aware of the concentration of data coverage that diffuses towards the edge of the scan section, particularly on sloped surfaces; this can be minimised if equal survey legs (the baseline distance between the scan positions) are maintained.
	Variability in ambient conditions during the data collection should be minimised in order to generate a consistent survey.	
Data processing and handling	Point cloud data produced from TLS are typically collected in arbitrary coordinate systems that calculate the distance between the target surface and the optical centre of the instrument.	Separate point clouds need to be related or registered to a common data space to form a complete, georeferenced survey.
		First task is to identify gross errors or blunders within the point cloud data.
	High demands on data structure, storage and archiving.	The higher the degree of data processing the greater the potential to incorporate error. Careful consideration should be given to the effects of data cleaning and thinning algorithms.

Georeferencing

Despite the ease with which a detailed survey can be produced, tying it in to geoids and local datums remains a thorny issue. Despite the manufacturers claims with regard to differential GPS it is not a simple matter to get precise *x,y,z* information for control points and to easily tie these to individual data points in a terrain model. The base station has to be set in a grid of RINEX stations, that is GPS receivers operating for years and post-processed to achieve such precise elevations. Many of us have experience of working in very remote locations without closely-knitted reference stations: to precisely set a benchmark in *x,y,z* in such locations is an extremely challenging task – and one conveniently glossed over by many earth scientists as they disseminate their science to a wider audience.

THE FUTURE

We began this book with two purposes in mind: first to introduce LiDAR to an interested readership and second to stimulate research into the use of LiDAR in the environmental sciences. LiDAR, both terrestrial and airborne, has the potential to help us understand the natural systems which have shaped, and continue to shape, our world. The experiences of the contributors to this volume reinforce our confidence in that potential, although all of us recognise that LiDAR is not a universal panacea in environmental data gathering. To ignore that potential would be to our ultimate disadvantage – never has our need to understand the physical world been greater, and we are beholden to use the best technologies to help us in our research. LiDAR, as we have collectively demonstrated, is one of those technologies.

Taking these points into account, the sheer ability of laser scanning systems to gather data provides a mechanism for circumventing many of the problems that have long limited alternative methods for recording, increasing our understanding of, and ability to predict, change in natural systems.

Training

The necessity for training should not be underestimated. Each manufacturer's system has a different setup and operational procedures. Whilst training can be purchased along with new equipment, the training material is often generic and not specifically oriented towards research in the environmental sciences. Early progress is likely to be slow, and skills acquired during frequent practice sessions will be invaluable in often unhelpful field conditions. Once acquired, the data require processing – while there are generic operations, for instance, meshing overlapping surveys, every piece of software is different, and developing the confidence that one's efforts have produced reliable results takes time. Finally, creating the surface models from the acquired data is also not for the unwary – should one create a TIN or use geostatistical software to create kriged estimates of the elevation? There is also the final researcher's nightmare – having perhaps travelled great distances to collect data for which return is difficult, and due to a combination of unfortunate circumstances, the results are not as expected; has one got the skills and expertise to achieve the desired 'silk purse' when presented with the 'sow's ear' of data problems? The editors have personal experience of this; lugging heavy TLS equipment together with rugged laptop, over-sized batteries and tripod and wires (pre Bluetooth) to map in centimetric detail an area covering more than 4 km^2 of unstable moraine in southeast Iceland in 2007. Subsequent data analysis revealed severe meshing problems over such a large area, caused by corrupted differential GPS locations of the reflectors (not something hinted at in the manufacturer's sales pitch!).

FINAL THOUGHTS

The technologies of LiDAR are maturing to the point of being operational in serving many mapping needs. Both airborne and terrestrial laser scanning are now well established methods for the acquisition of precise and reliable 3D geoinformation. Beyond the primary tasks in digital terrain model generation, airborne laser scanning has also proven to be a very suitable tool for general 3D modelling tasks and landscape analysis. At the same time, terrestrial laser scanning has been successfully used to acquire highly detailed surface models of objects such as building facades, and industrial installations. Use is now expanding rapidly in the natural environment and over the last few years, ground-based laser scanning has been rapidly adopted around the world for capturing three-dimensional survey data in a variety of environmental applications (Barber *et al.*, 2005). Sales of 3D terrestrial laser scanning hardware, software and services reached a total of $86.2 million in 2003, a growth of 22% over 2002, according to Spar Point Research LLC (www.sparllc.com, 2008). Driven by the need for faster, safer, more accurate and efficient ways to model existing conditions of built assets and civil infrastructure than traditional manual methods, the market was forecast to reach $293.7 million in 2008, a five-year compound annual growth of 28%. Thus the rapid take-up continues, with a particular interest in ensuring that data gathered is fit for purpose and value for money. However, there are issues in that different scanner manufacturers publish different technical specifications, which have the effect of complicating comparisons between different machines. There is a need therefore to publish defined standards based on rigorous benchmark tests.

It will be clear to the reader from the approaches described by all contributors to this volume that both airborne LiDAR and terrestrial laser scanning (TLS) offer distinct advantages to the environmental scientist. It must also be remembered that there are also potential pitfalls if the technology is applied inappropriately. Both advantages and disadvantages are summarised below.

Advantages

- Very high speed data collection – up to 100,000 points per second for airborne LiDAR with each data return having georeferenced *x-y-z* positional information, reflection intensity and colour provided by a calibrated high resolution digital camera.
- Information rich datasets which, once collected, may be interrogated at a later date for information which was not the focus of the original project.
- Massively improved data coverage. LiDAR data is collected at a much greater spatial density than previous survey methods, reducing the need for interpolation and allowing the data to be interrogated at a late stage to look for other features which may have initially been missed in the field.
- Straightforward procedures for the rapid collection of accurate spatial data can be easily implemented.
- The wealth of data collected with the new technology offers the ability to 'see' through obstacles formerly the bane of theodolite surveyors, for example tree cover can be 'removed' using computing algorithms and a combination of first and last-return data.

Disadvantages

- High entry costs, currently in the order of £100,000 in the UK for the hardware and associated software, although costs vary widely and academic users often qualify for educational discount from manufacturers.
- Lack of coverage in important areas, for example the problem of reflection from water surfaces when using eye safe infrared laser sources, causing problems in bathymetric applications.
- Poor machine calibration can result in interference causing loss of resolution, blank spots, corrugations and 'ripple effects' in the data.
- Inability to manipulate large volumes of data often associated with lower specification hardware (see Hetherington, Chapter 6) and poorly developed early generation software.
- Limited holistic application due to restricted laser frequency of operation (see Danson *et al.*, Chapter 13).

• Numerous instrumentation issues are inherent with current instruments that will affect range accuracy and feature delineation (see Heritage & Large, Chapter 2).

• There are a number of issues relating to training researchers in the use of TLS equipment, data acquisition, processing and modelling which slow initial take-up.

In conclusion, there are many clear trends evolving in the 3D terrain mapping marketplace. Airborne and ground-based LiDAR sensors are improving and can cover terrain faster and at higher resolution with the result being that more and more data are available to more and more end-users. Accompanying this trend is that of the more widespread distribution and availability of LiDAR. Over the past decade, the technology has moved from the realm of specialists towards much more widespread use, while many government entities are making government-funded survey data available to other agencies, Universities and non-governmental organisations. The software available to store and to analyse datasets also continue to improve. The result is that both airborne and terrestrial LiDAR instruments offer opportunities to research the natural environment across spatial and temporal ranges previously unachievable (and in some cases, unimaginable). We are looking at a potential revolution in our understanding of Earth systems provided that the technology is used correctly, data limitations are acknowledged and handled competently and data analyses are conducted in a rigorous and scientific manner. Publication of potentially paradigm shifting research is bound to follow.

REFERENCES

Abshire B, Einaudi F, Gentry BM, Schwemer GK. 2004. *LiDAR: Past, Present and Future in NASA's Earth and Space Programs*. National Aeronautics and Space Administration, Greenbelt, MD.

Barber DM, Mills JP, Bryan PG. 2005. Maintaining momentum in terrestrial laser scanning: a UK case study. ISPRS WG III/3, III/4, V/3 Workshop, *Laser Scanning 2005*, Enschede, the Netherlands, September 12–14, 2005.

Blair JBB, Hofton M, Luthcke SB. 2001. Wide- swath imaging LiDAR development for airborne and space borne applications. *International Archives of Photogrammetry and remote Sensing*, Volume XXXIV-3/W4 Annapolis, MD, 22–24 October 2001.

Crosby CJ, Arrowsmith JR, Jaeger-Frank E, Nandigam, Kim HS, Conner J, Memon A, Alex N, Baru C. 2007. Enabling access to high-resolution LiDAR topography through cyberinfrastructure-based data distribution and processing. *GEON Databases, Tools and Application Development* 61–63.

Goetz AF, Vane HG, Solomon J, Rock BN. 1985. Imaging spectrometry for earth remote sensing. *Science* **228**: 1147–1153.

Harding DJ, Lefsky MA, Parker GG, Blair JB. 2001. Laser altimetry canopy height profiles: methods and validation for closed canopy, broadleaf forests. *Remote Sensing of Environment* **76**: 283–297.

Heritage GL, Hetherington D. 2007. Towards a protocol for laser scanning in fluvial geomorphology. *Earth Surface Processes and Landforms* **32**: 66–74.

Hoff RM, Bösenberg J, Pappalardo G. 2008. The GAW Aerosol LiDAR Observation Network (GALION). http://www.igarss08.org/Abstracts/pdfs/1775.pdf

Lichti DD, Gordon SJ, Tipdecho T. 2005. Error models and propagation in directly georeferenced terrestrial laser scanner networks. *Journal of Survey Engineering* **131**: 135–142.

Lefsky MA, Cohen WB, Parker GG, Harding DJ. 2002. LiDAR remote sensing for ecosystem studies. *Bioscience* **52**: 19–30.

McCaffrey KJW, Jones RR, Holdsworth RE, Wilson RW, Clegg P, Imber J, Holliman N, Trinks I. 2005. Unlocking the spatial dimension: digital technologies and the future of geoscience fieldwork. *Journal of the Geological Society* **162**: 927–938.

Index

Note: Page numbers in *italics* refer to figures, those in **bold** refer to tables.